Industrial heat exchangers

a basic guide
second edition

G. Walker
University of Calgary,
Alberta, Canada

○HEMISPHERE PUBLISHING CORPORATION
A member of the Taylor & Francis Group

New York   Washington   Philadelphia   London

**Industrial
heat
exchangers:**
A basic guide, second edition

 2 3 4 5 6 7 8 9 0   E B E B   9 8 7 6 5 4 3 2 1 0

This book was set in Times Roman by Hemisphere
Publishing Corporation. The editors were Edward
Millman, Michael S. Schwartz, and John P. Rowan; the
designer was Sharon Martin DePass; the production
supervisor was Peggy M. Rote; and the typesetters were
Shirley J. McNett and Laurie Strickland. Cover design
by Sharon M. DePass.
Printing and binding by Edwards Brothers, Inc.

A CIP catalog record for this book is available from the
British Library.

Library of Congress Cataloging-in-Publication Data

Walker, G. (Graham), date.
    Industrial heat exchangers: a basic guide /
      G. Walker. — 2nd ed.
      p. cm.
    Includes bibliographical references.

    1. Heat exchangers. I. Title.
TJ263.W254 1990
621.402′5—dc20   90-4185
ISBN 0-89116-230-5   CIP

Contents

Preface

This elementary book on heat exchangers was conceived in the late 1970s, at the height of the frenzy of interest in all aspects of energy, including conservation.

At that time, many well-meaning but uninformed energy conservation efforts were proposed and implemented. Some involved expenditures of energy and money far beyond any savings that could possibly be achieved. Others required, for successful implementation, a contravention of the Second Law of Thermodynamics, a formidable challenge indeed.

The frenzy of interest in things energetic was dramatically subdued in the 1980s, principally by three events unrelated but all equally important. First, the change of leadership in Washington from President Carter to President Reagan changed the focus of U.S. Government concern from a preoccupation with energy and the environment to one of rebuilding and expanding the military capability, including "Star Wars," at the expense of nearly everything else.

This occurred at a time when the high oil prices achieved by the concerted actions of the Organization of Petroleum Exporting Countries (OPEC) stimulated oil production to record levels and, consequently, produced a surplus of oil so that the markets were flooded. All this was coupled with a small, but significant, reduction in world energy consumption arising mainly from the need to economize and perhaps a little from serious energy conservation efforts.

Finally, countries that were not members of OPEC expanded their production because of the high oil prices. In this they were greatly encouraged by the principal consumers, mainly the United States and Japan, seeking to diversify their sources of supply and thus break the OPEC stranglehold on their economic jugular.

This endeavor was remarkably successful, for in the early 1980s the oil production of the non-OPEC countries soon exceeded, by a substantial margin, that of OPEC.

As these market forces came into play, the price of oil plummeted, and the coherent structure of OPEC fell apart, thus further exacerbating the plunge of oil prices.

One consequence of these dramatic developments was a gross reduction of interest in energy conservation. The world moved on to other interests—resuming, without thought for the future, the profligate dissipation of those priceless chemical feedstocks represented by our crude oil supplies.

Moreover, the unbelievable quantity of natural gas discovered in recent times virtually guarantees that the party will continue for at least another century. Even more lavish reserves of coal are available in readily developed deposits widely dispersed throughout the world.

Token efforts at energy conservation are maintained by a few government departments and by the larger industrial conglomerates. However, these appear to be principally public relations exercises or the remnants of the momentum of departmental empires built up in the halcyon days of the energy frenzy. The funds available are largely derisory. Politicians at all levels have come to realize that "there ain't no more votes in energy."

Energy conservation does not, therefore, have the same sexy, growth-industry image possessed by computers and space. It is not likely to attract the brightest and best of our young graduating engineers.

It was, therefore, with surprise and, I confess, gratification, that I received from Ms. Florence Padgett, of Hemisphere Publishing Corporation, the invitation to undertake preparation of a second edition of this book.

Although the architects and others have now lost interest, it appears that the book has come to serve the useful purpose of introducing recent graduates to the real world of the thermal specialist. Thus equipped, these graduates are better prepared to proceed to the many and varied advanced texts in the field and to become familiar with the proprietary computer software now available at all the consultants' offices.

It was gratifying that the interest was sufficient to exhaust the publishers' stock in only a few years, with sufficient promise of future sales to justify the commission of a revised edition. However, I must confess the invitation left me somewhat perplexed: none of my other books has proceeded to a second edition, so I had no experience to guide my efforts.

What I did was to carefully read through and correct the original text in its entirety. Some readers of the first book were kind enough to write to tell me of some errors and imperfections. Fortunately, I maintained a file of all these letters, though I little expected to have the opportunity to implement the corrections. I much appreciate the thoughtfulness of all those who wrote to me in this way, and I hope that I have been able to make the correction or to implement their advice to their satisfaction.

Perusing the text after a lengthy interval, I came across some patches of purple prose I found somewhat embarrassing. In the end I left these alone, thinking that my best efforts would likely change only the shade of purple rather than eliminate it.

I found no paragraphs of the original text that I could easily discard. I always find editing to be a painful process, and because Ms. Padgett imposed no limitations of length on me, it was easy to persuade myself, in the end, that every paragraph was as valid now as in the earlier version.

In doing this I recalled that I had used several chapters of the original book for another work (Walker, G.: "Heat Exchangers: Handbook of Applied Thermal Design," E. C. Guyer [ed.], McGraw-Hill, New York, 1988), for which stringent limitations on the overall length and number of illustrations had been imposed. Editing these chapters to less than a third the size of the original was one of the most difficult and painful experiences I have had.

It was much easier and more pleasant to add to the book—a little here, a little there, amplifying the original text as time, experience, and the benefit of hindsight and second thoughts permitted.

It is probably true to say my additions and extensions exemplify the development of my own interests in particular areas rather than a significant advance in the technology. In a book so general as this, it is not possible to present comprehensively the leading edge technology.

The more substantial changes and additions include extra material on heat pipes in Chapter 3, some consideration in Chapter 2 of the advantageous physical properties for fluids used in heat transfer, and a new chapter, Chapter 13, on heat sinks, sources, and storage. I also expanded Chapter 7, on high- and low-temperature systems, to include some consideration of very large and very small systems.

In the course of my revisions I received a special issue of the *Journal of Heat Transfer,* published by the American Society of Mechanical Engineers in New York. This special issue is a collection of extended invited contributions by distinguished heat transfer specialists to commemorate the 50th anniversary of the Heat Transfer Division of ASME. I was once again surprised to recall just how young the science and technology of heat transfer is. At the time I was born, nearly 60 years ago, there were no special college classes or texts in heat transfer. Such treatment as there was of heat transfer was included in thermodynamics courses, and at that time Nusselt and his contemporaries were still establishing the scientific foundations of the technology.

In revising this text I took the opportunity to invite my former colleague and fellow author, Andrew Pollard of Queen's University, Kingston, Ontario, to take a second look at his original contribution of Chapter 12, Computer Analysis of Heat Exchangers. This he has done. I am grateful for his effort and glad to include his revised chapter.

As before, I am most grateful to my secretary, Karen Undseth, for her careful and tireless efforts to "get it right." She continues to simplify and smooth my path in so many ways.

I owe much to Burt Unterberger and Ann Penny of the University of Calgary for their assistance with the illustrations and photographs.

I completed the second edition while on a sabbatical from my normal teaching, research, and administrative duties at the University of Calgary. I am grateful for the support of the University of Calgary in countless ways, by the award of a Sabbatical Leave Fellowship and by the provision of facilities to ease the burden of the creation and birth of a new, or at least revised, text. My research effort at the university, including this present work, is supported financially by the National Science and Engineering Research Council of Canada, for which I am most grateful.

I have benefited from the advice and assistance of Graham Reader, head of the Department of Mechanical Engineering at the University of Calgary, and of my colleagues in that department. I am cognizant of the fact their labors were in no way eased by my enjoyment of a sabbatical fellowship as they toiled on.

Finally, I owe my greatest thanks to my wife, Ann. For more years than it seems reasonable for any man to expect, she has been my wise counselor, helpmate, and good friend.

Kihei, Maui   *G. Walker*
1989

Preface
to the First
Edition

This book about heat exchangers is different from all others I know. It is for buyers and users of heat exchangers and for young engineers, who, having taken a college course on heat transfer, find that they still know little about heat exchangers. The book will be of less interest to designers and builders of heat exchangers, but their needs are well served by the many excellent books already available.

Heat exchangers are used in every aspect of industrial, commercial, and domestic life concerned with energy flows. Large numbers of people are engaged in designing, making, selling, and using heat exchangers, and there is an extensive technical literature on the subject. The literature is principally directed to engineering and scientific specialists in the field. So far as I know, there is no small, low-cost user guide to heat exchangers for nonspecialists and newcomers. This is the niche that I set out to fill with this book.

The need for such a book became urgent with the onset of the energy crisis. The time to take energy conservation seriously is finally at hand, and, increasingly, the future will see many attempts to capitalize on the thermal-effluent streams now thrown away so unthinkingly. For these efforts an elementary understanding of heat exchangers will be mandatory to ensure, at the very minimum, that energy savings can be realized without contravention of the second law of thermodynamics. It is by no means unusual, on analysis of the energy flow and economics, to find that well-meaning but uninformed conservation efforts result in expenditures of energy and money beyond any savings that could possibly be achieved.

A requirement was therefore perceived for a book about heat exchangers, for present and potential users, describing the various types available, their advantages and disadvantages, some factors in their selection, the special fields of application, and how to go about buying, operating, and maintaining them for long life and optimal use. These are the areas I have included here.

The preparation of the book was facilitated by the great volume of existing literature contributed by many workers in the field of heat transfer engineering. I have tried to acknowledge all the sources from which I have reproduced the figures, tables, and verbatim quotes, and have sought the necessary permissions where appropriate. If omissions have been made, I offer my sincere apologies. Many manufacturers of heat-exchange equipment responded to my inquiries and supplied substantial useful and informative material. They are all acknowledged by inclusion in the directory or directly in the text.

The incomplete text was used as the notes for short courses on heat exchangers in 1980 at the University of California, Los Angeles, at Bath in the United Kingdom, and at the University of Calgary. To all who attended these classes and responded to my invitation to criticize the text, I offer my thanks. At the UCLA course I benefited greatly from the contributions of my associate lecturers, Dr. W. Armstrong of the Nooter Corp., St. Louis; Mr. Frank Rubin of Houston; Mr. T. Chase of Struthers-Wells Corp., Warren, Pa.; and Dr. J. Mondt of the General Motors Technical Center, Warren, Mich.

Dr. Andrew Pollard, a colleague at the University of Calgary, contributed Chapter 12, a review of computer simulation of heat exchangers. I am most grateful to Dr. Pollard for this timely and pertinent review. Mr. R. Smallwood and Mr. W. G. Ashbaugh contributed the "rogues gallery" of corrosion failures appended to Chapter 8. I am most grateful for their interest and help. The corrosion section of Chapter 8 is largely a condensation of Chapter 3 of Dr. Mars Fontana's excellent book on corrosion engineering. I am grateful for his permission to use the material and appreciate the help he gave me in its preparation.

My secretary, Karen Undseth, contributed greatly by her tireless efforts to sort and classify the material and finally to put the text into readable form. I am grateful also to Bert Unterberger and his assistants for their help in the preparation and execution of the diagrams. Marlene Stewart checked the text and prepared the index in her customary exemplary fashion and thereby relieved me entirely of this demanding and vital last lap.

I completed the text in the course of my normal duties as a professor at the University of Calgary. I am grateful to the university for support in countless ways, particularly the financial support of the University Grants Committee, using funds provided by the National Scientific and Engineering Research Council of Canada.

My family and my students were denied much of my time and interest that they could rightfully expect. I thank them for their tolerance and understanding.

G. Walker

1 Introduction

Heat exchangers are devices to enhance or facilitate the flow of heat. Countless examples are found in everyday life. Every living thing is equipped in some way or another with heat exchangers. Mammals have complicated heat exchangers. The primary unit is the lungs, which not only cool the body by saturating the expelled air with water vapor but also serve as complex mass exchangers, taking oxygen from air and carbon dioxide from the blood. The skin acts as a supplementary heat exchanger, changing its character to promote or inhibit the transfer of heat from the body depending on the temperature, humidity, and velocity of the air.

For reasons known only to God, mammals (with the notable exception of camels) have a specific operating temperature; even a few degrees of variation, up or down, result in severe physiological disturbance. An adequate, well-controlled heat-exchange process is therefore vital, literally a matter of life and death.

Industrial heat exchangers are fortunately much less complicated, although the trappings of technological civilization make them no less vital. Manufactured heat exchangers are found in every facet of our life. Boiling water or frying an egg requires a heat exchanger. Refrigerators operate on a vapor-compression cycle using two heat exchangers-one to cool the freezer compartment, and the other to transfer to the air heat from the freezer compartment plus the work required to drive the system. Automobiles are equipped with heat exchangers, miscalled radiators, and all electrical and electronic equipment must be provided with heat exchangers for cooling.

Heating and cooling systems for buildings involve the use of heat exchangers. Electricity is generated in base-load electric power stations that depend on heat exchangers for the generating and conducting of the steam used to drive the turbine-powered alternators. Oil refineries and chemical processing plants use

1

many different heat exchangers. Food processing, baking, brewing, mixing, and freezing all involve the use of heat exchangers of one sort or another.

It is evident that heat exchangers exist in great numbers and are widely used throughout industry and commerce. They are found in an enormous range of heat-transfer capacities. The smallest (<1 W) are included in miniature cryocoolers for infrared thermal imaging, heat-seeking missile guidance, or superconducting electronic applications. The largest (>1 GW) are the boilers, condensers, and condenser cooling-water air coolers in base-load electric power stations.

This vast range of capacities embraces a diversity of types, shapes, and arrangements of heat exchangers. The majority serve to facilitate the energy transfer between two fluid streams at different temperature levels. Others involve solid mass heating or cooling systems. One example is the convective cooling fins placed on electrical apparatus to dissipate to the ambient air the heat generated by internal resistance. In food processing, sides of beef are cooled for refrigerated preservation or, conversely, meat and bread are heated for cooking.

CLASSIFICATION OF HEAT EXCHANGERS

Heat exchangers used to transfer heat between two or more streams of fluid may be classified broadly in two groups:

1. Recuperative heat exchangers
2. Regenerative heat exchangers

A *recuperative* exchanger is equipped with separate flow conduits for each fluid. The fluids flow simultaneously through the heat exchanger in separate paths, and heat is transferred from the hot fluid to the cold fluid across the walls of the flow section.

A *regenerative* exchanger has only a single set of flow channels through a relatively massive solid matrix. The hot and cold fluids pass through the matrix alternately. When the hot fluid is passing through (called the "hot blow"), heat is transferred from the fluid to heat the matrix. Later, when the cold fluid passes through (called the "cold blow"), heat is transferred from the matrix to the fluid, and the matrix cools. When continuous flow of fluid is required, a duplicate matrix must be provided with quick-acting valves to switch the flows periodically. The matrices will then experience the cold blow and the hot blow successively, periodically, and alternately. In other cases a single matrix is used and the flow is switched cyclically by moving the matrix or the flow headers controlling the fluid flow to and from the matrix. Figure 1.1 shows some of the possible variations of regenerative heat exchangers.

Regenerators are presently used in only a few specialized applications, principally exhaust-air preheaters in large electric power stations, in steel plants, in gas turbines, and in industrial gas liquefaction processing. There is a family of regenerative machinery—Stirling, Ericsson, Vuileumier, and Gifford cycle engines—

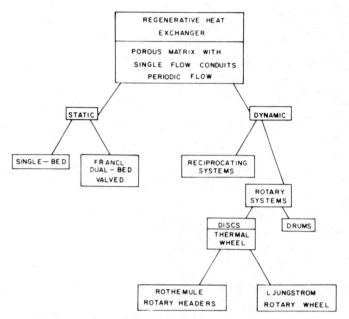

FIGURE 1.1

Design variants and classification of regenerative heat exchangers.

all of which embody a regenerative engine. Nevertheless, recuperative exchangers are preferred for the great majority of cases; however, many further applications of regenerative heat exchangers appear likely in the future, as attempts are made to utilize low-temperature thermal effluent streams. For these moderate-temperature applications, regenerative heat exchangers are attractive because they can be made low in cost, with gravel aggregate, paper and plastic honeycomb, or any finely divided material as the regenerative matrix.

Recuperative heat exchangers exist in many variations (some of which are shown in Fig. 1.2) but can be broadly classified into the tubular and plate types. Tubular heat exchangers have tubes (mostly but not always circular in cross section) for the flow passages for one or both fluids. Plate exchangers have stacks of die-formed plates in which the fluid flow passages are impressed. The stacks of plates may be brazed or welded in a permanent assembly (plate-fin types) or simply compacted in a frame by compression bolts with sealing gaskets between the plates (plate-and-frame type).

Tubular heat exchangers are found in the great majority of applications, primarily for historical and economic reasons. They have been in use for a long time and are familiar to many engineers. Standard models provided by well-established manufacturers are available in a wide range of capacities at reasonable cost, as are the custom-made units found in oil refineries and chemical process plants and in the power industry. Standards for design and procedures for use,

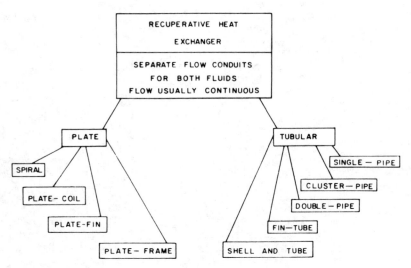

FIGURE 1.2
Design variants and classification of recuperative heat exchangers.

maintenance, and inspection are all well known and familiar. Furthermore, in many cases the fluid streams are substantially pressurized and the exchanger serves as a pressure vessel as well as to transfer energy. Circular tubes and shells are good shapes with which to contain pressures. The pressures and temperatures that can be handled in tubular exchangers are virtually unlimited. Similarly, there is no foreseeable limit to the range of flow capacities that can be accommodated.

Nevertheless in some applications, appreciable savings in cost, weight, and space can be gained with plate exchangers. They have an established and growing place in the technology. Increasing numbers of plate exchangers are being used.

Heat exchangers for very high temperatures or very low temperatures require special consideration. Other special cases arise with corrosive fluids, thick viscous liquids, solid-liquid slurries, and multicomponent vapors whose components condense at different temperatures.

Many heat exchangers are used for processes involving a phase change of the fluids from liquid to vapor or vice versa, generally described as *latent* heat transfers. Processes involving a change in temperature without a change of phase are called *sensible* heat transfers.

Exceptionally high rates of heat transfer may be attained in systems where the fluids boil or condense. The situation is further complicated because the fluids assume complex two-phase patterns that are difficult to analyze and predict.

To newcomers and nonspecialists in the field, one of the most surprising and disturbing aspects of heat-exchanger technology is the lack of precision in design and performance prediction. For new systems it is rarely possible to have confidence closer than ±20 percent in the predicted performance. Also, in use the internal surfaces become fouled, and the thermal performance progressively deteriorates to less than half the "as new" performance. In such cases the ease of cleaning the surfaces to restore performance is an important criterion.

Vibration in heat exchangers is another important area of concern, particularly in large, high-rating systems with high-velocity flows. The vibration may arise from the fluid motion, because of pressure pulses in the fluids, or it may be structure-borne vibration arising from mechanical vibration originating outside the exchanger. When the frequency of the vibration corresponds to the natural frequency or related harmonics of some part or assembly, the amplitude of vibration may build to dangerous levels, resulting in physical damage to the structure or leakage between the fluid streams.

Every process involving the transfer of heat from one fluid to another must include a net temperature drop. The hot fluid will cool to a lower temperature, and the cool fluid will heat to a higher temperature. However, the maximum temperature attained by the cool fluid will always be *less than* the maximum temperature of the hot fluid. Similarly, the minimum temperature of the hot fluid will always be *greater than* the minimum temperature of the cool fluid. There must be a temperature difference between the fluids to induce the transfer of energy from the hot fluid to the cold fluid. This is a requirement of the second law of thermodynamics.

Similarly energy will always be expended in causing fluids to flow through the exchanger, and this becomes manifest in the pressure difference at the inlets and outlets for the two fluids. Frictional degradation of pressure energy during the flow causes the fluid pressure at the outlet to be lower than that at the inlet.

Thermal performance is, in general, related to the frictional pressure drop. Improvements in thermal performance will likely be accompanied by increased pressure drop. With high-density fluids (liquids) the frictional power expenditure is generally small relative to the rate of heat transfer. With low-density fluids (gases) it is only too easy to effect an improvement in the heat-transfer rate that requires an equivalent or greater expenditure of mechanical energy in the form of compressor or fan work. Such improvement in heat transfer is dearly bought, for in most systems mechanical energy is "worth" up to 10 times its equivalent in low-grade thermal energy. The design and use of a heat exchanger always require careful consideration of both the heat-transfer rates attained and the mechanical power necessary to achieve them.

Careful selection of the proper materials to use with the fluids to be handled is another vital aspect of heat-exchanger technology. Corrosion, leading to failure of the heat exchanger to contain the fluids, occurs in many forms that must be recognized and sufficiently understood to provide a sound basis for logical and economic choices of materials.

ORGANIZATION OF THE BOOK

Chapter 2 is a review of basic heat transfer and fluid flow relevant to heat exchangers. Chapter 3 is concerned with types of tubular exchangers, and Chap. 4 with plate exchangers.

Some aspects of thermal design are discussed in Chap. 5, and mechanical design in Chap. 6. High- and low-temperature systems are considered in Chap. 7. Some of the factors responsible for degradation of performance in service—corrosion, fouling, and maldistribution of flow—are considered in Chap. 8.

Chapter 9 outlines significant aspects of the specifications for heat exchangers and the procedures for inspection, testing, and maintenance. Codes and standards relating to heat exchangers are considered in Chap. 10.

Regenerative heat exchangers are described in Chap. 11. Chapter 12, contributed by Dr. A. Pollard, contains a brief outline of future developments for the uncompromising simulation of heat exchangers using large digital computers, which may eventually revolutionize the design processes for heat exchangers.

Chapter 13 contains a brief survey of heat sinks, sources and storage systems.

Appendix A is a glossary, and App. B contains a directory of heat-exchange manufacturers. A specimen customer specification for tube-and-shell heat exchangers is included in App. C. The book ends with a brief review of the literature relating to heat exchangers, in App. D.

2

Thermodynamics, heat transfer, and fluid flow

THE FIRST AND SECOND LAWS OF THERMODYNAMICS

Thermodynamics is a science concerned with the behavior of liquids and gases resulting from changes in temperature or pressure. Many energy conversion systems for generating power and refrigeration are designed, based on the thermodynamic behavior of the fluid concerned, to satisfy the first and second laws of thermodynamics.

The first law is simply a statement of the principle of conservation of energy: Energy can neither be created nor destroyed. According to the first law it is impossible to have a "black box" from which energy can be extracted continuously without replenishment. Energy must be supplied to the box if work (a form of energy) is to be taken from it. A simple example is the car battery—a literal as well as figurative black box. Electrical energy can be supplied by the battery but only until it becomes exhausted and must be recharged by energy supplied to it.

The second law is more subtle. A simple statement of the second law is: Heat can flow from one body to another only when there is a difference in temperature, and the heat flows from the hotter body to the cooler. The significance of this statement to heat exchangers is self-evident. However, there are corollaries of the second law that are less obvious, and many wise and experienced engineers have been led to attempt a contravention of the law.

The second law applies to all thermal energy conversion systems. In power systems it limits the fraction of energy supplied at high temperature that is converted to work. The remainder, the difference between the energy supplied and the work done, must be rejected as waste heat at low temperature.

Of the heat supplied, the fraction that can be converted to work is the ratio

$$\text{Thermal efficiency} = \frac{\text{work done}}{\text{heat supplied}}$$

The largest possible thermal efficiency is the so-called Carnot efficiency η_{Carnot} for a particular thermodynamic cycle called the Carnot cycle:

$$\eta_{Carnot} = \frac{T_{max} - T_{min}}{T_{max}}$$

where T_{max} = maximum cycle temperature at which heat is supplied (at constant temperature)

T_{min} = minimum cycle temperature at which heat is rejected (at constant temperature)

The Carnot efficiency is a theoretical ideal that practical engines cannot achieve in practice. The best that can be done, in very large base-load steam plants generating electric power, is about 60 percent of the Carnot. Smaller plants, diesel engines, spark-ignition engines, and gas turbines all achieve a smaller fraction, 20 to 40 percent of the Carnot value. Actual thermal efficiencies can, therefore, be represented as

$$\eta_{act} = \frac{F(T_{max} - T_{min})}{T_{max}}$$

where F is characteristic of the system being represented and ranges from 0.01 to 0.6.

This simple equation illustrates some important facts about thermal power systems. Of the energy supplied, the fraction that can be converted to work depends principally on the temperatures at which the energy is supplied and rejected. These temperatures are the principal parameters affecting the efficiency of all heat engines.

Consider the system shown in Fig. 2.1. It consists of a fluid circuit including a coupled compressor and expander and two heat exchangers, one receiving heat from a high-temperature source at T_{max}, the other rejecting heat to a low-temperature sink at T_{min}. On leaving the high-temperature heat exchanger the fluid expands in the expansion engine, and work is produced. The expanded fluid then moves to the low-temperature heat exchanger and is cooled. Finally it is compressed to some high pressure, and it then passes to the heater to complete the circuit. If the work required to compress the fluid is less than that produced on expansion, a surplus of work is available at the shaft; this is the useful output of the system.

The energy transfers involved are

Q_{in} = heat supplied at temperature T_{max}

Q_{out} = heat rejected at temperature T_{min}

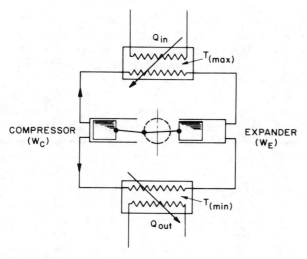

FIGURE 2.1
Closed-cycle thermodynamic system.

W_E = work produced in the expander

W_C = work consumed in the compressor

The net work output $W = W_E - W_C$ is the difference between Q_{in} and Q_{out}; that is, $W = Q_{in} - Q_{out}$. The thermal efficiency of the system is

$$\eta_{act} = \frac{W}{Q_{in}} = \frac{Q_{in} - Q_{out}}{Q_{in}} = 1 - \frac{Q_{out}}{Q_{in}}$$

Furthermore, $\eta_{act} = F\eta_{Carnot} = F(T_{max} - T_{min})/T_{max}$.

The manipulation of thermodynamic data along these lines permits elementary determination of the *quantities* of energy to be transferred to or from systems to accomplish the desired conversion of heat to work, etc. Thermodynamics takes no account of the limiting *rates* at which the energy may be transferred or the *size* of the equipment necessary to accomplish the transfer. This is the domain of the science of heat transfer.

THE NATURE OF HEAT

Newcomers, on reading technical journals or listening to specialists, will likely form a mental picture of heat as some weightless, invisible fluid that is somehow triggered to flow from one body to another by a difference in temperature. This model was the basis for the *phlogiston theory* much in evidence until about 1850, when Joule established that energy can be converted from one form to another and so laid the basis for the emergent science of thermodynamics.

A better simplistic model is postulated by the *kinetic theory of gases*. This holds that matter (molecules) is made up of "billiard balls" (atoms), either singly or in groups of two or three. The balls move about randomly with some velocity of translation. They also spin in their axes and, when coupled in groups of two or three, vibrate and rotate together as a group.

Temperature is a measure of the excitation of the system, the energy with which the balls move about, vibrate, or rotate. High-temperature systems contain highly excited billiard balls, rushing about and vibrating most energetically. Lower-temperature systems contain less energetic billiard balls. When a high-temperature system is brought adjacent to a low-temperature system, the highly excited, fast-moving, vibrating, rotating balls meet and collide with the cold, sluggish balls, which then become excited and increase in temperature. The original high-temperature system exhibits a less excited state and decreases in temperature, having imparted energy to the lower-temperature system.

For gases, particularly at low pressures, this model is fairly representative of the actual situation. The billiard balls are the molecules of gas, which may be a single atom, monatomic, diatomic, or triatomic. For gases at high pressure and for liquids, the kinetic-theory model is less representative. The molecules of matter are in close proximity, and other forces become significant. For solid materials it is even less applicable. For metals having many free electrons, one can stretch the model to embrace the concept of an electron gas. For nonmetals with few free electrons, the primary vehicle for heat transmission is lattice vibration, and the concepts of the kinetic theory do not hold.

MODES OF HEAT TRANSFER

There are three modes of heat transfer: conduction, convection, and radiation. For engineering calculations it is convenient to consider the three modes separately and then combine the results into an overall solution.

Conduction Heat Transfer

Conduction is the mode of heat transfer through solids. A metal bar heated at one end will soon increase in temperature at the other end because of conduction heat transfer along the bar.

The equation defining the rate of heat transfer, called the Fourier equation, is

$$\dot{q} = -kA\frac{dT}{dx}$$

where $\dot{q}$ = rate of heat transfer in x direction
dT/dx = temperature gradient at point being considered
A = cross-sectional area perpendicular to x direction
k = thermal conductivity of solid

The thermal conductivity is characteristic of the material and must be determined experimentally if the material is unknown or from tables of data if the material is known. Metals have high conductivities compared with nonmetals. Some typical values for metals used in heat exchangers are given in Table 2.1. The units for conductivity are heat flow per unit area per unit thickness per unit temperature; that is, $Btu/(ft^2 \cdot ft \cdot °F)$ or $W/(m \cdot K)$.

The heat flux through a plane wall of thickness x with steady-state surface temperatures of T_A on one face and T_B on the other is

$$\dot{q} = \frac{-kA(T_A - T_B)}{x}$$

This can also be written as $\dot{q} = (T_A - T_B)/R_w$, where $R_w = x/kA$, the thermal resistance. This form of the equation corresponds to the well-known Ohm's law of electrical engineering,

$$I = \frac{E}{R}$$

where I = current flow (equivalent to $\dot{q}$, the rate of heat transfer)
E = potential difference (equivalent to $T_A - T_B$, the temperature potential)
R = electrical resistance (equivalent to R_w, the thermal resistance)

The equation for heat transfer is clearly equivalent in form to Ohm's law. Many complicated problems in heat transfer are handled as electrical analogies, using the theory of networks devised for resolving electrical-engineering problems.

The radial heat flux through a circular tube of length L may be determined in a similar way. Let the inside radius and wall temperature of the tube be r_i and T_i, respectively, and the outside radius and wall temperature be r_o and T_o, respectively. In the Ohm's-law format,

$$\dot{q} = \frac{T_i - T_o}{R_w}$$

TABLE 2.1

Typical Values of Thermal Conductivity for Metals Used in Heat Exchangers

Metal	Thermal conductivity, $kW/(m \cdot °C)$			
	$-100°C$	$0°C$	$100°C$	$400°C$
Aluminum alloy	120	160	180	–
Steel, carbon	–	43	42	36
Stainless steel	16	17	19	19
Copper	407	380	374	350
Brass	88	100	120	140
Nickel/chrome	12	13	16	18

where $R_w = \ln (r_o/r_i)/2\pi kL$.

For multilayered pipes or walls the same thermal-resistance concept may be used. Consider, for example, the composite three-layer wall shown in Fig. 2.2. The heat flow in the direction shown will be

$$\dot{q} = \frac{T_1 - T_4}{x_A/k_A A + x_B/k_B A + x_C/k_C A}$$

and the equivalent electric series circuit will be as shown in Fig. 2.2.

The multiple-wall pipe shown in Fig. 2.3 can be similarly treated. For the three-layer system shown, the heat flux will be

$$\dot{q} = \frac{2\pi L(T_1 - T_4)}{\ln (r_2/r_1)/k_A + \ln (r_3/r_2)/k_B + \ln (r_4/r_3)/k_C}$$

For thermally insulated systems where, for example k_B is very small compared with the thermal conductivities of a thin-wall steel pipe (A) and a thin-wall aluminum jacket (C), it is possible to neglect the thermal resistance of the metals and reduce the equation to

$$\dot{q} = \frac{2\pi L(T_1 - T_4)}{\ln (r_3/r_2)/k_B}$$

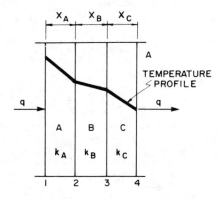

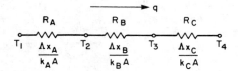

FIGURE 2.2
Conduction heat transfer in multilayer wall.

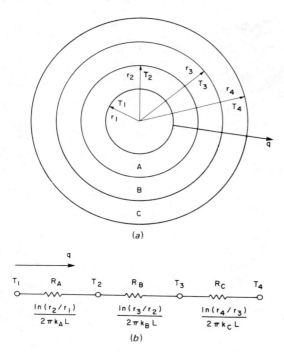

FIGURE 2.3

Conduction heat transfer in multiwall pipe.

Conduction in Fins

The amount of energy transferred from a solid wall to a fluid flowing past it depends on a number of factors, including the area of the surface exposed to the fluid. To increase the heat transfer one can, therefore, extend the area of fluid-surface interface by adding fins as shown in Fig. 2.4. The actual design of a fin is a complicated exercise in economic optimization, beyond our level of interest here. For our purpose it is sufficient to note that the temperature distribution of the fin depends on the thermal conductivity of the fin material. All the heat to be dissipated by convection from the fin must enter the fin at the root and pass by conduction along the fin for eventual dissipation by convective heat transfer from the surface to the fluid. The temperature diminishes along the fin from a maximum at the root.

The addition of fins to a bare tube will, of course, increase the heat transferred. It is important to remember that the addition of the fins "shadows" a fraction of the bare pipe and thereby reduces its effectiveness. Furthermore, it is vital to provide a secure bond to the pipe to eliminate interface thermal resistance between the pipe surface and the fins. A fin cut or cast from the solid material of the tube is likely superior to bare tubes to which fins are

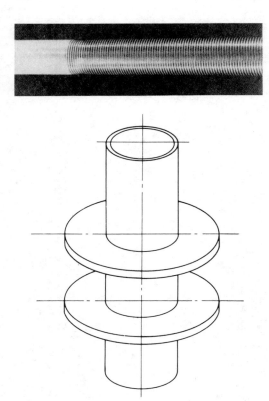

FIGURE 2.4
Finned surface.

subsequently attached. Joining the fins to the tube by means of a liquid-metal technique (welding, brazing, soldering, or dip-galvanizing) eliminates concern about the interface thermal resistance.

The addition of fins to a bare tube will not increase the heat transfer by the same proportion as the area is increased. The average temperature of the fins is lower than the bare-tube temperature, and the fluid flow will be impeded by the fins.

Interface Resistance

When conducting materials are placed in contact, heat is transferred, tending to equilibrate the temperature. Such contact is never as perfect and complete as that of surfaces that are fused and bonded to form an integral whole. Apparently flat surfaces in contact appear on a microscopic scale as opposing peaks and valleys. The peaks come in contact, and the valley interstices are voids filled with low-conductivity gas (air) or a squeezed metallic gasket of soft conducting material (lead, indium, copper, etc.).

The interruption in heat flow at the interface can be accounted for as an additional thermal resistance called *interface resistance* and defined by $R_i = \Delta T_i/(\dot{q}/A)$. For surfaces in perfect thermal contact the interface resistance is zero because the interface temperature difference ΔT_i is zero.

Convective Heat Transfer

Convective heat transfer takes place when heat is transferred from a solid surface to a fluid because of relative motion between the fluid and solid. Conduction also occurs between the fluid and the solid wall, for which no relative motion is necessary.

The cooling equation proposed by Isaac Newton defines convective heat transfer:

$$\dot{q} = hA(T_F - T_W)$$

where $\dot{q}$ = rate of heat transfer
A = area for heat transfer
T_F = fluid temperature
T_W = wall temperature
h = heat-transfer coefficient

The value of h, the heat-transfer coefficient, varies widely. Typical values are given in Table 2.2. Note that *forced* convective heat-transfer coefficients are much greater than *free* convective heat-transfer coefficients, and that liquids have higher coefficients than gases (primarily the effect of density). Fluids changing phase from liquid to vapor (boiling) or vapor to liquid (condensing) have exceptionally high rates of heat transfer compared with non-phase-change systems.

It is not possible to predict the value of h with great certainty in a new, untried configuration. It is normally calculated from empirical correlations determined from experimental work. The value of h depends on the properties of the flowing fluid, the geometry of the flow path, and the fluid velocity. For a

TABLE 2.2

Range of Values for the Heat-Transfer Coefficient h

	h	
Mode	Btu/(h·ft²·°F)	W/(m²·K)
Free convection (air)	1–5	5–25
Forced convection (air)	2–100	10–500
Forced convection (water)	20–3,000	100–15,000
Boiling water	500–5,000	2,500–25,000
Condensing water vapor	1,000–20,000	5,000–100,000

given fluid and flow path, the heat-transfer coefficient is principally dependent on the fluid velocity.

In forced convective heat transfer the fluid is pumped at velocities greater than those arising from natural convection buoyancy forces. The dependence of h on fluid velocity is, then, approximately as follows:

Flow inside tubes (laminar): $h \propto V^{1/3}$

Flow inside tubes (turbulent): $h \propto V^{3/4}$

Flow outside tubes: $h \propto V^{2/3}$

When heat transfer is primarily due to natural convection, the fluid motion arises because of fluid density changes resulting from temperature effects. The magnitude of h depends on the fluid-wall temperature differential approximately as follows:

High temperature differences: $h \propto \Delta T^{1/3}$

Low temperature differences: $h \propto \Delta T^{1/4}$

It is convenient to extend the notion of thermal resistance to convective heat transfer. The equation $\dot{q} = hA (T_F - T_W)$ can then be expressed as $\dot{q} = (T_F - T_W)/R_c$, where R_c is the thermal resistance in convective heat transfer, defined as $R_c = 1/hA$.

Fouling Resistance

After some period of use as a heat exchanger, the amount of convective heat transferred for a given temperature drop will decrease. This occurs because deposits accumulate on the heat-exchange surface and interpose an additional thermal barrier to heat flow. The deposits include mineral salts precipitated from water, carbonaceous deposits resulting from the polymerization of organic compounds on the overheated surface, sedimentation from contaminated liquors, rust and scale formed by corrosive action, and other miscellaneous depositions.

The deposits are generally termed *fouling,* and to account for the phenomenon a *fouling resistance* $R_f = \Delta T_f/(\dot{q}/A)$ is sometimes defined. The fouling resistance is so variable it is difficult to determine precisely. Nevertheless, the accumulated experience of manufacturers and users has resulted in the preparation of tables of "fouling factors" by the Tubular Equipment Manufactures Association [TEMA, 1978]. Values corresponding to R_f above range from 0.0009 $m^2 \cdot K/W$ for surfaces exposed to alcohol vapors to 0.0004 $m^2 \cdot K/W$ for industrial air.

Radiation Heat Transfer

In radiation heat transfer, energy is transported in the form of electromagnetic waves traveling at the speed of light. Radiation can occur in a perfect vacuum, in contrast to conduction and convection, which occur only when a transfer medium is available.

The sun provides the most familiar example of radiant energy, the result of a large-scale exothermic fusion reaction occurring as a relatively steady process. We see the sun as a huge, fiery sphere having a temperature of 12,000 K. This glowing, superheated sphere emits intense thermal radiation that travels across the 150 Gm of virtually total vacuum separating earth and sun and reaches us in 8 min. It is the sole support of all life on the planet. No living thing would survive very long if the sun were suddenly to disappear or stop emitting radiation.

The quantity of radiant energy leaving a body is

$$\dot{q}_{\mathrm{rad}} = \epsilon \sigma A T^4$$

where $\dot{q}_{\mathrm{rad}}$ = rate of radiant energy emission

ϵ = emissivity, a characteristic of the surface; $0 \leqslant \epsilon \leqslant 1$

A = surface area of the body

T = temperature of the body surface

σ = Stefan-Boltzman constant, 5.7×10^{-8} W/(m$^2 \cdot$K^4)

Note the rate of energy emission is a function of the fourth power of the temperature.

Radiation heat transfer can normally be ignored unless temperatures are very high, as in flames, combustion systems, or solids heated to red heat. Radiation is also very important in low-temprature (cryogenic) systems where the cold bodies are isolated in high-vacuum enclosures to eliminate convection and conduction effects.

When a body emitting the radiation is completely enclosed, for example, a flame or fire in a boiler firebox, the rate of heat transfer from the enclosure surfaces is $\dot{q}_r = \epsilon_1 \sigma A_1 (T_1^4 - T_2^4)$, where the subscripts 1 and 2 refer to the emitter and enclosure, respectively.

When two bodies 1 and 2 are adjacent but body 2 does not completely surround and enclose body 1, the rate of radiant heat transfer from the hot body to the cold body is

$$\dot{q}_{r(1-2)} = \alpha_2 A_2 F_{(1-2)} \, \epsilon_1 A_1 \sigma(T_1^4 - T_2^4)$$

where α_2 = absorptivity, a characteristic of the surface of the absorber;
$0 \leqslant \alpha_2 \leqslant 1$

A_2 = surface area of absorbing body

$F_{(1-2)}$ = view factor, indicating the proportion of the surface of body 2 that is covered by the optical image of body 1; $0 \leqslant F_{1-2} \leqslant 1$

Radiation heat transfer is combined in nearly all engineering work with the other modes of heat transfer, conduction and convection. It is, therefore, convenient to extend the notion of equivalent thermal resistance to radiation heat transfer.

The rate of heat transfer by radiation may be written as

$$\dot{q}_{rad} = \frac{T_1 - T_2}{R_r}$$

where $\dot{q}_{rad}$ = rate of radiation heat transfer
$\quad\quad T_1$ = temperature of hot body
$\quad\quad T_2$ = temperature of cold body
$\quad\quad R_r$ = thermal resistance for radiation heat transfer
R_r is defined by the equation

$$R_r = \frac{T_1 - T_2}{\alpha_2 A_2 F_{(1-2)} \epsilon_1 A_1 \sigma(T_1^4 - T_2^4)}$$

Combined Heat Transfer

It is rarely necessary to become involved in detailed consideration of the separate heat-transfer processes taking place within a heat exchanger. Instead, the various processes are all combined into an overall heat-transfer coefficient U defined so that the total heat transfer in the heat exchanger may be determined from the simple equation $\dot{q} = UA\ \Delta T$. To demonstrate, consider the situation shown in Fig. 2.5. Fluid A, at some high temperature, is separated by a metal wall from fluid B at a lower temperature. Heat will be transferred from fluid A to fluid B. There are three heat-transfer processes taking place: convective heat transfer from

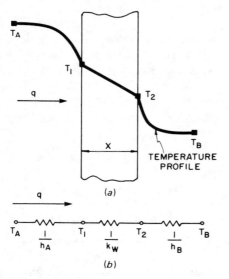

FIGURE 2.5
Combined heat transfer.

fluid A to the wall, conduction heat transfer through the wall, and convective heat transfer from the wall to fluid B. The temperature distribution will be as shown in Fig. 2.5. Fluid A near the wall will decrease in temperature from T_A to the wall temperature T_1. There will be a decrease in wall temperature from T_1 to T_2 through the wall. Fluid B close to the wall will experience a further decrease in temperature down to the temperature T_B of the principal mass of the fluid.

Assuming the direction of heat flow to be in one direction only, we can represent the rate of heat transfer per unit area of wall as

$$\dot{q}_{AB} = h_A(T_A - T_1) = \frac{T_A - T_1}{R_A}$$

or

$$\dot{q}_{AB} = \frac{k_w(T_1 - T_2)}{x} = \frac{T_1 - T_2}{R_w}$$

or

$$\dot{q}_{AB} = h_B(T_2 - T_B) = \frac{T_2 - T_B}{R_B}$$

We can write this as

$$\dot{q}_{AB} = \frac{T_A - T_B}{1/h_A + x/k_w + 1/h_B} = \frac{T_A - T_B}{R_A + R_w + R_B}$$

where R_A, R_w, and R_B are the equivalent thermal resistances $1/h_A$, x/k_2, and $1/h_B$. We prefer to use the simpler equation $\dot{q}_{AB} = U(T_A - T_B)$ and can do so if $U = 1/(R_A + R_w + R_B)$.

Extending the concept further, we can include other thermal resistances discussed earlier: the interface thermal resistance R_i, the fouling thermal resistances R_{fA} and R_{fB} for the walls adjacent to liquids A and B, respectively, and perhaps the radiation thermal resistance R_r. The overall heat-transfer coefficient U therefore embraces a large number of terms and complicated processes:

$$U = \frac{1}{R_A + R_i + R_{fA} + R_w + R_{fB} + R_B + R_r}$$

In a heat exchanger various areas may be selected for use as the reference area. The surface area of tubes is commonly used, but in high-pressure systems with thick tube walls there may be an appreciable difference in the internal and external surface areas of the tubes. When the area ratio A_A/A_B is significantly greater than unity, it is necessary to specify the reference area A adopted for $\dot{q}_{AB} = UA\,\Delta T$ and to adjust the individual fouling resistances so as to relate to a common reference area; that is, $R'_A = (A_B/A_A)R_A$ and $R'_{fA} = (A_B/A_A)R_A$ when A_B is the common reference area.

Typical values of U are listed below:

System	U, W/(m²·K)
Gas/gas (free convection)	1–2
Gas/gas (forced convection	10–30
Liquid/liquid (free convection)	25–500
Liquid/liquid (forced convection)	200–2500
Condensers	300–4000
Boilers	300–6000

Given this range of possible values, it is clearly unwise to select a value of U arbitrarily for preliminary sizing or evaluation. Experience with a similar unit is the best indicator of the likely value. In the absence of direct experience, one may profit from the experience of others, as distilled into the composite chart reproduced in Fig. 2.6. To use the chart, select first, on the upper and lower scales, the fluids of interest. Connect these two selections (as shown by the dashed line for the example). The intersection of this line with the middle scale indicates the approximate value of the overall heat-transfer coefficient U. Figure 2.6 is in British units, and 1 Btu/(h·ft²·°F) is equal to 5.7 (say 6) W/(m²·K). Therefore, to obtain values for U in SI units of watts per square meter per kelvin, multiply by 6 the numerical values on the middle scale of Fig. 2.6.

Fluid Flow

Fluid in motion represents a form of energy. Work is required to compress the fluid, to accelerate it to some velocity u, and to move it about from one place to another. The fluid can be made to produce work as it falls from one level to another, as it slows down, or as it expands with decrease in pressure. Common examples are water wheels and hydroelectric plants driven by water falling under gravity in a river or lake. The *potential* energy of the water is converted to work. In a similar way the wind causes a windmill to rotate and produce electricity, grind corn, or pump water. Here the *kinetic* energy, the velocity energy of the wind, is converted into work. Similarly, the *pressure* energy of a fluid can be converted to work as the fluid is expanded to a lower pressure. An inflated rubber balloon will accelerate when released because the air it contained expands through the orifice. Jet aircraft fly because of the forces induced when high-pressure streams expand in the engine nozzle or jet.

Fluids in motion experience frictional effects, so the amount of work that can be recovered *from* a fluid stream is always less than the work or energy

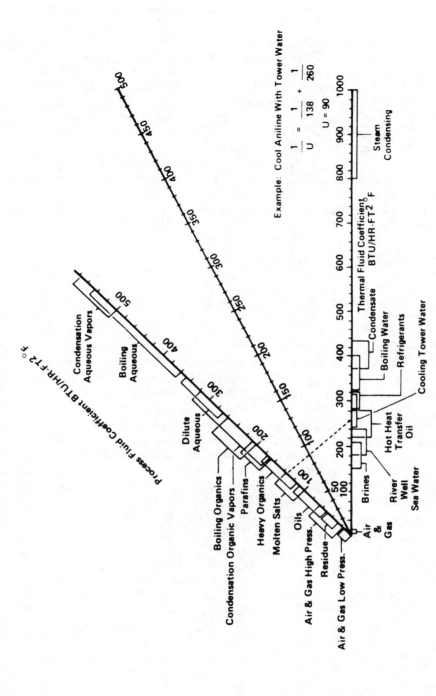

FIGURE 2.6
Combination chart for finding the overall heat-transfer coefficient U. (*After Frank [1978].*)

21

added to it. The frictional dissipation of energy arises because of the fluid viscosity. It results in some frictional heating and pressure drop in the fluid.

Fluids can be classified as incompressible or compressible, depending on the change in fluid density with pressure. If there is no significant increase in density as the pressure increases, the fluid is considered incompressible. Liquids are usually assumed to be incompressible, whereas gases and vapors are compressible. The density of a liquid does, of course, increase with an increase in pressure, but the effect is sufficiently small to be ignored in many flow situations. In a similar way, when the velocity of motion is relatively low and pressure and temperature changes are small, the flow of gases and vapors can be treated as incompressible to simplify matters.

Incompressible fluid flows can be represented by the well-known Bernoulli equation:

$$\frac{p}{\rho} + \frac{u^2}{2g_c} = \text{const}$$

where p = pressure at a particular point in the flow
     ρ = fluid density
     u = fluid velocity
     g_c = gravitational constant

The Bernoulli equation indicates that the sum of the pressure and kinetic and potential energy of a fluid stream remains constant unless some work is done or heat is transferred. The relative significance of the various terms can be changed. The pressure can be reduced but, to compensate, the velocity of the fluid must be increased to provide an increased kinetic-energy term. In many cases the various energy terms are specified in terms of the *head* or height of a column of water or mercury that the flow would support if all motion ceased and the energy was converted to potential energy.

Compressible flow is, in principle, exactly the same but becomes more involved in the details because of the dependence of fluid properties on changes in pressure, temperature, and density. For one-dimensional compressible flow, the steady-flow energy equation applies:

$$i + \frac{u^2}{2g_c} + \dot{q} + W = \text{const}$$

where i = enthalpy $e + p/\rho$
     e = internal energy
     p = pressure
     ρ = fluid density
     u = fluid velocity
     $\dot{q}$ = heat transferred
     W = external work done

Fluid Friction

Fluid friction effects are generally accounted for in terms of a *friction factor f* or *skin friction coefficient* C_f. The pressure drop in a tube may be represented by an equation of the form

$$\Delta p = \frac{fL\rho u^2}{2g_c d}$$

where Δp = pressure drop along pipe
f = friction factor
L = length of pipe
ρ = fluid density
u = mean fluid velocity
d = diameter of pipe

The value of the friction factor f is not constant but is a function of the Reynolds number $\mathrm{Re} = \rho u d/\mu$ (see the next section for a discussion of the Reynolds number). For laminar flow ($\mathrm{Re} < 2400$), $f = 64/\mathrm{Re}$. For turbulent flow in smooth tubes, $\mathrm{Re} < 2 \times 10^5$, and $f = 0.3/\mathrm{Re}^{1/4}$ approximately.

Similar expressions are used to assess the fluid friction factors and pressure drops in other flow situations—crossflow over tube banks, in stacks of corrugated tubes, finplates, and other assemblies of interest with respect to heat exchangers. In all cases the pressure drop is a function of the *square* of the fluid velocity and directly proportional to the fluid density. The friction factor f decreases with an increase in the Reynolds number, that is, an increase in the density or fluid viscosity.

An extensive compilation of fluid-friction data relevant to heat exchangers is contained in the classic text by Kays and London [1964].

THE CONVECTIVE HEAT-TRANSFER COEFFICIENT

The introduction to convective heat transfer given above is adequate for those content with a superficial acquaintance. Others wishing to go further but without the benefit of a university course in fluid mechanics and heat transfer may find the literature confusing and difficult to assimilate. It abounds with unfamiliar dimensionless groupings or parameters that are not readily appreciated by the newcomer. This book would therefore be seriously deficient if an attempt were not included to make these understandable. It is, however, recommended that this section be omitted at the first reading.

Laminar and Turbulent Flow

There are, in general, two types of fluid flow, laminar and turbulent. Laminar flow, characterized by low velocities, is an easy, languid flow with no mixing of

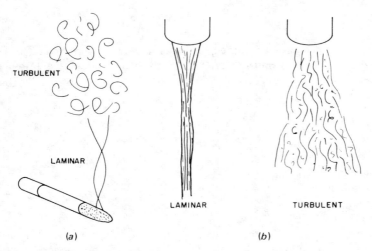

TURBULENT

LAMINAR

LAMINAR

TURBULENT

(a)   (b)

FIGURE 2.7
Laminar and turbulent flow: (a) cigarette, (b) water faucet.

the streams. Turbulent flow occurs at higher velocities and is accompanied by vigorous intermixing of the streams; most flows are turbulent. The difference can be simply observed with cigarette smoke or a water faucet, as illustrated in Fig. 2.7.

In the nineteenth century, Osborne Reynolds, a professor at the Manchester College of Science and Technology, studied the flow of fluids and the transition from laminar to turbulent flow. Reynolds found that the transition varied among different fluids and depended on the type of flow path and the velocity of the flow, and he deduced a grouping of parameters to characterize the situation. This dimensionless grouping is the so-called Reynolds number,

$$\mathrm{Re} = \frac{\rho u d}{\mu}$$

where ρ = fluid density
u = fluid velocity
μ = fluid viscosity
d = some characteristic dimension of the flow path

Two flows of particular interest are flow in a circular tube and flow over a flat plate.

For tube flow, the characteristic dimension d is the tube diameter. Reynolds found that the flow in tubes changes from laminar to turbulent when the Reynolds number exceeds a critical value of 2400. Within reasonably close limits, say 2200 to 2600, the transition occurs whether the fluid is water, glycerine, air, hydrogen, gasoline, or some other gas, liquid, or vapor.

The characteristic dimension for plate flow is the distance from the leading edge of the plate. For local values of the Reynolds number at some distance x downstream of the leading edge, the characteristic dimension is simply x. The Reynolds number increases as one moves away from the leading edge of a plate, and it reaches a maximum at the trailing edge. The critical Reynolds number for plate flow to change from laminar to turbulent flow is

$$Re_{crit} > 5 \times 10^5$$

Noncircular Ducts

When the duct through which the fluid is flowing is not circular in section, it is usually adequate to calculate the equivalent hydraulic diameter $D_H = 4A/P$, where A is the cross-sectional area of flow, and P is the wetted perimeter:

1. For a circular tube: $A = \pi D^2/4$ and $P = \pi D$, so that $D_H = (4\pi D^2/4)(1/\pi D) = D$.
2. For a square duct of side L: $A = L^2$ and $P = 4L$, so that $D_H = 4L^2/4L = L$.
3. For a rectangular duct of height H and length of side L: $A = LH$ and $P = 2L + 2W$, so that $D_H = 4LH/2(L + H) = 2LH/(L + W)$.

Viscous Boundary Layer

Consider fluid flowing with a velocity u past a flat surface. Because of viscosity effects the fluid near the wall will move more slowly than fluid unaffected by the presence of the wall. The fluid flowing past the wall will exert a drag force on the wall, tending to cause it to move in the same direction as the fluid.

The situation is represented diagramatically in Fig. 2.8. Fluid near the wall, moving with a velocity lower than the free-stream velocity, is enclosed by the dashed line. This region of fluid is called the *viscous boundary layer*. The boundary starts at the leading edge of the plate and increases in thickness in moving along the plate.

The characteristic dimension x for calculating the Reynolds number for

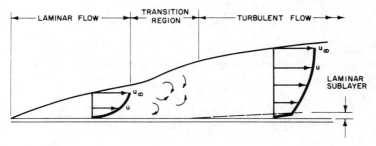

FIGURE 2.8
Viscous boundary layer for a flat plate.

plate flow is measured as the distance from the leading edge. At the leading edge $x = 0$ and, therefore, $Re = 0$, since $Re = \rho u d/\mu$. As we move along the plate, x increases, Re increases, and the thickness of the boundary layer, the region of disturbed fluid, increases.

For the first part of the boundary layer the flow is laminar, even though the mainstream flow may be turbulent. At a sufficient distance from the leading edge, for the Reynolds number to attain the critical value 5×10^5, the flow changes from laminar to turbulent. As a result, the thickness of the boundary layer is substantially increased, and the intense agitation of fluid in the turbulent boundary layer increases the drag force on the plate. The velocity distribution in the laminar and turbulent boundary layers is shown in Fig. 2.8.

An analogous situation exists when fluid flows in a tube, as shown in Fig. 2.9. The boundary layer forms at the leading edge, the entrance of the tube, and the flow is initially laminar. The thickness of the boundary layer increases downstream, and eventually the boundary layers meet at the center of the tube. The flow is then said to be *fully developed*. The initial part, before the flow becomes fully developed, is said to be the *entrance region*.

Fully developed flow can be either laminar or turbulent in character. The difference is principally in the velocity distribution across the flow section. Typical fluid velocity profiles for laminar and turbulent flow are shown in Fig. 2.9.

Thermal Boundary Layer

If the temperature of the wall is different from the temperature of the fluid, heat transfer will take place with energy passing from the high-temperature to the low-temperature system. Near the wall the fluid temperature will be intermediate between the fluid stream temperature T_∞ and the wall temperature T_w. There is a thermal boundary layer similar to the viscous boundary layer. It is enclosed within an imaginary envelope drawn where fluid at the free-stream temperature is closest to the wall, i.e., unaffected by the wall. Just as there is a velocity profile in the momentum boundary layer, so there is a temperature profile in the thermal boundary layer.

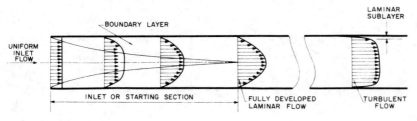

FIGURE 2.9
Velocity profiles in tube flow.

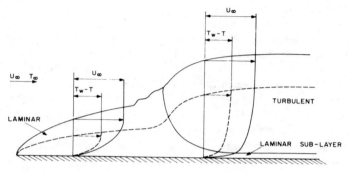

FIGURE 2.10
Boundary-layer temperature and velocity profiles for a flat plate.

The velocity and temperature profiles are juxtaposed in Fig. 2.10 for both laminar and turbulent flow on a flat plate. Similar effects for tube flow are shown in Fig. 2.11.

Relative Thickness of the Thermal and Momentum Boundary Layers

The thicknesses of the momentum and thermal boundary layers are not necessarily the same. For many fluids the viscous boundary layer is thicker than the thermal boundary layer. Liquid metals are notable exceptions to this general rule.

The relative thicknesses of the momentum and thermal boundary layer depend on the properties of the fluid concerned, particularly the *transport* properties. The relationship is generally expressed in a dimensionless group called the Prandtl number after the great German physicist Ludwig Prandtl.

The Prandtl number Pr is defined simply as the ratio

$$\text{Pr} = \frac{\nu}{\alpha}$$

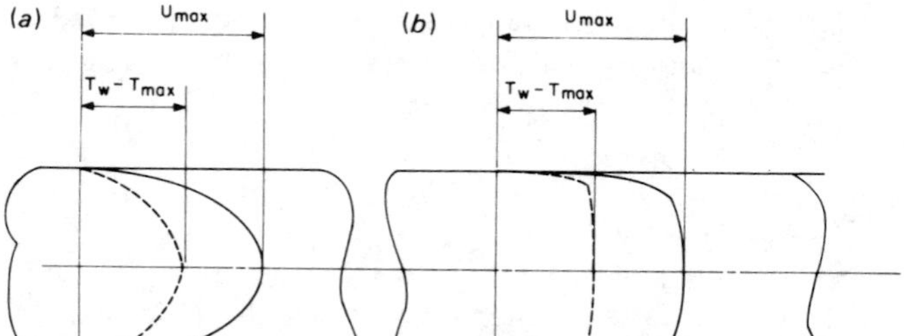

FIGURE 2.11
Temperature and velocity profiles in tube flow: (*a*) laminar, (*b*) turbulent.

where ν = kinematic viscosity of fluid
$\quad\alpha$ = thermal diffusivity of fluid

The kinematic viscosity is indicative of the rate at which momentum diffuses through a fluid because of molecular motion. The thermal diffusivity is indicative of the rate of diffusion of heat in the fluid. The ratio of these quantities is therefore a measure of the relative magnitudes of diffusion of momentum and heat in the fluid. These diffusion rates are precisely the quantities that determine how thick the boundary layers will be for a given flow: Large diffusivities mean that viscous or temperature effects are experienced further out in the flow.

The Prandtl number is therefore the connection between the velocity field and the temperature field. Now

$$\nu = \frac{\mu}{\rho} \quad \text{and} \quad \alpha = \frac{k}{\rho C_p}$$

where μ = dynamic viscosity of fluid
$\quad\rho$ = density of fluid
$\quad k$ = thermal conductivity of fluid
$\quad C_p$ = specific heat at constant pressure of fluid

Then

$$\text{Pr} = \frac{\nu}{\alpha} = \frac{C_p \mu}{k}$$

It is important that consistent units be used in computing the Prandtl number. In SI units, μ would be in kilograms per second per meter, C_p in kilojoules per kilogram per Celsius degree, and k in kilowatts per meter per Celsius degree. In British units μ would be in pounds mass per hour per foot, C_p in Btu per pound per mass per Fahrenheit degree, and k in Btu per hour per foot per Fahrenheit degree.

When consistent units are used, the Prandtl number is a dimensionless number. Typical values, at normal temperature, are given in Table 2.3.

The common values of $0.7 < \text{Pr} < 1$ for many gases is striking. Also many liquids, apparently dissimilar, have a Pr in the range 2 to 4. However, the need to remain cautious about this is demonstrated by observing the high value for engine oil ($\text{Pr} > 10,000$) and the very low value for mercury ($\text{Pr} < 0.03$).

Heat-Transfer Coefficient

Heat will be transferred when a fluid moves with velocity u_∞ relative to a solid wall and there is a significant temperature difference between the fluid and the wall. The rate of heat transferred $\dot{q}$ can be calculated very simply from the equation

TABLE 2.3

Typical Prandtl Numbers

Substance	Prandtl number
Gases (at atmospheric pressure)	
Helium	0.71
Hydrogen	0.69
Oxygen	0.70
Nitrogen	0.71
Carbon dioxide	0.75
Ammonia	0.87
Water vapor	1.06
Liquids (saturated)	
Ammonia	2.02
Carbon dioxide	4.10
Sulfur dioxide	2.00
Dichlorodifluoromethane	3.5
Glycerin	12.5
Ethylene glycol	204
Engine oil	10,400
Mercury	0.025

$$\dot{q} = hA(T_w - T_\infty)$$

where $\dot{q}$ = rate of heat transferred
$\quad h$ = convective heat-transfer coefficient
$\quad A$ = contact area for heat transfer between fluid and wall
$\quad T_w$ = wall temperature
$\quad T_\infty$ = fluid free-stream temperature

The convective heat-transfer coefficient varies widely, over several orders of magnitude, and depends principally on the fluid velocity, the characteristics of the fluid, and, very importantly, on whether the fluid is experiencing a change of phase.

It is both convenient and conventional to express the heat transfer in terms of yet another dimensionless group, the Nusselt number Nu, after Wilhelm Nusselt, another noted German engineer and scientist active in developing the science of heat transfer in the 1930s. The Nusselt number is composed of three elements:

$$\mathrm{Nu}_x = \frac{hx}{k}$$

where Nu_x = Nusselt number

h = convective heat-transfer coefficient

x = some characteristic dimension

k = thermal conductivity of fluid

For a flat plate the characteristic dimension x is the distance from the leading edge for local values of Nu_x or the total length L for the overall coefficient $\overline{\mathrm{Nu}}_L$. For a tube the characteristic dimension is the tube diameter d.

The heat transfer, fluid velocity, and fluid properties may be related through equations of the form $\mathrm{Nu} = C\,\mathrm{Re}^m\,\mathrm{Pr}^n$, where C, m, and n are constants and indices characteristic of the nature and circumstances of the flow.

The heat-transfer coefficient h is contained in the Nusselt number Nu. The fluid velocity u_∞, is contained in the Reynolds number Re. The fluid properties $(\mu C_p/k)$ are contained in the Prandtl number Pr. Parenthetically, it may be noted that for many gases $\mathrm{Pr} \cong 1$ and for many situations $n \cong 0.5$, so that $\mathrm{Pr}^n \cong 1$.

A few idealized situations, including laminar flow of a fluid over a flat plate or in tube flow, can be analyzed theoretically, and the values of the constants C, m, and n predicted. Most practical situations are too complicated for theoretical analysis of the heat transfer. Instead, the results of experimental measurements are processed to obtain generalized heat-transfer correlations. These are then presented in the technical literature in the form

$$\mathrm{Nu} = C\,\mathrm{Re}^m\,\mathrm{Pr}^n$$

with the specific values of C, m, and n given for recommended ranges of applicability of Re and, perhaps, Pr.

Some well-known relations are:

1. For laminar flow over a flat plate:

$$\mathrm{Nu}_x = 0.332\,\mathrm{Pr}^{1/3}\,\mathrm{Re}_x^{1/2} \quad \text{and} \quad \overline{\mathrm{Nu}}_L = 2\,\mathrm{Nu}_{x=L}$$

The overall heat-transfer coefficient $\overline{\mathrm{Nu}}_L$, applicable to the whole plate, has a value corresponding to twice the local heat-transfer coefficient calculated for $x = L$, the length of the plate.

2. For turbulent flow over a flat plate with $5 \times 10^5 < \mathrm{Re}_L < 10^7$:

$$\overline{\mathrm{Nu}}_L = \frac{hL}{k} = \mathrm{Pr}^{1/3}\,(0.037\,\mathrm{Re}_L^{0.8} - 850)$$

Different correlations are recommended for $\mathrm{Re}_L > 10^7$.

3. For laminar flow in a tube:

$$\mathrm{Nu}_d = \frac{hd_0}{k} = 4.36$$

Provided the flow is laminar in the tube, the Nusselt number is constant and independent of the Reynolds or Prandtl number. It is the only common flow situation where this is so.

4. For turbulent flow in a smooth tube:

$$\mathrm{Nu}_d = 0.023 \ \mathrm{Re}_d^{0.8} \ \mathrm{Pr}^n$$

where n is 0.4 for heating and 0.3 for cooling.

These are only a few of the many combinations of Nu, Pr, and Re found in the literature to cover various flow situations and wall surfaces. As noted, only the simplest of flow situations are susceptible to theoretical analysis. The remainder require empirical relationships formulated by processing experimental data.

The advantage of dimensionless-group relationships is their generality. Experiments can be carried out with air or water in, say, steel tubes. Now, provided the correct dimensionless groups are used, the results obtained can be used to predict the heat transfer of glycerin or helium flowing in brass or carbon-fiber tubes. This is very fortunate; imagine the confusion that would prevail if separate account had to be made of each possible liquid and gas in combination with all the different metal tubes, plates, etc.

Relation between Fluid Friction and Heat Transfer

It was noted above how the thermal and viscous boundary layers are related, the connection being the Prandtl number. It is not surprising that there is a strong connection between heat-transfer and fluid-friction effects. The effects of fluid friction are manifest in the viscous drag forces acting on a body moving through a fluid or in the pressure drop of a fluid flowing through a pipe.

Attempts to improve the heat transfer between a wall and fluid at different temperatures will, in general, always be accompanied by an increase in friction effects. It is easy to see why. For a given fluid and wall surface combination at given temperatures there is little that can be done to increase the heat transfer apart from

1. Increasing the velocity to increase the Reynolds number
2. Changing the character of the surface by adding fins or a rough, irregular surface

It is almost self-evident that an increase in the flow velocity or the surface area for contact between the wall and the fluid will increase the heat transfer, but only at the expense of an increase in flow friction.

The connection between heat transfer and fluid friction is in fact so precise that reasonably good estimates of heat transfer can be made by measuring the friction effects. This technique is widely used, for it is easier to measure pressure drop or drag force than to make the careful and tedious measurements necessary for heat-transfer experiments. It is particularly useful for new, untested flow situations in tubes with rough internal surfaces or ducts with irregular flow sections. The relationship is conventionally expressed as $St_x\ Pr^{2/3} = C_{fx}/2$, where St_x is the Stanton number, a new dimensionless group defined as

$$St_x = \frac{Nu_x}{Re_x\ Pr} = \frac{hx/k}{(\rho u_\infty x/\mu)(C_p \mu/k)} = h\rho C_p u_\infty$$

and C_f is the friction coefficient in the defining equation

$$\tau_w = C_f\ \frac{\rho u_\infty^2}{2}$$

where τ_w = viscous shear stress interface between fluid and wall
  ρ = fluid density
  u_∞ = free-stream fluid velocity
The above equations may not be familiar to readers, and we shall not explore them further here. The significant feature to note is that measurements of fluid friction can, in principle, be translated into good estimates of heat transfer.

To illustrate, consider the heat transfer for flow in pipes having rough walls. Established correlations are generally sparse. Therefore, in a new flow situation, it is recommended that a simple measurement be made of the pressure drop along a length of the tube. The flow friction coefficient f may then be determined from the equation

$$\Delta p = f\ \frac{L}{d}\ \rho\ \frac{u_m^2}{2g_c}$$

where Δp = pressure drop
  f = friction coefficient
  L = length of flow section
  d = diameter of flow section
  ρ = fluid density
  u_m = mean fluid velocity
It has been found that for rough tubes

$$St\ Pr^{2/3} = \frac{f}{8}$$

Having determined f from the pressure-drop measurement, one may then proceed to determine the Stanton number and the heat-transfer coefficient h.

To those unfamiliar with the field this will seem a terribly complicated and tedious procedure. Indeed it is. The justification for it all is that experience shows that the system works to allow reasonable predictions of heat transfer.

Emphasis should be placed on the word reasonable. Given the complicated thermal and viscous situations described above in the fluid adjacent to the solid wall, it is very difficult indeed to be absolutely precise in estimating the proper value of the convective heat-transfer coefficient to be used. In a new situation, it is rarely possible to have confidence in one's prediction better than ±30 percent. This will come as a staggering lack of exactitude to accountants and others used to meticulous accounting of the last penny. This lack of precision implies that the design of heat-exchange apparatus is a black art: One piece of apparatus designed conservatively with low values of h will be twice the size of another piece designed optimistically with high values of h.

Experience is the key to success. If one is building a piece similar to another built previously, it is possible to be more confident of its performance than that of a new, untried design.

Some Further Complications

We have noted how the value of the heat-transfer coefficient h varies from one flow situation to another. Even for a given flow situation, the value of h varies widely in different locations. As an example, consider once more the simple case of flow of a fluid over a flat plate. Assume the plate is hotter than the fluid and that heat is being transferred from the plate to the fluid. We know from the previous discussion that thermal and viscous boundary layers will be established as shown in Fig. 2.12, with regions of laminar and turbulent flow. The local values of the heat-transfer coefficient, measured experimentally or determined using the empirically based relationships found in the literature, will vary as shown by the upper diagram. At the leading edge, because the boundary layer is so thin, the heat transfer will be very effective and h will be very high, tending to infinity at the actual leading edge. Further along the plate, the value of h will decline rapidly as the boundary layer thickens until, at a Reynolds number of approximately 5×10^5, the flow will change from laminar to turbulent. The turbulent boundary layer will be substantially thicker than the laminar boundary layer and, owing to the intense fluid agitation, the heat-transfer coefficient h will be very substantially increased above that for the laminar region at the transition point. With further progression along the plate, the heat-transfer coefficient h will decline, but much less acutely than in the laminar region.

The local heat-transfer coefficient h is a measure of the rate of heat transfer from a unit area of the plate per degree of temperature difference between the fluid and the plate. The high value of h at the leading edge indicates a great loss of heat from the plate to the fluid in that region. Because of high rates of heat transfer, the plate may become cooler in those areas where h is high. Whether or not the plate temperature does change depends on how the plate is heated and the thermal conductivity of the wall material. If it is heated by a concentrated heat source such as the gas flame in the center of the plate shown in Fig. 2.12, the temperature distribution in the plate will surely be

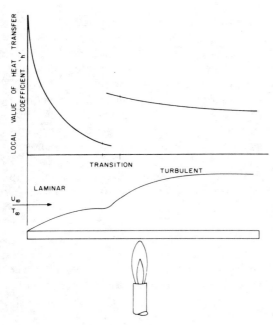

FIGURE 2.12
Variation of convective heat-transfer coefficient on a flat plate.

uneven. If the plate is very thin or is made of material having a low thermal conductivity, the uneven distribution will be emphasized.

As the fluid passes over the plate, heat is transferred to it from the plate. The fluid temperature, therefore, increases from a low value at the leading edge of the plate to a maximum value at the trailing edge of the plate.

In this situation we are confronted with the dilemma of what temperatures to use in the simple equation $\dot{q} = hA(T_w - T_\infty)$. We have seen that h varies widely over the plate and that the wall temperature T_w varies from point to point. Similarly, the fluid temperature is not constant because of heat transfer from the plate, and it increases as the fluid traverses the plate. Furthermore, local temperature of the fluid near the wall varies in the thermal boundary layer over the range T_w to T_∞.

For the purpose of estimating heat transfer, we generally assume some average value of h applied over the whole plate. We assume also either a constant wall temperature or a constant heat flux from the wall and, finally, we assume some kind of average fluid temperature. In many calculations, the procedure is to guess some temperatures that might apply and then use these in calculations to demonstrate that the guessed temperatures are in fact reasonable estimates. If the guessed temperatures are not compatible with the calculated heat balances, refined estimates are made and the computations repeated until an adequate level of agreement is obtained. Such a procedure is called "cut and try" or an "iterative" method.

Properties of the fluid such as density, thermal conductivity, viscosity, and specific heat all vary significantly with the temperature. Tables of such data are presented as functions of temperature in the various design handbooks. These data are used to calculate the Reynolds and Prandtl numbers. A question facing the investigator is what fluid temperature to use since, in the thermal boundary layer, it varies from T_w near the wall to T_∞. It is customary to take the simple arithmetic average temperature $T_f = (T_w + T_\infty)/2$. This is called the *film temperature*. Such a gross simplification has limited validity but, fortunately, the change in properties with temprature is sufficiently small for the approximation to be used with no appreciable error.

Similar difficulties arise with fluid flow in a tube. At the tube inlet the fluid enters with a temperature T_∞. Initially, the flow is not fully developed, and there is a thin laminar boundary layer. Eventually the boundary layer around the internal circumference may become sufficiently thick to equal the radius of the tube. The boundary layers then join to form fully developed flow. The flow changes from laminar to turbulent at a Reynolds number of 2400.

All the while, heat is being transferred from the hot tube wall to the fluid, and the temperature of the fluid across the tube section is not constant. The average temperature across the flow section, called the *bulk* or *mixing-cup* temperature, increases as the fluid moves down the tube and is appreciably greater at exit than at inlet. For calculation purposes, it is sometimes assumed that the tube wall temperature remains constant or that the heat flux from the tube wall to the fluid is constant. In practice the situation is probably different from either of these assumptions.

When fluids flow on the outsides of tubes, there are two important cases to consider. The first is *parallel* flow, where the fluid velocity is parallel to the axis of the tube. This can be treated simply as flow over a flat plate having a depth $P = \pi D$, the circumference of the tube, where D is the external diameter of the tube. Fluid effects arising from the curvature of the tube are generally discounted.

The other important case is fluid flowing normal to the axis of the tube. This is called *crossflow*. The flow situation around a tube in crossflow is indicated in Fig. 2.13. At low fluid velocities, the situation corresponds to that represented in Fig. 2.13*a*, where the streamlines simply part at the tube, are squeezed together in passage over the tube, and then join again downstream of the tube.

At higher fluid velocities, the fluid boundary layer separates from the wall as shown in Fig. 2.13*b*. Immediately downstream of the tube there is a region of intensely agitated fluid, with reverse circulation of the fluid as shown.

Under certain conditions the reverse circulation flow can take the form of vortices that are shed alternately from one side of the tube and the other as

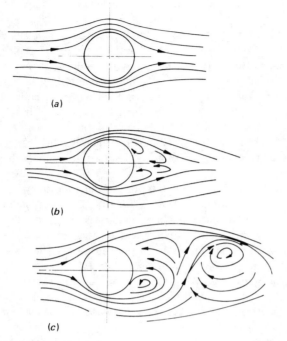

FIGURE 2.13
Crossflow past a tube: (*a*) laminar, (*b*) turbulent, boundary layer separation, (*c*) vortex shedding.

shown in Fig. 2.13*c*. These downstream vortices are known as *Karman vortex streets*. The phenomenon is one of the principal sources of vibration problems in heat exchangers and is considered in more detail in Chap. 6.

The local values of the heat-transfer coefficient *h* vary in an extremely complicated fashion around the circumference of a tube in crossflow. Experimental correlations for average overall heat-transfer coefficients may be found in the literature in the customary form:

$$\text{Nu} = C \, \text{Re}^n \, \text{Pr}^{1/3}$$

C and *n* vary with the Reynolds number, ranging from $C = 0.98$ and $n = 0.33$ at $0.4 < \text{Re} < 4$ to $C = 0.026$ and $n = 0.80$ at $4 \times 10^4 < \text{Re} < 4 \times 10^5$.

Natural and Forced Convection

In our consideration of convective heat transfer between a surface and a fluid we assumed the fluid was moving relative to the surface with a velocity u_∞. This is called *forced* convective heat transfer. The relative velocity may be established by using a fan to blow air over the surface or using a pump to push liquid along a tube, or alternatively by moving the surface through a stagnant fluid, e.g., the motion of air through a car radiator.

There is an important area of convective heat transfer where the fluid is not physically forced to move over the surface by some external means such as a fan or pump. The fluid still moves over the surface, however, due to buoyancy forces established when heat transfer between the surface and the fluid causes changes in the fluid density. As fluid is heated it becomes lighter and rises away from the surface like a hot-air balloon. If the fluid is cooled, it becomes heavier and falls downward. Such action is called *natural* convection. It is the principal method of cooling some electrical equipment and the condenser cooling water of many large-base-load power stations.

The fluid velocities in natural convection are very much lower than in forced convection. As a consequence, the values of the heat-transfer coefficient h for natural convection are, for air, customarily in the range 5 to 25 W/(m$^2 \cdot$°C). This compares with values of 10 to 500 W/(m$^2 \cdot$°C) with forced-convection air coolers.

Natural convection is attractive because no power supply or moving parts are involved, with consequent savings in operating costs. However, because of the low values of h, the surface area for heat transfer must be increased by 2 to 10 times that required for forced-convection systems. This increased area is most likely to increase the capital cost of the equipment.

With natural-convection systems the fluid has no free-stream velocity u_∞ so of course the Reynolds number $\text{Re} = \rho u_\infty x / \mu$ is always zero. Another dimensionless group is used to take the place of the Reynolds number. This new group is called the Grashof number, defined as

$$\text{Gr}_x = \frac{\beta g_c (T_w - T_\infty) x^3}{\nu^2}$$

where Gr_x = Grashof number

$\quad \beta$ = volume coefficient of expansion (a fluid property)

$\quad T_w$ = wall temperature

$\quad T_\infty$ = fluid temperature remote from the wall

$\quad x$ = some characteristic dimension (distance from top or bottom of wall)

$\quad \nu$ = fluid viscosity

Empirical correlations for different natural convective heat-transfer situations may be found in the technical literature in the now familiar form $\text{Nu}_x = C \, \text{Pr}^m \, \text{Gr}^n$, where C is a constant (for a given range of Grashof and Prandtl numbers), and m and n are indexes of the Prandtl and Grashof numbers, respectively. Frequently the values of n and m are sufficiently close for a common value to be used. In such cases it is customary to combine the Grashof and Prandtl numbers into a new group called the Rayleigh number $\text{Ra} = \text{Gr} \, \text{Pr}$, so that $\text{Nu} = C \, \text{Ra}^m$.

Combined Free and Forced Convection

Many practical situations involve convective heat transfer that is partly forced and partly natural. These circumstances arise when the fluid is forced over the surface but at a low enough velocity so that the natural-convection buoyancy forces are significant. There are two situations to consider. When the forced and free convection currents are in the same direction, the situation is called *ordinary combined convection.* When the currents are opposed, it is called *opposed combined convection.* It is customary to evaluate the heat transfer using charts available in the literature. Alternatively, the free-convection and forced-convection heat transfers can be separately estimated and then combined. A useful criterion of the significance of natural-convection effects is that when $Gr/Re^2 > 1.0$ natural-convection effects will dominate the heat-transfer processes. This can arise either as a result of extremely low forced-convection fluid velocities or a very high temperature difference between the fluid and the wall surface.

BOILING AND CONDENSING

Many important convective heat-transfer processes involve a change of phase of the fluid from vapor to liquid (condensation) on a cold plate or from liquid to vapor (boiling) on a hot plate. The processes of heat transfer are considerably complicated by a change in the phase of the fluid. The physical characteristics and properties of the liquid and the vapor are usually so different that they behave as two quite different fluids, but with the sometimes unfortunate characteristic of one changing to the other at particular levels of fluid pressure and temperature.

Relatively large quantities of heat must be transferred to evaporate liquids or condense vapors. With a single-component fluid the process of evaporation or condensation occurs at constant pressure and temperature. This and the associated large heat transfers are the fundamental reasons why water-steam systems are used for base-load electric power generation, and vapor compression cycles for refrigeration systems. Both approximate operation on the thermodynamic Carnot cycle and have a higher thermal efficiency or coefficient of performance than is possible in systems where the fluid does not change phase.

A line diagram for a steam power generation system is shown in Fig. 2.14, and one for a vapor compression refrigeration plant is shown in Fig. 2.15. In both cases the fluid flows in a closed loop consisting of two heat exchangers, the boiler or evaporator and the condenser. Between the heat exchangers the fluid is compressed to the high pressure or expanded to the low pressure of the cycle. In the power generator the fluid is water, and boiling occurs at *high* pressure. In the refrigerator the fluid, one of the fluorocarbon Freons, boils at *low* pressure.

It is clear that in both systems the heat exchangers are vital components. In fact, they are invariably the largest and most expensive elements in the

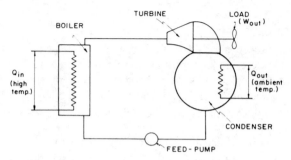

FIGURE 2.14
Line diagram for a steam power generation system.

system. In the steam power generating plant, the specially treated water vapor from the turbine is condensed, in the condenser, by cooling water circulating in the condenser tubes. The cooling water may be taken from a lake, river, or coastal estuary, passed through the condenser, and then returned. However, the quantities of heat involved have become so large as the size of power stations increases that there are few sites where this simple bypass cooling can be tolerated without harmful environmental effects. The alternative is to add an additional heat exchanger, a cooling tower, to transfer heat from the condenser cooling water to the atmosphere.

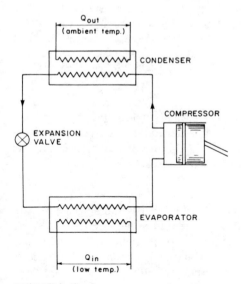

FIGURE 2.15
Line diagram for vapor compression refrigeration system.

Condensing

When fluid in the vapor phase approaches a cold surface, heat is transferred from the vapor to the surface. The temperature of the fluid decreases and the density increases. If the temperature of the wall is less than the *saturation* temperature of the vapor, then the vapor condenses on the surface. This is a matter of common everyday experience; condensation of the water vapor in one's breath on a cold window is one example.

The saturation temperature at which the change from vapor to liquid occurs depends on the pressure of the vapor and increases as the pressure increases. In the presence of other gases or vapor, the temperature at which a particular fraction condenses depends primarily on the *partial pressure* of that component but is affected also by the other constituents.

The process of condensation occurs in one of two ways: as dropwise condensation or as filmwise condensation. In dropwise condenstation, the liquid condensate coagulates into droplets, leaving a substantial fraction of the surface bare of liquid. The droplets drain from the surface under gravity. In film condensation, the liquid condensate "wets" the entire surface and therefore isolates (or insulates) the surface from the incoming vapor. Because part of the cold surface remains exposed, dropwise condensation is much more effective than filmwise condensation (up to 10 times more effective). Great efforts are made to develop or prepare surfaces that promote dropwise condensation. Unfortunately, after a relatively short interval of use, these surfaces invariably become contaminated or degrade to filmwise condensation. Sometimes minor amounts of additive in the condensing fluid are helpful; small amounts of oil added to water often promote dropwise condensation. However, the effect is rarely permanent and dependable so that condensers are invariably designed for filmwise condensation.

Boiling

Boiling heat transfer or evaporation is a complicated subject, far more so than the inverse process, condensing. Yet boiling is a vital process used in many modern plants. Boiling is characterized by very high rates of heat transfer, orders of magnitude greater than generally obtained with convective heat transfer.

Pool Boiling

The simplest boiling process occurs when a heated wall is submerged in a still pool of fluid. The heat-transfer coefficient h or the rate of heat transfer $\dot{q}/A$ may be represented, as shown in Fig. 2.16, as a function of the temperature difference $T_w - T_f$ between the wall and the fluid. When the temperature difference is small, fluid near or in contact with the hot surface is heated by conduction from the wall. Owing to the consequent reduction in the density of the heated fluid, it rises away from the heated wall. A natural-convection motion of the fluid is thus established, and the pool is heated by heat transfer from the submerged heater.

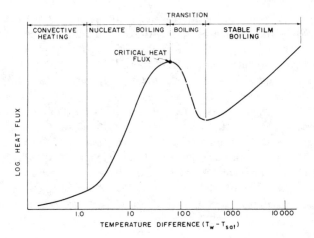

FIGURE 2.16
Pool-boiling heat-transfer coefficient as a function of temperature difference between wall
surface and fluid.

The convection currents may become so vigorous that when the average
temperature of the pool approaches the saturation temperature, the surface of
the pool becomes visibly disturbed and evaporation of the liquid occurs from the
surface.

At higher temperature differences between the hot wall and the pool fluid,
bubbles of vapor may grow on the hot wall. These form at *bubble nucleation
sites* found at small pits or irregularities on the heated wall. There are very large
differences in the density of vapor and liquid, and eventually the buoyancy
forces of the bubble overcome those due to surface tension. The bubble detaches
and ascends toward the liquid surface. If the pool temperature is below the
saturation temperature, the bubble will collapse at a short distance from the
heated wall as the vapor in the bubble cools and condenses.

Eventually, as the pool becomes heated and approaches the saturation
temperature, the bubbles ascend further, eventually breaking through the surface
and releasing their vapor to the space above the pool surface.

As the temperature difference between the fluid and the heated wall is
further increased, the bubble population density and the frequency of bubble
formation increases, with enhanced values of the heat-transfer coefficient and
consequently the heat-transfer rate. This process is called *nucleate boiling*. It is
the process for which equipment for pool boiling is normally designed.

Eventually the temperature differential becomes so great that so many
bubbles are formed and their agitation is so intense that the bubbles coalesce to
form a vapor film separating the liquid from the hot surface. The interposition of
the vapor film between the liquid and the hot wall causes a reduction in the rate
of heat transfer from the hot wall. It therefore becomes even hotter, which
aggravates the situation and results in a further decrease in the transfer of heat.

This unstable *film boiling* process continues until eventually the heater surface is covered entirely by vapor. When the surface becomes so hot that radiation heat transfer begins to play a part, the rate of heat transfer rises once more. However, the surface temperature necessary to increase the heat transfer back to the peak of the nucleate-boiling curve is usually well above the temperature at which common engineering materials can be safely used.

The peak of the nucleate-boiling curve is called the *critical* or *burn-out heat flux*. At temperature differences less than that corresponding to the critical value, the boiling process is a stable nucleate-boiling process. Above that temperature difference, the unstable film-boiling regime is incipient. The development of film boiling is usually violent and accompanied by the sudden periodic collapse and regrowth of the vapor film, so that the hot surface experiences sudden changes in temperature and substantial hydraulic forces that may excite dangerous vibrations.

Boiling with Forced Convection

The process of boiling is further complicated when it occurs during the forced circulatory flow of liquid. The complications arise because of the different flow regimes that may prevail in the presence of a two-phase fluid, liquid and vapor, depending on the circumstances of the flow (principally the geometry), the proportions of liquid and vapor, and the rate of mass flow.

Some of the different flow regimes in two-phase flow are represented in Fig. 2.17. The figure is arranged in four separate columns, two relating to tube flow and two relating to the crossflow of fluid over tubes. In a boiling system, the categories would correspond to

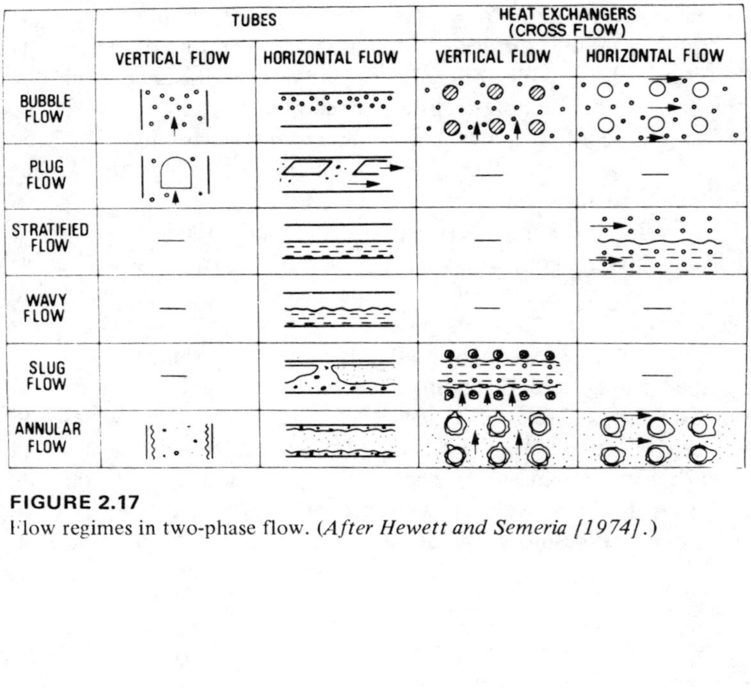

	TUBES		HEAT EXCHANGERS (CROSS FLOW)	
	VERTICAL FLOW	HORIZONTAL FLOW	VERTICAL FLOW	HORIZONTAL FLOW
BUBBLE FLOW				
PLUG FLOW			—	—
STRATIFIED FLOW	—		—	
WAVY FLOW	—		—	—
SLUG FLOW	—			—
ANNULAR FLOW				

FIGURE 2.17
Flow regimes in two-phase flow. (*After Hewett and Semeria [1974].*)

1. Flow in a tube with hot walls
2. Flow in a heat-exchanger shell with heated tubes or fuel-rod elements

Separate columns for vertical flow and for horizontal flow are provided for both tube flow and crossflow.

At least six different flow regimes are recognized by specialists in two-phase flow systems; these are designated bubble, plug, stratified, wavy, slug, and annular flow. Not all regimes exist in all four cases included in Fig. 2.17.

Consider the situation of a fluid flowing in a tube. If the tube walls are hotter than the fluid, heat will be transferred from the tube to the fluid, causing the temperature of the fluid to increase. Initially, the process will occur by simple convection. The temperature of the fluid adjacent to the tube wall will be higher than that of the fluid in the center of the tube.

As the fluid proceeds down the tube, its temperature progressively increases. Eventually the fluid near the wall reaches the saturation temperature, and boiling begins. Bubbles grow from nucleation sites on the wall, but when still small they are torn from the wall by viscous forces of the moving fluid. The bubbles flow with the fluid and, in the superheated boundary, may actually grow in size. When they move into the mainstream of the fluid, still below the saturation temperature, the bubbles collapse and disappear.

Eventually all the fluid attains the saturation temperature, and so the bubbles remain and continue to grow. At this stage the fluid at the wall is sufficiently heated for vigorous nucleate boiling to take place, and large numbers of bubbles are produced. This is called *bubble flow*.

As the bubbles grow in the supersaturated fluid, they coalesce and form large bullet-shaped bubbles moving along the tube, separated at fairly regular intervals by liquid that probably contains a high proportion of bubbles. This is called *plug flow*. Further along the tube the flow changes to *stratified flow* when the vapor and liquid separate, owing to the great difference in density. The vapor at the top of the tube moves along faster than the liquid at the bottom of the tube. In this mode they behave virtually as independent fluids, although of course the evaporation of liquid to the vapor phase continues. Heat transfer to liquid in the bottom part of the tube occurs at a higher rate than heat transfer to the vapor in the upper part.

The high-velocity vapor flowing over the slower liquid causes waves to be established on the surface of the liquid, giving rise to *wavy flow*. Eventually some waves become so great that the crests reach the top of the tube to initiate *slug flow*. In this regime occasional but regular slugs of liquid, filling the whole tube, move down the tube with a high velocity corresponding to the vapor velocity and well above the mean liquid velocity.

Eventually, the proportion of vapor becomes so large that a relatively stable system known as *annular flow* becomes established. The remaining liquid is squeezed in a thin film against the wall, and the vapor flows at high velocity through the central core.

As further evaporation occurs, the liquid film is sufficiently depleted so that it can no longer wet the wall completely, and ultimately a single-phase vapor flows in a tube with dry walls.

The process of boiling liquid in a hot tube is substantially the same as the pool-boiling process, and the rates of heat transfer are not very different so long as liquid wets the wall of the tube. In general, the rate of heat transfer is enhanced by the convective action of the flow over and above the rate in pool boiling. With dry walls, of course, the process of heat transfer changes dramatically to the single-phase convective process, the rates of heat transfer are very much lower, and the tube walls increase very much in temperature unless the sudden drop in the rate of heat transfer can be compensated for by moderating the rate at which heat is supplied to the walls.

The above is a grossly oversimplified and glib explanation of an extraordinarily complicated phenomenon. Those wishing a more comprehensive and substantial treatment are referred to the excellent review by Hewett and Semeria [1974].

HEAT TRANSFER FLUIDS

In most cases where heat exchangers are used to transfer heat from one fluid to another, the fluids can be broadly classified as (1) the process fluid or (2) the coolant.

The fluids may be simple, one component (only one constituent gas or liquid), and a single phase (either a gas or a liquid). They may also be more complex fluids comprising a mixture of different gases or liquids, perhaps in different phases (i.e., solid fragments in a liquid with some gas or vapor with perhaps some melting, precipitation, solidification, boiling, and condensation going on).

In most cases, the heat exchanger user has little or no control over the fluids that will pass through the heat exchanger. The job is simply to manipulate and control the fluids that others have prescribed.

The process fluid is usually specified at the outset, and, since its treatment (cooling or heating) is the *raison d'etre* for the heat exchanger, there is little point in wishing it were possible to change it.

In many cases, the process fluid is to be cooled from a temperature that is greater than the ambient atmospheric value. Normally, the coolants available are water or air. We have seen from Table 2.2 that water cooling is many times more effective than air cooling and is invariably the system of choice if a choice were available. However, as pressures on the environment grow, sources of uncontaminated water that are readily accessible for cooling purposes are becoming harder to find. To an increasing extent, we are driven to use air as the cooling medium for many industrial processes.

It may not always be possible or convenient to use the air directly for cooling. Sometimes it is better to use an intermediate fluid (liquid) loop to extract heat from the process fluid or other heat source and convey it to an air-cooled heat exchanger for dissipation to the air. A very common example of an intermediate fluid loop is the water-cooled internal combustion automobile engine. The hot water from the engine is transferred to the radiator (actually a transverse flow, finned, tubular convective heat exchanger) where the water is cooled and is ready for return to the engine while the heat is "dumped" to the environment by heating air passing through the radiator.

For most cooling situations, water is an excellent heat transfer medium. Relatively speaking, it is cheap, available in nearly all parts of the world, non-toxic, and, in its pure state, relatively noncorrosive or otherwise not harmful to many materials.

Water combines the attractive physical properties of high latent high of vaporization and freezing, high thermal conductivity, high specific heat, and high density with moderate viscosity. This combination allows a large amount of heat to be carried off at high rates of heat transfer with only moderate pumping work required. This reduces the size and cost of equipment necessary to handle a given heat load when compared with other heat transfer media.

Water does dissolve some salts and minerals and absorbs them in solution to become actively corrosive to other materials (i.e., sea water is highly corrosive to mild steel). Water is also susceptible to contamination by microbiological organisms requiring the use of chlorine or fungicide.

However, one of the biggest difficulties with water as a heat transfer fluid is the limited range of temperature over which it can be used. Water freezes at 0 °C and, unlike most other fluids, expands as it solidifies. This unusual characteristic is the reason ice floats in water rather than sinking to the bottom. For instance, as lakes freeze, there is an insulating layer of ice formed on the surface, thus preventing the whole lake from freezing solid to the obvious detriment of lake and pond life.

However, in engineering systems, the slight expansion of water on freezing can, in a closed container, result in the generation of enormous forces sufficient to crack the pipe or severely strain the metal wall of the containment.

The use of anti-freeze (ethylene glycol) mixed with the water in the radiator is mandatory for use in areas subject to freezing temperatures (Canada and the northern United States).

In other applications involving refrigeration and the transfer of heat at temperatures below the freezing point of water, it is customary to use low freezing point liquids known collectively as *brines*. The term was originally applied to aqueous solutions containing inorganic salts such as calcium chloride and sodium chloride. It is now generally applied also to nonsaline aqueous solutions containing ethylene or propylene glycols and the alcohols.

Generally, the freezing point of the brine is depressed as the percentage of the additive increases. Thus, a solution containing 12% percent sodium chloride freezes at $-8\,°C$. This decreases to $-22\,°C$ in a 21 percent solution.

At the other end of the scale, water boils at 100 °C (at atmospheric temperature). While this modest boiling temperature is a boon to coffee and tea drinkers, it does impose a severe constraint on the range of temperature over which water is useful as a coolant.

It is possible to elevate the boiling temperature by increasing the pressure of the water. However, the required pressure increases quite rapidly. For example, to increase the boiling temperature to 200 °C, it is necessary to increase the pressure to over 4 atm. (4 bar, 60 lb per sq in). One cannot go far in this direction before it is necessary to become involved in pressure vessels with all the costs and cares involved in their manufacture, acquisition, and use.

Steam engines, of course, do involve very high pressures and temperatures, but then special equipment, thick walls, and great care are used to ensure safe operation. Steam is used to about 250 °C for process heating and to over 700 °C (and 70 bar) for power production in steam turbines.

Gases are sometimes used as heat transfer fluids. Air is the most common heat transfer fluid after water and, by far, the most commonly used gaseous heat transfer fluid. Compared with water, air is a light, low density fluid with a low specific heat capacity and low thermal conductivity. It is therefore much less effective than water in removing heat from a body and, per unit of heat transferred, requires much more energy, in fan work, to move the air than the equivalent pump work for water.

Gases have the advantage over water and other liquids in that they do not change phase (except at very low temperatures) and so may be used over the widest range of temperatures. Water vapor, present at low pressure in all natural air streams, sometimes causes problems when the air is cooled (air conditioning) and becomes saturated with moisture causing some of the vapor to be precipitated as liquid water.

For special situations, gaseous heat transfer fluids may be specified in closed systems carrying heat from a source for dumping to an air- or water-cooled heat exchanger at near ambient temperatures. In such closed systems, any suitable gas may be used. The preferred gases are those with a low molecular weight, such as hydrogen and helium. Hydrogen is two to three times better than air; helium slightly less so. However, hydrogen is very difficult to contain, is very highly inflammable, and is damaging in various wide ranges to metals and other materials. For these reasons, helium is often selected for use in closed systems with a gaseous heat transfer medium.

SUMMARY

In this chapter we have reviewed, in somewhat cavalier fashion, the basic tenets of thermodynamics, the first and second laws, the different modes of heat transfer, and the use of electrical analogies for heat-transfer calculations. We reviewed also the elementary concepts of fluid flow and the inevitable degradation of energy as frictional heat. We then went on to consider the complexities of convective heat transfer accompanied by a phase change, either from vapor to liquid (condensation) or from liquid to vapor (evaporation) and some of the desirable properties of heat transfer fluids.

Our coverage is sufficient to indicate to nonspecialists some of the reasons why heat transfer is so imprecise and to illustrate some of the problems confronting those who design heat exchangers. Readers wishing to explore fundamental heat transfer beyond the level covered here are referred to the many excellent texts available, including that by Kreith and Black [1980].

REFERENCES

Frank, O.: Simplified Design Procedures for Tubular Heat Exchangers, pp. 1–25 in "Practical Aspects of Heat Transfer," American Institute of Chemical Engineers, New York, 1978.

Hewitt, G. F., and R. Semeria: Aspects of Two-Phase Gas-Liquid Flow, chap. 12, in N. Afgan and E. U. Schlunder (eds.), "Heat Exchangers: Design and Theory Sourcebook," Hemisphere, Washington, D.C., 1974.

Kays, W., and A. L. London: "Compact Heat Exchangers," McGraw Hill, New York, 1964.

Kreith, F., and W. Z. Black: "Basic Heat Transfer," Harper & Row, New York, 1980.

3

Tubular heat exchangers

INTRODUCTION

Tubular exchangers are used in great numbers, far more than any other type of exchanger. They are made in a wide variety of sizes and styles, ranging from the tiny units used in miniature cryocoolers to giant installations containing thousands of tubes and used as condensers in base-load power stations.

Tubular exchangers are so widely used because the technology is well established for making precision metal tubes capable of containing high pressures in a variety of materials. There is virtually no limit to the range of pressures and temperatures that can be accommodated. Furthermore, the holes for tubes in the tube sheets, the end plates to which the tubes are attached, are easily made by drilling. This ease of manufacturing established the use of tubular exchangers historically and led to the growth of major manufacturers and the development of industry standards for manufacture. The widespread use of tubular exchangers, their relatively low cost, their familiarity to users, and the many years of satisfactory service routinely obtained lead to their being the first choice in new installations, a situation containing all the elements of a self-perpetuating dynasty: "We are because we are."

Tubular exchangers can be broadly classified by the character of the fluids they are designed to handle:

1. *Liquid/liquid.* This is by far the most common application of tubular exchangers. Typically, cooling water on one side is used to cool a hot effluent stream. Both fluids are pumped through the exchanger so that the principal mode of heat transfer is forced convective heat transfer. The relatively high density of the liquids results in very high rates of heat transfer, so there is little incentive in conventional situations to use fins or other devices to enhance the heat transfer.

2. *Liquid/gas.* This is also very common, typically for the air cooling of a hot liquid effluent. The liquid is pumped through the tubes with very high rates of convective heat transfer. The air, in crossflow over the tubes, may be in the forced or free convective mode. Heat-transfer coefficients on the air side are low compared with those on the liquid side. Fins are usually added on the outsides (air side) of the tubes to compensate.

3. *Gas/gas.* This type of exchanger is found in the exhaust-gas/air preheating recuperators of gas-turbine systems, steel furnaces, and cryogenic gas-liquefaction systems. In many cases one gas is compressed so the density is high, while the other is at low pressure with a low density. Normally the high-density fluid flows inside the tubes. Internal and external fins are sometimes provided to enhance the heat transfer.

4. *Condensers.* These may be liquid (water) or gas (air) cooled. Alternatively, the heat from condensing streams may be used to heat another processing fluid. Normally the condensing fluid is routed in either of two ways:

   a. Outside the tubes with liquid cooling, i.e., a water-cooled stream condenser. Fins are not necessary either inside or outside the tubes.

   b. Inside the tubes with gas cooling, i.e., an air-cooled Freon condenser heat pump or refrigerating system. Fins are normally provided on the outsides of the tubes. The tubes may be flattened to obtain a better ratio of surface area to flow cross section.

5. *Evaporators and boilers.* This important group of tubular heat exchangers can be subdivided into two classes:

   a. *Fired systems,* in which the products of the combustion of fossil fuels (coal, oil, natural gas) at very high temperatures but ambient pressure (and hence low density) generate steam under pressure. Normally the water and/or steam is contained inside the tubes (water tube boiler), and pin fins, studs, or some other form of extended surface is applied to the tubes to enhance the heat transfer. In smaller systems the combustion products may pass through the tubes (fire tube boilers). Locomotive boilers and small low-pressure boilers (Lancashire boilers) are of this type.

   b. *Unfired systems.* These embrace a great variety of different boiler and evaporator heat exchangers, extending over a broad temperature range from high-temperature nuclear steam generators to very low temperature cryogenic gasifiers for liquid-natural-gas evaporation. Many chemical and food-processing applications involve the use of steam or electric heating to evaporate solvents, concentrate solutions, distill liquors, or dehydrate compounds.

TYPES OF TUBULAR EXCHANGERS

Single-Pipe Exchanger

The simplest conceivable tubular exchanger is the single bare tube through which fluid passes. Heat is exchanged between it and the environment surrounding the tube. Many building heating systems are of this type. The perimeter heating pipe is installed horizontally at ground level, and hot water or steam is circulated through the tube. Air flows (in crossflow) over the pipe by natural convection and heats the room. The heat transfer from the bare pipe may be enhanced by incorporating fins on the external surface. It is commonplace nowadays to find finned perimeter heating tubes comprised of rectangular aluminum fins mounted on a small-bore thin-wall copper pipe. The fin is squeezed onto the pipe during manufacture but is not otherwise secured to the pipe. Routine handling and differential thermal expansion with repeated heating and cooling of the pipe in service loosens the fins, with the result that they contribute essentially nothing to the heat-transfer process. If fins are necessary, it is vital that there be good thermal continuity between the metal and the pipe. Use of the same metal for the pipe wall and the fins and joining with soldered, brazed, or conducting epoxy joint cement resolves many problems of this nature.

Any pipe, whether intended as a heat exchanger or not, will lose heat by natural convection to the environment and by conduction along the pipe supports. If the temperature difference between the pipe and its surroundings is high or, with external pipes, the wind is blowing strongly, the heat loss (or gain) can become appreciable. To reduce heat transfer, many pipes are routinely insulated and supported on low-conductivity supports.

Clustered-Pipe Exchanger

A development of the single-pipe heat exchanger is shown in Fig. 3.1. Two or more tubes are joined by a thermally conducting medium so that heat is transferred between fluids flowing in the tubes. Sometimes a cluster of tubes is arranged around a central core tube. High-density fluid passes through the core tube. The return stream of the low-density fluid passes through the multiple tubes arranged around the core tube. This construction is favored in some small cryogenic counterflow heat exchangers. The conducting medium connecting the tubes is deposited by electroplating and is periodically interrupted to reduce axial conduction along the tubes.

Double-Pipe Exchanger

A conceptually similar system is the double-pipe heat exchanger shown in Fig. 3.2, which consists simply of a central tube contained within a larger tube. It was at one time widely used for small-capacity systems. It was relatively cheap and easy to manufacture and offered great flexibility, since it could be made in any length and the tubes could be selected from a wide range of available sizes and materials. It was customary to operate with the high-pressure, high-temperature, high-density, or corrosive fluid in the small inner tube, with the less-demanding fluid (water) in the outer tube. This minimized the requirement for thick-walled expensive alloy materials to contain the former fluid.

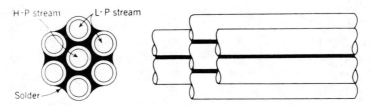

FIGURE 3.1
Cluster-type tubular exchanger.

Stacks of double-pipe heat exchangers were commonly found, as shown in Fig. 3.3, arranged in either parallel or series arrangements. It is very unfortunate that the standard fittings used for double-pipe heat exchangers are no longer as widely available as previously. As a consequence, the practice of fabricating double-pipe heat exchangers has been largely superseded by the use of custom or standard shell-and-tube exchangers.

Tube-and-Shell Exchangers

Sometimes, to increase capacity or reduce the required length, more than one internal tube is incorporated within the outer tube enclosure, as shown in Fig. 3.4. The most common form of multitubular heat exchanger is the tube-and-shell exchanger shown in Fig. 3.5. This is very widely used for liquid/liquid heat transfer, for liquid/condensing-vapor applications, and for liquid/evaporating-vapor systems.

The great number of tube-and-shell heat exchangers in use have led to the emergence of industry standards that have been widely adopted. The best-known standards for tubular heat exchangers are the *TEMA Standards* of the Tubular Exchanger Manufacturers Association, which include a basic nomenclature and classification scheme for tube-and-shell heat exchangers.

The tube-and-shell exchanger is comprised of four principal subassemblies:

1. Front end
2. Rear end
3. Tube bundle
4. Shell

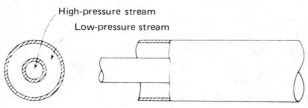

FIGURE 3.2
Double-pipe heat exchanger.

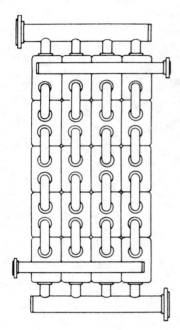

FIGURE 3.3
Multiple stack of double-pipe heat exchangers.

These subassemblies, of a variety of types, can be arranged in different combinations. Figure 3.6 shows a number of these subassemblies, which are designated by alphabetic characters. The figure shows

1. Five types of front-end stationary-head subassemblies, types A, B, C, N, and D.
2. Seven shell types designated E, F, G, H, J, K, and X.
3. Eight types of rear-end head subassemblies, designated L, M, N, P, S, T, U, and W.

The subassemblies can be combined in a variety of ways. The resulting exchanger is designated by a three-letter combination characterizing, in order, the stationary front head, the shell, and the rear head, e.g., type AES.

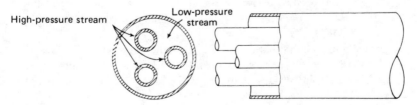

FIGURE 3.4
Multiple-tube heat exchanger.

FIGURE 3.5
Tube-and-shell heat exchanger.

The complete designation also includes the size of the exchanger. This is a combination of two numbers indicating first the shell-and-tube bundle diameter and second the tube length. The nominal diameter is the inside diameter of the shell in inches, rounded to the nearest integer. The nominal length is the tube length in inches. For U tubes the length is taken as the straight length from the end of the tube to the tangent of the bend.

For the shell type represented in Fig. 3.6 as type K, kettle reboiler, the size designation is a three-part number describing

1. The port or tube-bundle diameter, rounded, in inches
2. The shell diameter, rounded, in inches
3. The tube length

Some typical size and type designations are given below. The standard terminology presented in Table 3.1 identifies the various numbered parts and components in Figs. 3.7 through 3.12.

Figure 3.7 shows a combination of subassemblies described as a split-ring floating-head exchanger with removable channel and cover, two tube passes, single-pass shell, of $23\frac{1}{4}$-in inside diameter with tubes 16 ft long. The full designation is size 23-192 type AES.

Figure 3.8 shows a fixed-tube-sheet, single-pass shell with expansion joint,

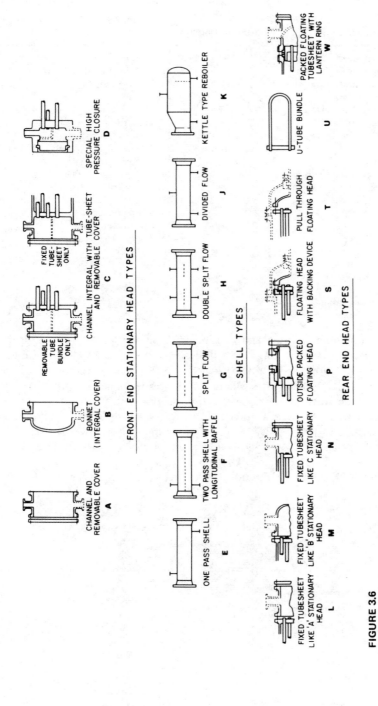

FRONT END STATIONARY HEAD TYPES

A — CHANNEL AND REMOVABLE COVER

B — BONNET (INTEGRAL COVER)

C — CHANNEL INTEGRAL WITH TUBE-SHEET AND REMOVABLE COVER (REMOVABLE TUBE BUNDLE ONLY / FIXED TUBE-SHEET ONLY)

D — SPECIAL HIGH PRESSURE CLOSURE

SHELL TYPES

E — ONE PASS SHELL

F — TWO PASS SHELL WITH LONGITUDINAL BAFFLE

G — SPLIT FLOW

H — DOUBLE SPLIT FLOW

J — DIVIDED FLOW

K — KETTLE TYPE REBOILER

REAR END HEAD TYPES

L — FIXED TUBESHEET LIKE 'A' STATIONARY HEAD

M — FIXED TUBESHEET LIKE 'B' STATIONARY HEAD

N — FIXED TUBESHEET LIKE C STATIONARY HEAD

P — OUTSIDE PACKED FLOATING HEAD

S — FLOATING HEAD WITH BACKING DEVICE

T — PULL THROUGH FLOATING HEAD

U — U-TUBE BUNDLE

W — PACKED FLOATING TUBESHEET WITH LANTERN RING

FIGURE 3.6

Subassemblies of tube-and-shell heat exchangers.

TABLE 3.1

Standard Nomenclature of Heat Exchanger Components

1. Stationary head—channel	21. Floating head cover—external
2. Stationary head—bonnet	22. Floating tubesheet skirt
3. Stationary head flange—channel or bonnet	23. Packing box
4. Channel cover	24. Packing
5. Stationary head nozzle	25. Packing gland
6. Stationary tubesheet	26. Lantern ring
7. Tubes	27. Tierods and spacers
8. Shell	28. Transverse baffles or support plates
9. Shell cover	29. Impingement plate
10. Shell flange—stationary head end	30. Longitudinal baffle
11. Shell flange—rear head end	31. Pass partition
12. Shell nozzle	32. Vent connection
13. Shell cover flange	33. Drain connection
14. Expansion joint	34. Instrument connection
15. Floating tubesheet	35. Support saddle
16. Floating head cover	36. Lifting lug
17. Floating head flange	37. Support bracket
18. Floating head backing device	38. Weir
19. Split shear ring	39. Liquid level connection
20. Slip-on backing flange	

stationary head-bonnet front end, stationary head-nozzle rear, of $23\frac{1}{4}$-in inside diameter, with tubes 16 ft long. The full designation is size 23-192 type BEM.

Figure 3.9 shows an outside packed floating-head exchanger with removable channel and cover, single-pass shell, of $23\frac{1}{4}$-in inside diameter, with tubes 16 ft long. The full designation is size 23-192 type AEP.

Figure 3.10 shows a U-tube-bundle heat exchanger having the channel integral with the tube sheet and a removable cover, two-pass shell with longitudinal baffle, of $23\frac{1}{4}$-in inside diameter, with tubes 16 ft long. The full designation is size 23-192 type CFU.

Figure 3.11 shows a pull-through, floating head, kettle-type reboiler, having

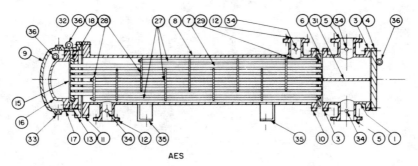

AES

FIGURE 3.7
Tubular exchanger.

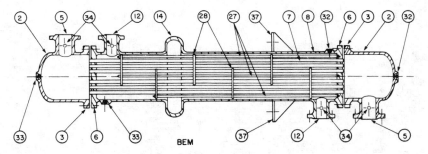

FIGURE 3.8
Tubular exchanger.

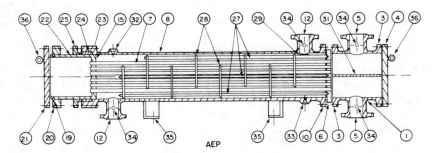

FIGURE 3.9
Tubular exchanger.

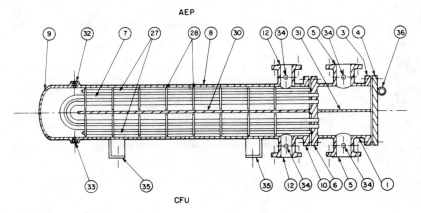

FIGURE 3.10
Tubular exchanger.

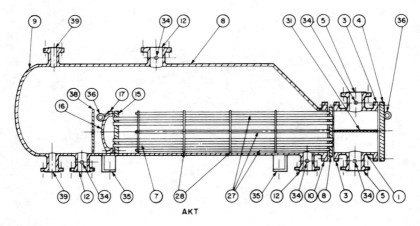

FIGURE 3.11
Tubular exchanger.

removable channel and cover, of $23\frac{1}{4}$-in port diameter and 37-in inside diameter, with tubes 16 ft long. The full designation is size 23/37-192 type AKT.

Figure 3.12 shows an externally sealed floating-tube-sheet heat exchanger with removable channel and cover, divided-flow shell, of $23\frac{1}{4}$-in inside diameter, with tubes 16 ft long. The full designation is size 23-192 type AJW.

The possible variations are virtually endless. For special designs not embraced by the variations shown in Fig. 3.6, descriptions to suit the manufacturers are recommended. For example, a single-tube-pass, fixed-tube-sheet exchanger with conical ends might be described as type BEM with conical heads.

Differences among the various types of heat exchangers are principally design or construction features incorporated to facilitate cleaning of the inside or outside tube surfaces and to accommodate the differential thermal expansion arising from large temperature differences in the fluids. The type AES heat exchanger shown in Fig. 3.7 is of relatively complicated construction and can be

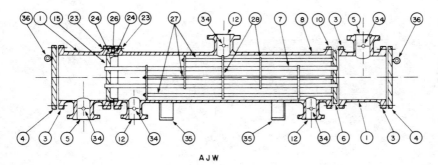

FIGURE 3.12
Tubular exchanger.

4. Cleanliness of fluids. Processes with above-average requirements for cleanliness may require the use of special materials. It is less costly to provide special tubes than to provide both special tubes and special shell.
5. Hazard or expense of fluids. Leakage of fluid is less likely from tube side than from shell side in most types of exchangers.
6. Pressure drop. The pressure drop inside the tubes can be more accurately forecast than that in the shell. Wide deviations from theoretical values arise as a result of fabrication tolerances in the shell leakage clearances. Where the fluid pressure drop is critical and must be accurately predicted, fluids should pass through the tubes.
7. Fluid viscosity. To maximize heat transfer, both fluids should be in turbulent flow. Fluids of high viscosity may be laminar in the tubes but turbulent in the shell (depending on the clearance space between the tubes). If the flow is laminar in both shell and tubes, the viscous fluid should be in the tubes, for which more reliable heat-transfer and flow-distribution predictions can be made.
8. Mass flow. In general it is better to put the fluid having the lower mass flow on the shell side. Turbulent flow is obtained at lower Reynolds numbers on the shell side. Furthermore the complexity of multipass construction may be avoided, with a consequent improvement in exchanger effectiveness.
9. Cleaning. The shell and tube outer surfaces are more difficult to clean than the tube internal surface. Therefore, the cleaner fluid should pass through the shell.

Flow Arrangements

A variety of shell-side and tube-side flow arrangements are used in tubular heat exchangers. The arrangement chosen for any particular application generally represents a compromise among a number of factors, some conflicting. These include the maximum temperature difference between the fluids (LMTD; see Chap. 5), manufacturing techiques and cost, pressure drop, friction effects, and fouling of heat-exchanger surfaces by the precipitation of sediments or polymerization.

Axial counterflow is the ideal configuration from a thermal point of view (see Chap. 5). Fluid on the shell side flows longitudinally over the tubes in the direction opposite the flow of fluid inside the tubes. In this arrangement the temperature differential between the fluids is a minimum, and very high thermal effectiveness can be achieved.

The single-tube-pass, single-shell-pass arrangement shown in Fig. 3.14 approximates the ideal. However, it is difficult to establish reasonably uniform flow distribution on the shell side. To improve the situation, baffles are provided to guide and control the shell-side flow, causing it to pass through the tube bundle in an advantageous manner. The baffles also serve to space and support the tubes. Support is required about every 30 to 40 tube diameters along the length of the tube, and uniform spacing of the tube assures that flow and

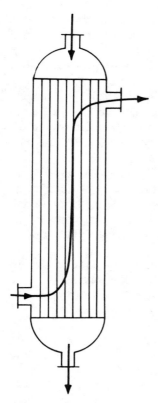

FIGURE 3.14
Single-tube-pass, single-shell-pass counterflow unbaffled heat exchanger.

heat-transfer distributions in the tube bundle will be reasonably balanced. The baffles should have a hole pattern identical to that of the end tube sheets.

Baffles. Various arrangements of baffles may be used, four of which are illustrated in Fig. 3.15: segmental, strip, disk and doughnut, and orifice.

The segmental baffle is simply a disk with a segment removed. It is characterized by the *percent baffle cut,* the ratio of the depth of cut to the baffle diameter, expressed as a percentage. Segmental baffles are arranged so the cut edges of alternate baffles are at 180° to each other. This causes the shell-side fluid to move perpendicularly back and forth in progressing along the shell. The flow over the tubes is therefore partly crossflow and partly axial flow.

Segmental baffles are the most common type of baffle. They are associated with good heat transfer rates for the pressure drop and power consumed. Every baffle supports the tubes, and their longitudinal spacing and the depth of baffle cut can be varied over a wide range to suit the duty of the exchanger.

The double-segmental or strip baffle is similar to the segmental baffle, with the advantage that large flows can be handled. Alternate baffles consist of the large center section, a disk with upper and lower segments removed, succeeded downstream by upper and lower sections, a disk with the center section removed.

The disk-and-doughnut baffle consists of a central disk somewhat larger in diameter than half the shell diameter, followed by a disk of the full shell diameter but with a central hole somewhat smaller than half the shell diameter. This type of baffle overcomes the partial fluid bypassing of the tube bundle that occurs with segmental baffles. However, it has the disadvantages that heat-exchanger tie rods must all be located in the tube bundle, and that support for the inner tubes is provided only indirectly by the tubes in overlap between the disk and the doughnut.

Orifice baffles consist of full disks with oversize holes through which the tubes pass. The shell fluid flows through the orifice between the hole and the tube, and the resulting turbulence of the flow enhances the heat transfer. They

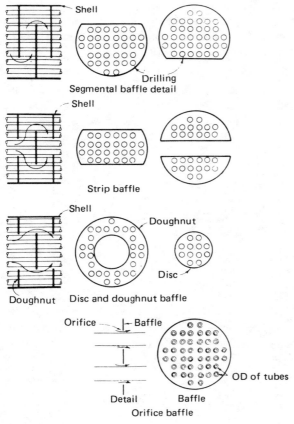

FIGURE 3.15
Baffle arrangements in heat exchangers. (*After Mueller [1973].*)

are rarely used, for they have been found less efficient than other types. They are difficult to clean and provide no support for the tubes, which are then subject to vibration.

Multipass flow arrangements. In addition to the simple single-tube-pass, single-shell-pass arrangement with baffle supports, alternative multipass flow arrangements are possible. One of the most popular is the "hairpin" or U-tube bundle (Fig. 3.16), giving two tube passes in shells with one pass or with two passes. The U-tube bundle with two shell passes is the short, fat equivalent of a single-tube-pass, single-shell-pass counterflow exchanger with the advantage that all the fluid connections are located at one end of the exchanger. It provides a very compact and convenient installation; however it is necessary to provide sufficient clearance to remove the tube bundle for cleaning.

Larger exchangers with up to four tube passes are made with single- or two-pass shells, as shown in Fig. 3.17. It is not unusual to find more than four tube passes. Six or even eight tube passes are sometimes used, but the large temperature variations make it difficult to achieve the equivalent thermal effectiveness of two coupled exchangers in either the series, the parallel, or the series-tube, parallel-shell connection shown in Fig. 3.18.

Crossflow arrangements. In some exchangers (particularly large condensers), the shell-side fluid passes in crossflow over the tube matrix as shown in Fig. 3.19. Exchangers having a single entry and single exit for the shell-side fluid include longitudinal baffles at inlet and outlet to distribute the flow as well as the

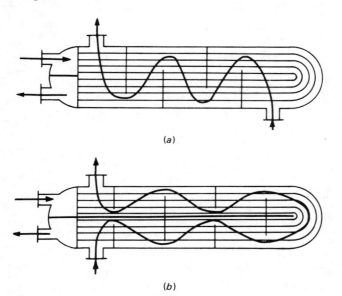

(a)

(b)

FIGURE 3.16
U-tube, two-tube-pass heat exchanger with (a) one and (b) two shell passes.

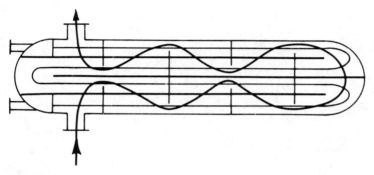

FIGURE 3.17
Four-tube-pass, two-shell-pass heat exchanger.

normal tube support baffles. Alternatively, to gain better flow distribution, multiple entry and exit connections may be provided on the shell side from headers.

Tube Patterns

The tubes may be arranged in a variety of ways, of which four have come to be recognized as standard. These are shown in Fig. 3.20. The corresponding hole patterns in the tube header sheets are shown in Fig. 3.21. The equilateral triangular arrangement gives the strongest header sheet for a given shell-side flow passage area and is, therefore, favored for applications with a large pressure differential between the fluids. The triangular arrangement is not as easy to clean as the square pattern and is not recommended where mechanical shell-side cleaning is anticipated.

Extended-Surface Tubes

Most shell-and-tube exchangers are fabricated with bare tubes, but where the circumstances warrant extended-surface tubes are used. The extended fin surface may be added to a bare tube by welding or some form of mechanical rolling. Alternatively, the tube may be extruded to produce an integral continuous fin.

Finned tubes for shell-and-tube exchangers sometimes have the longitudinal fins shown in Fig. 3.22 rather than the radial fins favored for crossflow air heaters and coolers. A typical two-tube-pass, single-shell-pass exchanger with longitudinal fins is shown in Fig. 3.23.

Finned-tube exchangers are particularly suitable in applications where there are large differences in the heat-transfer coefficients of the shell-side and tube-side fluids. One example is steam condensing inside the tubes with very high rates of heat transfer to heat a thick, viscous oil moving rather slowly on the shell side. In these cases the additional area for heat transfer on the shell side compensates for the difference in heat-transfer coefficients.

Finned heaters are designed for straight-through flow without baffles and

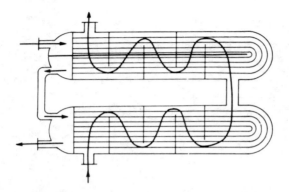

SERIES COUPLING, TUBESIDE AND SHELLSIDE

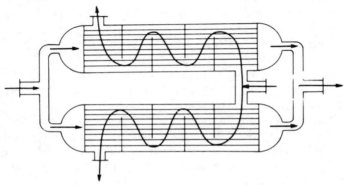

PARALLEL COUPLING, TUBESIDE AND SHELLSIDE

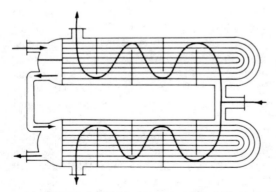

SERIES TUBESIDE, PARALLEL SHELLSIDE COUPLING

FIGURE 3.18
Series and parallel coupling of heat exchangers.

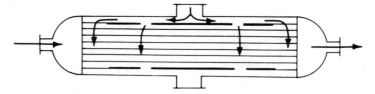

SINGLE ENTRY CROSSFLOW WITH INTERNAL
FLOW DISTRIBUTION BAFFLES

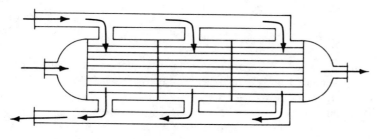

MANIFOLD INLET AND EXIT WITH
INTERNAL TRANSVERSE BAFFLES

FIGURE 3.19
Shell-side crossflow arrangements.

support plates, and manufacturers claim this minimizes the accumulation of sediment and reduces corrosion. Furthermore, tube vibration problems are reduced because of the added stiffness of the tube, and the possibility of the tubes being damaged by contacting each other and the baffles or supports is reduced.

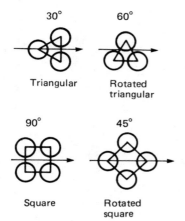

30°   60°

Triangular   Rotated
              triangular

90°   45°

Square   Rotated
            square

FIGURE 3.20
Standard tube patterns. Note: flow arrows are normal to the baffle cut edge.

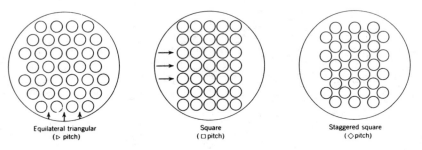

Equilateral triangular   Square   Staggered square
(▷ pitch)   (□ pitch)   (◇ pitch)

FIGURE 3.21
Tube header-sheet hole patterns. (*After Fraas and Ozisik [1965].*)

The principal advantage of finned tubes is that the effects of fouling are very much reduced, compared with bare tubes. Fouling is the generic description for the additional thermal resistance that arises in service when deposits accumulate on the heat-transfer surfaces. Experience has shown that finned-tube exchangers can operate for very long periods (10 years or more) without maintenance or cleaning, with only a moderate decrease in capacity (25 percent), in situations where heavy fouling and severe deterioration of bare-tube exchangers is experienced.

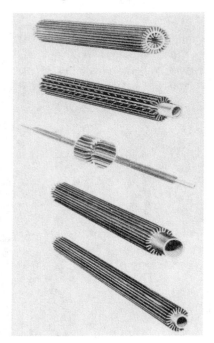

FIGURE 3.22
Longitudinal finned tube. (*Courtesy Brown Fintube Co., Tulsa, OK.*)

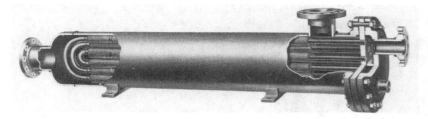

FIGURE 3.23
Two-tube-pass, single-shell-pass line heater with longitudinal finned tubes. (*Courtesy Brown Fintube Co., Tulsa, OK.*)

Internally finned tubes are occasionally used, with the fins attached to the tubes by welding or brazing. Alternatively, the tube may be extruded with internal and external fins integral with the tube, like the double-fluted-tube configuration shown in Fig. 3.24. Another possibility for enhancing the heat transfer of plain tubes is the spirally grooved tube shown in Fig. 3.25. Finally, in place of internal finning, it is sometimes advantageous to use the helical ribbon insert shown in Fig. 3.26. This is most useful in evaporators, where the ribbon causes the liquid droplets to impinge by centrifugal action on the wall of the tube, thereby promoting the boiling process.

FIGURE 3.24
Double-fluted-tube configuration.

FIGURE 3.25
Spirally grooved tube.

Tube-and-Fin Exchangers

Many applications for heat exchangers involve the transfer of heat between a liquid or vapor and a gas. In most cases the gas is atmospheric air used to cool the liquid or vapor stream, or itself being heated or cooled for air conditioning. The liquid varies, depending on the application, and may be water, refrigerant, or an acid, oil, or alcohol. Condensing vapors include steam, refrigerants, and hydrocarbons.

Tube-and-fin heat exchangers are well suited to this duty. They consist simply of a bank of finned tubes, perhaps a single row but customarily several rows deep with manifold or U-tube connections at the ends of the tube sections. A typical example is shown in Fig. 3.27. The liquids flow through the tubes, and air flows in crossflow over the tubes.

The heat-transfer coefficient for liquid flow in the tubes is very much higher (orders of magnitude greater) than for airflow over the tubes. To compensate for this difference, the area for heat transfer on the air side is increased by providing fins. Great ingenuity has been exercised to maximize heat transfer on the air side in compact arrangements by varying the geometry and nature of the fins, corrugating, crimping, etc. Such efforts rarely meet with success, for the increased heat transfer is invariably accompanied by increased aerodynamic friction. The resulting pressure drop may require the investment of more work than is gained by the improved heat transfer.

In cases where space and volume are not at a premium, fin design is largely guided by economic considerations. Many air-cooled systems utilize aluminum fins wrapped or extruded over the base tube.

Applications

A familiar application of tube-and-fin heat exchangers is the automotive radiator. Water cools the engine, and the heat transferred must be dissipated to the atmosphere. The water is circulated by a pump driven by the engine, and air is drawn over the fins of the heat exchanger by a fan that is also driven from the engine.

Another familiar application is the air-cooled tube-and-fin heat exchanger found at the rear of domestic refrigerators. In this heat exchanger, refrigerant vapor at high pressure cools and condenses en route to expansion in the

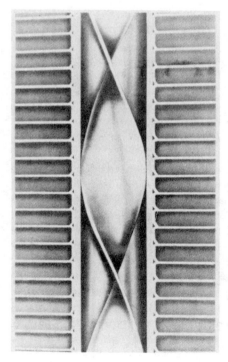

FIGURE 3.26
Helical ribbon insert.

FIGURE 3.27
Tube-and-fin heat exchanger for air-conditioning system. (*Courtesy Custom Coils Ltd., Southhampton, U.K.*)

Joule-Thomson (JT) valve and subsequent cold chamber, where the low-pressure fluid evaporates in a plate-and-fin exchanger, producing refrigeration. The low-pressure vapor is then compressed before it arrives back at the high-pressure condenser.

The same system is used in larger-scale refrigerating and air-conditioning plants and in heat pumps, where refrigerant is either condensed or evaporated and air is either heated or cooled, depending on the use. Similarly, hot-water and steam-heating or chilled-water and brine-cooling systems are used to heat or cool air for comfort systems in buildings, planes, ships, vehicles, etc. Air coolers are used for lubricant oil-cooling systems on aircraft, ships, vehicles, and stationary gas-turbine or diesel-engine power generators.

In much larger sizes, air coolers are used in petroleum refineries, in the process industries, and in all manner of commercial enterprises for dissipating waste heat to the atmosphere. Condensers for steam turbines are sometimes directly air-cooled. In large base-load power stations, an intermediate condenser cooling-water circuit is used, wherein the heat transferred from the steam in the condenser is ultimately transferred to the atmosphere in a "wet" or "dry" cooling tower. A wet cooling tower involves partial evaporation of the condenser cooling water to saturate the air passing through the cooling tower. This is becoming increasingly unacceptable because of the loss of water, which may be in short supply, and the impact of large volumes of saturated air released into the atmosphere. Air coolers are the preferred alternatives.

Natural and Forced Convection

The tubes in the tube bank may be arranged horizontally so that natural convective air motion is established. However, the low heat-transfer coefficients that are characteristic of natural-convection systems demand an exchanger of such great size that it is rarely feasible to depend on this method. The alternative is a forced-convection system with the airflow driven by fans over the tubes. This greatly improves the heat transfer. It allows the use of a smaller exchanger with consequent savings in capital cost and relaxes the need for tubes to be mounted horizontally. The provision of a fan and a motor to drive it partially offsets the saving in capital cost, for work is required to drive the fan, and regular maintenance is necessary. Operating costs are therefore increased.

The drive fan can be mounted upstream or downstream of the tube bank, as shown in Fig. 3.28. When the fan is mounted upstream the arrangement is called a *forced-draft* system, and when mounted downstream it is an *induced-draft* system. There is little difference in the two arrangements, but a slight improvement in fan efficiency results from mounting it on the cold-air side; i.e., for forced draft if the coil is heating the air, and for induced draft if it is cooling.

Fans are almost invariably of the axial-flow type, with large-volume, low-pressure characteristics, selected to impart an air velocity of 1 to 5 m/s through the tube bundle. They usually have four to six blades and a diameter somewhat smaller than the width or length of the tube bank.

Attempts to economize on capital cost by operating a smaller fan at high

FIGURE 3.28
Induced-flow tube-and-fin heat exchanger for air conditioning. (*Courtesy Custom Coils Ltd., Southampton, U.K.*)

speeds can result in fan noise becoming a problem and will almost certainly result in reduced fan efficiency. Fan noise level is a very strong function (sixth power) of the fan-blade tip velocity and is a prime factor in establishing the fan speed. Fans are commonly driven by electric motors, but steam-turbine drives are sometimes used for the largest sizes.

Construction

Air-cooled heat exchangers can be generally categorized into two types:

1. Single tube pass
2. Multiple tube pass

In the single-tube-pass type shown in Fig. 3.29, the finned tubes are located at the tube ends in header tanks or manifolds. The tube ends are secured by the tube rolling in heavy-wall manifolds, or by brazing and soldering in header tanks for less arduous duties.

Multiple-tube-pass heat exchangers (Fig. 3.30) have the tubes interconnected at the ends by a U connection so the liquid in the tube makes two or more traverses of the tube bank. This type of exchanger is customarily called a coil, even though the tubes are straight. It is widely favored for the condensing and evaporating systems in steam heating or refrigerating and air conditioning.

FIGURE 3.29
Single-tube-pass air cooler. (*Courtesy Brown Fintube Co., Tulsa, OK.*)

FIGURE 3.30
Multiple-tube-pass air-cooled heat exchanger. (*Courtesy Brown Fintube Co., Tulsa, OK.*)

The tubes used in tube-and-fin heat exchangers range from the small-diameter (5-mm) tubes found in compact aircraft oil coolers or air-conditioning systems to the 5-cm-diameter tubes found in viscous oil and flue gas service. Standard tube lengths range up to 10 m, with tube bundles up to 4 m wide and as many as 30 tubes deep. Fin heights vary up to 2 cm with up to four fins per centimeter of tube length. The ratio of extended surface area to bare tube area ranges from 6 to 20.

Aluminum is an economical fin material for temperatures up to 400°C on the liquid side. For low-temperature work up to 120°C, the aluminum fin may be mechanically wrapped under tension to the roughened surface of the base tube. This provides an economical tube suitable for use in low-vibration situations. For higher temperatures, the fin may be wrapped under tension in a groove machined in the base-tube wall, or the fins may be produced by a high-pressure cold extrusion process. In corrosive conditions, a base tube other than aluminum may be used.

For the smaller air-conditioning and steam-heating coils and automotive radiators, thin-wall copper or brass tubing with soldered or brazed joints is widely used. The fins are a flat helically coiled copper or aluminum strip or wire coil wound into the tube and secured by dip-brazing or soldering.

Tank Heaters and Coolers

Tubular exchangers are often used for heating and cooling the contents of tanks. The contents may require heating to reduce the viscosity and facilitate pumping, to prevent solidification of waxes or other solids, or to melt the contents prior to processing. The contents may require cooling to prevent the buildup of excessive vapor pressure, to slow down a reaction, or to induce the selective deposition of solids. In batch processing operations, the contents are frequently heated or cooled for evaporation, concentration, etc.

There are, in general, two types of tank heating systems: gross contents heating and suction heating. For gross heating a tubular coil is either located within the tank or secured externally to the wall of a thermally insulated tank.

Internal tank coils are made in a wide variety of configurations—flat spirals, helical coils, hairpin tube bundles, and ring headers as illustrated in Fig. 3.31. To take advantage of natural convection, heating coils tend to be located at the bottom of the tank, and cooling coils at the top. Heating may be accomplished by electrical resistance heating, by steam condensing, or by the circulation of hot water or some other medium. Similarly, the cooling fluid may be evaporating refrigerant fluid or chilled water or brine.

Frequently the heat-transfer coefficient inside the tube is so much greater than that outside the tube that the use of external fins is justified. For suction heating a hairpin tube bundle is located at the exit from the tank, as shown in Fig. 3.32. The heating process is preferential and confined almost entirely to fluid leaving the tank. The tube bundle may actually be inserted in the tank or incorporated in the suction line leaving the tank. In this latter form it is called a *line heater* and, of course, may be used in any application where the heating of

FIGURE 3.31
Types of tubular tank heaters. (*Courtesy Brown Fintube Co., Tulsa, OK.*)

FIGURE 3.32
Suction heater. (*Courtesy Brown Fintube Co., Tulsa, OK.*)

pipeline contents is required. Sometimes large tanks are equipped with one or more external circulation loops, with line heaters incorporated in the circulating loop.

Fired Heaters

A wide variety of types and sizes of tubular exchangers are available in which heating is produced by the hot products of the combustion of gas, oil, or coal. These systems are used for raising steam, for superheating steam, and for heating oils and vapors generally. They are called *fired heaters,* and we shall consider them in more detail in Chap. 7.

Evaporators

The process of evaporation is widely used in the chemical and refining industries. There are many different types of evaporators, and most incorporate some form of tubular heat exchangers. An excellent and comprehensive survey of evaporators has been given by Rubin et al. [1977].

   In the great majority of cases the heating medium is steam located in the tubes, with evaporating fluid passing over the tube bundle in the shell. Sometimes a series of evaporator systems, called a *multiple-effect* evaporator, is used to gain economy in steam consumption via a sequential decrease in the steam temperature and pressure from one evaporator to the next.

Scraped-Surface Exchanger

Scraped-surface heat exchangers are used for processes likely to result in the substantial deposition of solids on the exchanger surface. Such processes include the precipitation of crystals and of waxes in chilling applications. The use of a scraped-surface exchanger prevents a significant buildup of the solids.

Scraped-surface heat exchangers are most commonly of double-pipe construction, with the process fluid in the inner pipe and the cooling or heating fluid (steam or water) in the annulus. A rotating element is contained within the tube and is equipped with spring-loaded blades scraping the inside tube walls, as shown in Fig. 3.33. The rotating element is driven at slow speed by an electric motor outside the exchanger. Normally, several exchangers are gang connected, either in series or in parallel, with a chain drive coupling all the exchangers to the drive motor.

NONMETALLIC EXCHANGERS

Most heat exchangers of all types are fabricated from metals. Metals are available in a wide range of types and forms with various strength characteristics. They are easily worked and readily joined by welding, brazing, or soldering. Economic considerations dictate the use of the metal of lowest cost that is compatible with the fluids to be used in the heat exchanger. The most significant characteristic of a metal from the aspect of fluid compatibility is its resistance to corrosion.

Corrosive liquids and vapors such as hydrochloric, sulfuric and phosphoric acids require the use of such exotic metals as titanium, tantalum, zirconium, and Hastelloy and other alloys. Certain fluids are so corrosive that no metal alloys can be found with suitable corrosion resistance, within the bounds of economic feasibility.

Plastics

The alternative is to use nonmetallic systems. Plastic materials, particularly polytetrafluoroethylene or PTFE (Teflon or Fluon), are particularly favored because of their exceptional resistance to chemical attack and degradation. Very effective heat exchangers are constructed from clusters of long, thin, flexible tubes of PTFE, secured at the tube ends to headers of the same material. The long, flexible tube clusters are coiled or draped in open cooling tanks containing cooling water. Another technique is to plate with PTFE the interior surfaces of the tubes and bonnet covers and any other internal surfaces exposed to the corrosive fluids.

Graphite

Graphite is also very widely used in heat exchangers for highly corrosive fluids. Graphite has a very high resistance to corrosion, is a good conductor of heat, and

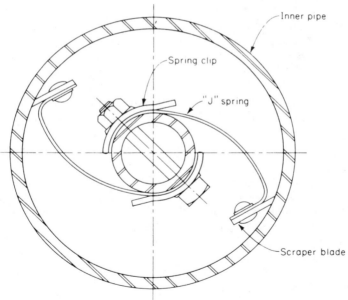

FIGURE 3.33
Scraped-surface heat exchanger.

has a low coefficient of thermal expansion (so it is resistant to thermal shock). It is readily machinable but cannot be welded and has low tensile strength. Equipment design is therefore predicated toward loading the graphite in compression. The temperature and pressure limits for graphite exchangers are generally $350°F$ and 250 lb/in^2. The use of graphite as an engineering material has a long history but has been greatly expanded in the process industries over the last quarter century, following the development of techniques to render the graphite nonporous by impregnating it under pressure with suitable resins. The expansion of its use coincided with the general expansion of the chemical industry, particularly with regard to chlorinated and fluorinated organic compounds.

The difficulty of low tensile strength can be overcome by ingenious design whereby the graphite components are subject to compressive rather than tensile stress. One approach is to use metal tie rods or metal face plates in tension to maintain graphite components in compression loading, even when the internal pressure is applied. Another approach is to enclose the graphite within a steel shell. The close-fitting graphite components are installed at elevated temperatures so that on cooling the metal shell contracts and preloads the graphite components in compression.

For higher-pressure applications the cross-bore technique is often used, wherein rows of parallel holes are drilled in solid blocks of impervious graphite. Two sets of mutually perpendicular holes are drilled, perhaps of different diameters, one set for the corrosive tubeside fluid and the other for the noncorrosive shellside fluid. The incorporation of this cross-bore technique in tube-and-shell and cubic exchangers is illustrated in Figs. 3.34 and 3.35.

Although it is resistant to corrosion by most fluids, graphite is not recommended for use with bromine, nitric and chromic acids, fluorine, iodine, red phosphorous, and chromium-plating solutions. Before use with other highly reactive chemicals, the graphite manufacturer should be consulted at an early stage to ensure compatibility of the resin impregnants used to seal the graphite.

Glass

Glass is another material that is used for heat exchangers in special situations. It is particularly attractive for use with acids and other corrosive fluids. Glass is highly resistant to corrosion by virtually all fluids. Glass is also favored in the pharmaceutical and food-processing industries, where purity of the product is essential. If the processed materials contact only glass and Teflon, the possibility of contamination is greatly reduced.

The processing of chlorinated and brominated organic fluids is often faciliated in impervious borosilicate glass vessels, to overcome the problem of permeation existing with many other materials.

Heat exchangers containing glass are commonly found in the two types of construction illustrated in Figs. 3.36 and 3.37. Figure 3.36 shows a conventional

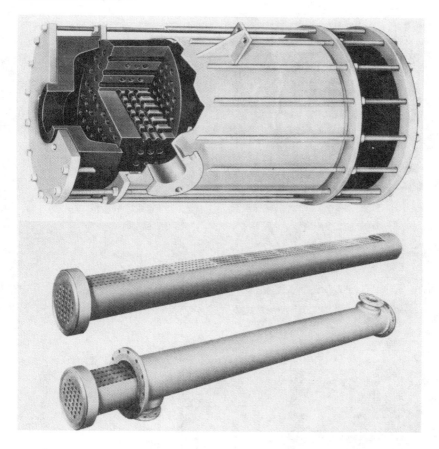

FIGURE 3.34
Cross-bore graphite tube-and-shell heat exchanger. (*Courtesy Carborundum Co., Solon, OH.*)

tube-and-shell heat exchanger in which the tubes are made of glass with PTFE tube sheets and a PTFE threaded fitting securing each tube in the tube sheet. The shell may also be of glass if both the shell-side and tube-side fluids are corrosive. In such cases the permissible operating pressure and temperature are limited to about 20 lb/in^2 and 350°F. (Higher pressures, up to 75 lb/in^2, and temperatures of 359°F are possible with carbon-steel shells. For corrosive fluids the shells may be lined with rubber or Teflon.)

Figure 3.37 shows coil condenser and coil boiler elements in glass. In one case the coolant water circulates in the coil. In the other case steam or some other heating fluid circulates in the coil. The elements can be stacked one upon the other in a vertical column as part of a circulating loop.

Glass has such a low coefficient of thermal conductivity [1 W/(m·°C)] compared with metals [40 to 60 W/(m·°C)] as to appear quite unsuitable for

FIGURE 3.35
Cross-bore graphite cubic heat exchanger. (*Courtesy Carborundum Co., Solon, OH.*)

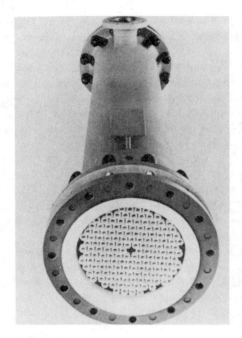

FIGURE 3.36
Tube-and-shell heat exchanger with glass tubes and Teflon tube sheet and tube seals. (*Courtesy Corning Glass Co.*)

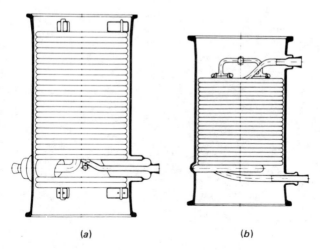

(a)   (b)

FIGURE 3.37
Glass coil. (*a*) condenser, (*b*) boiler. (*Courtesy Corning Glass Co.*)

heat-exchanger use. Fortunately, the disadvantage is not so great as it first appears, for the smooth surface of glass resists the growth of fouling deposits. Many times the fouling resistance is the predominant thermal resistance (see Chap. 2), and the conductivity of the tube material is relatively insignificant. Glass is unlikely to become a common heat-transfer material but advantages of extreme corrosion resistance, lack of contamination, and impermeability combined with ready availability in a variety of forms at relatively low cost create opportunities for its continued use in a significant sector of the process, pharmaceutical, and food and drink industries.

NONCIRCULAR TUBES

Normally the tubes used in tubular heat exchangers are circular in cross section. In the few exceptions to this general rule, the tubes are flattened to an elliptic shape or a rectangular form with semicircular ends.

One example of the use of noncircular tubes is the lamella heat exchanger made by the Swedish Alfa-Laval Co. and illustrated in Figs. 3.38 and 3.39. The lamella heat exchanger is considered a variant of the conventional tube-and-shell exchanger and is designed specifically for the longitudinal countercurrent "tube-side" flow of both the tube and shell fluids.

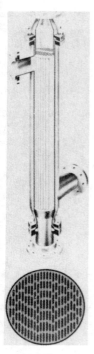

FIGURE 3.38
Cross section of lamella bundles in shell. (*Courtesy Alfa-Laval, AG.*)

FIGURE 3.39
Section of lamella heat exchanger. (*Courtesy Alfa-Laval, AG.*)

The flattened tubes, called lamellas, are made of two strips of plates having a profile impressed in them and spot- or seam-welded together. The various sections are assembled into a matrix with spaces between the lamellas. Both ends of each lamella are joined by peripheral welds to the channel cover or tube sheet. One channel cover is fixed and the other floats, to allow for differential thermal expansion.

The manufacturer claims that because of high turbulence levels, uniformly distributed flow, and smooth internal and external surfaces the lamellas do not foul easily. They are used in general heating and cooling duties and for heat recovery in pulp and paper mills, in alcohol plants, and in petrochemical and other chemical industries.

Noncircular tubes are also found occasionally in liquid/gas heat exchangers. The flattened tubes provide a more favorable ratio of perimeter to cross-sectional area to enhance the heat transfer on the external finned air side and to diminish heat transfer and increase pressure drop on the liquid side, where there is plenty of margin for adjustment.

HEAT PIPES

Heat pipes are a form of tubular heat exchanger quite different from any of the types discussed above. The elements of a heat-pipe heat exchanger are shown in Fig. 3.40. The exchanger consists simply of a metal tube closed at both ends and about one-quarter full of liquid. One end of the tube is heated and the other end is cooled. Liquid in the tube evaporates at the heated end and condenses at the cooled end. There is therefore a flow of vapor from the heated end to the cooled end, and a flow of condensate liquid back to the cooled end. A liquid has a higher density than its vapor; to ensure return of the liquid to the heated end, it is therefore advantageous to operate the heat pipe vertically or inclined with the

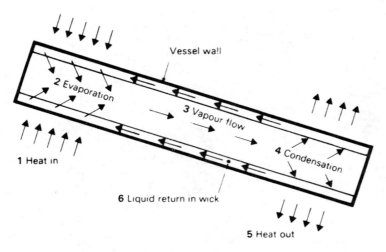

FIGURE 3.40
Heat-pipe heat exchanger.

cooled end uppermost, utilizing gravity to sustain the liquid and vapor flows. In space applications with a near-zero gravity environment, it is necessary to include a wick of wire screen or a similar material around the internal periphery of the tube. Surface-tension forces and capillary action between the large surface area of the wick and the liquid act to "pump" the liquid back to the evaporative regions.

By using a heat pipe, very large amounts of energy can be transferred over relatively long distances (several meters) with only a degree or two of temperature difference between the heated and cooled ends. The effect is the same as would be attained with a solid metal bar having superlative thermal conductivity, orders of magnitude greater than the best thermal conductors known—pure silver, copper, and aluminum.

Heat pipes are used in an increasing variety of heat-exchanger applications, over a very wide spectrum of temperatures ranging from cryogenic to very high. Of course the liquid used depends very much on the operating temperature. In use, the pressure in the heat pipe is the saturation pressure of the fluid, corresponding to the temperature of the fluid that is evaporating and condensing. At high temperatures liquid metals are used to keep the operating pressures at moderate values. At ambient temperatures water is often used. For lower temperatures the halogen refrigerants are preferred.

Figure 3.41 shows a multiple-element heat-pipe heat exchanger suitable for waste-heat recovery in an inlet-air preheater or similar application. There is a very large literature concerning heat pipes. Reay [1979] has provided a good summary of information about design, applications, and availability and has included an extensive bibliography.

So far, heat pipes have tended to be relatively small, a few kilowatts at most and usually far less. However, large heat pipes are in prospect. They are being considered for use with large coal-fired Stirling engines (Walker [1983]) to transport heat from a fluidized bed coal combustor to the heater head of a Stirling engine.

It is anticipated that large coal-fired Stirling engines could find application in the power ranges 0.5–5 MW for stationary power, marine propulsion, cogeneration plants, and the population engines for railway locomotives and the large off-highway vehicles used for mining, forestry, construction, and agriculture. Given this power range and the realistic engine thermal efficiencies of 20–30 percent, it would be necessary to convey large quantities of heat in the range 1.5–25 MW.

In most studies of the possibilities, sodium heat pipes are favored. They operate satisfactorily at 700°–800°C. This is a temperature range where the fluidized bed coal burner operate effectively with virtually no nitrogen oxide emission and thus with little impact on the environment. Limestone is supplied to the bed along with coal and sulphur in the coal combustion with the limestone so that no sulphurous oxides (acid rain) are produced.

Also, 700°–800°C is also a good temperature at which to operate stirling engines. The engines would be highly pressurized, but, at this modest temperature, nothing more exotic than stainless steel is required for the hot parts.

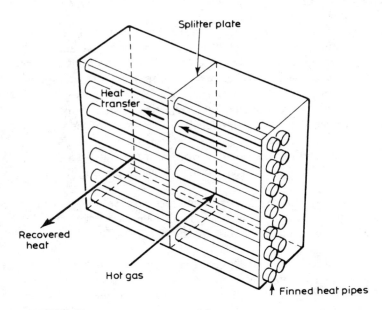

FIGURE 3.41
Arrangement of multiple-element heat-pipe heat exchanger for waste-heat recovery. (*After Reay [1979].*)

Many of the studies for these large heat pipes favor the use of two stage systems. In this scheme, many identical first-stage heat pipes are located in the hostile environment of the fluidized bed. The sodium vapor produced in all these separate first-stage heat pipes rises above the bed to the condensing section. The condensing sections are grouped so that many feed their heat to a common second-stage system, which is actually connected to one of the heater heads of the large multi-cylinder Stirling engine.

The two-stage heat pipe provides the possibility for several of the first-stage systems to be punctured in the violent and convective motion of the fluidized bed and lose their sodium charge and become ineffective.

However, a failure of some of the first-stage units will not cause the entire system to lose the whole sodium charge as would be the case with a single stage heat pipe developing a leak in one of the many "fingers" located in the bed.

Gas-balasted heat pipes, also called variable conductance heat pipes, are sometimes used. These are heat pipes with an additional section added above the customary section, itself located, of course, above the evaporator section. The gas-ballasted heat pipe contains two fluids, the normal heat transfer medium, say, sodium, and an inert gas of low density and low thermal conductivity, say, argon.

In normal use, the sodium performs its thermal function, evaporating in the evaporator section and rising up in the heat pipe as vapor to the condensing section where it gives up its heat, condenses, and returns by gravity or capillary pumping in the wick to the evaporator section.

The argon plays virtually no part in this process but migrates to the additional upper section above the condenser. There is a well-defined and stable sodium-argon interface at some level near the top end of the condensing section. The argon does not change phase, and, although there may be come convective motion established, the amount of heat transferred by the argon is negligible.

Any change in the pattern of heat addition or rejection to or from the evaporator and condenser section immediately produces an adjustment of the interface position.

Thus, for example, consider the situation when heat transfer from the condensing section is cut off for some reason, perhaps because the Stirling engine receiving the heat siezes up.

Heat is still being supplied to the evaporator section even though none is being removed from the condenser. The result is an increase in the operating temperature of the heat pipe fluid and a corresponding increase in the saturation pressure.

This rise in pressure compresses the argon, causing the interface between the sodium and the argon to rise into the region previously occupied only by the argon. That space may be designed to be air cooled rather than thermally insulated like the rest of the heat pipe between the evaporating and condensing sections. As the sodium enters the air-cooled section the sodium immediately condenses, losing heat and limiting the increase in pressure and temperature of the heat pipe.

The gas-ballasted heat pipe is thus an effective fail-safe and fool-proof automatic thermal "dump" that comes into play when the normal thermal load is removed.

Another practical advantage of the argon-filled region is that it provides refuge for the gaseous corrosion products generated within the heat during normal operation with, in many cases, deleterious effects on thermal performance if there is no place for it to migrate to.

CARTRIDGE HEAT EXCHANGER

Thomas Margittai [1980] invented the cartridge heater shown in Fig. 3.42. It consists of a series of double-walled cylindrical cartridges telescoped together. The heating or cooling medium circulates within the annuli of the double-walled cartridges, flowing from the largest to the smallest. The cartridge flow conduits are coupled by U connectors as shown. The process fluid passes in the annuli formed between the outside of one cartridge and the inside of the adjacent larger cartridge. Spiral guide vanes are welded on the external diameters to cause the process to follow an extended circuitous flow path. Sealing of the flanges is accomplished with 0 rings.

This design results in remarkably compact tubular heat exchangers with some appreciable advantages for certain applications. There is a single channel for the process fluid, instead of the multiple channels of the conventional tube-and-shell system. Problems of maldistribution of flow therefore cannot arise, which is of great advantage for heat- or shear-sensitive fluids, emulsions, and slurries containing fibers and sediments. Any tendency toward fouling or sedimentation results in partial closure of the single flow channel, with the result that the flow velocity increases at that point and acts to clear the obstruction.

Cartridge heaters are available in a variety of metals and other materials, including glass and PTFE Teflon plastic linings. There appears to be no fundamental limit to the pressure and temperature ranges that may be accommodated. Further information on this interesting new development may be obtained from Dr. T. Margittai, 778 Cornwall Road, State College, PA 16801.

SUMMARY

In this chapter we have surveyed the broad field of tubular heat exchangers. Many applications involve liquid-to-liquid heat transfer for cooling, heating, condensing, and evaporating. By far the great majority of tubular heaters for these applications are the double-pipe and tube-and-shell types, which are found in a wide range of sizes, shapes, and flow configurations.

Another important class of tubular heat exchangers is concerned with liquid-to-gas heat transfer. These units are usually equipped with extended surfaces on the gas side to compensate for the relatively low heat-transfer coefficients on that side.

Fired heaters, systems for condensing and evaporating, and scraped-surface exchangers for precipitating solutions were briefly mentioned.

Some fluids are so corrosive as to require nonmetallic exchangers; the use of PTFE plastic materials, graphite, and glass as exchanger materials was briefly considered.

Finally some special applications involving the use of noncircular tubes were reviewed.

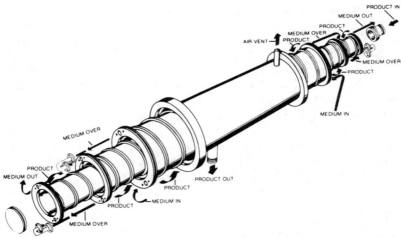

FIGURE 3.42
Margittai cartridge heater.

REFERENCES

Fraas, A. P., and M. N. Ozisik: "Heat Exchanger Design," Wiley, New York, 1965.

Margittai, T.: Private communication, 1980.

Mueller, C. A.: Heat Exchangers, chap. 18 in W. M. Rohsenow and J. P. Hartnett (eds.), "Handbook of Heat Transfer," McGraw-Hill, New York, 1973.

Reay, D. A.: "Heat Recovery Systems," E. Spon., London, Halstead/Wiley, New York, 1979.

Rubin, F. L., C. E. Ernst, A. D. Holt, M. S. Standiford, and D. Stuhlborg: Heat Transfer Equipment, chap. 11 in R. H. Perry and C. H. Chilton (eds.), "Chemical Engineers' Handbook," 5th ed., McGraw-Hill, New York, 1973.

Walker, G., O. R. Fauvel, and E. J. Johnson: Coal Fired Stirling Engines for Railway Locomotive and Stationary Power Applications, *Proc. I. Mech. Eng.*, vol. 197, no. 46, 1983.

4

Plate heat exchangers

INTRODUCTION

Plate heat exchangers are less widely used than tubular exchangers but have characteristics that make them the system of choice in some applications. Plate heat exchangers can be classified in four principal groups:

1. Plate-and-frame heat exchangers, used as an alternative to tube-and-shell exchangers for low- and medium-pressure liquid/liquid heat-transfer applications.
2. Spiral heat exchangers, used as an alternative to tube-and-shell exchangers for low- and medium-pressure applications of all kinds. They are of particular value where low maintenance is required, and with fluids tending to sludge or containing solids in suspension.
3. Plate-coil heat exchangers, made from previously embossed plates to form a conduit or coil for liquids coupled by fins.
4. Plate-fin heat exchangers, in which a stack of die-formed corrugated plates is welded or brazed to provide compact heat-exchanger surfaces for various gas/gas exchange processes. They are commonly used in cryogenic applications and in exhaust or inlet air preheating waste-heat recovery systems for buildings, gas turbines, etc.

PLATE-AND-FRAME HEAT EXCHANGERS

Construction

Plate-and-frame heat exchangers are made of an assembly of the pressed metal plates shown in Fig. 4.1. These are aligned on a frame and secured between covers by tie bars, as shown in Fig. 4.2. Gaskets are set in the grooves around

FIGURE 4.1
Die-formed pressed metal plates. (*Courtesy Alfa-Laval.*)

the periphery of each plate to contain the fluids and direct the flow distribution. Ports for the inlet and outlet of both hot and cold fluids are punched or blanked into the corners of each plate. When the plates are assembled, the ports are aligned to form the distribution headers for flow through the plate pack.

Fluid flow occurs in the fine passages between alternate plates. Except at the ends, therefore, there are hot fluids flowing on both sides of the two plates forming the conduit in which the cold fluids flow, and vice versa. The distribution of fluids to alternate plates is effected by selectively removing part of the circumferential gasket separating the plates, as shown at the left-hand ports in the specimen plates of Fig. 4.1. Intermixing of the two fluids in the event of gasket failure is prevented by leakage grooves in the gasket near the headers.

Plates having different fluid flow patterns impressed in them can be incorporated, depending on the fluids to be used and the application. Substantial flexibility in the heat-transfer and pressure-drop relationships can thereby be gained. Plates of two or three surface geometries, but of the same overall size, can be mixed in the same exchanger. This feature, called plate mixing, is illustrated in Fig. 4.3, which shows two adjacent plates; the front plate has the customary chevron patterns, whereas the rear plate has a horizontal pattern impressed in the surface. Various chevron patterns of different angles and pitch are used.

Plates may be made of any metal that can be dieformed and may be of virtually any shape. Plate heat exchangers are made in a very wide range of sizes. Two large units are shown in Fig. 4.4. The conventional flow arrangement is for the inlet and outlet connections to be mounted on the fixed end, the head, with the connections arranged to cause the fluids to move in countercurrent flow. However, many other arrangements are possible.

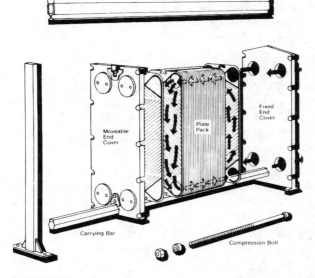

FIGURE 4.2
Construction elements of plate-and-frame heat exchanger. (*Courtesy APV.*)

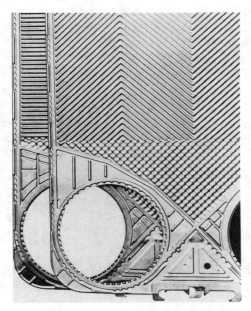

FIGURE 4.3
Plate mixing; adjacent plates have different compressed patterns. (*Courtesy APV.*)

Advantages

The principal advantage of plate-and-frame heat exchangers is the accessiblity of the heat-exchange surface. The follower may be rolled back along the frame, and the plates separated for inspection and cleaning. This facility for sanitation led to the substantial initial development of plate heat exchangers by Seligman [Dummett, 1978] for the dairy and food-processing industries in the 1920s, although the original invention dates back to the late nineteenth century. Other advantages now appear to be fostering a trend to its adoption generally in the process industries as an alternative to tube-and-shell heat exchangers.

The advantages claimed for plate-and-frame heat exchangers, as compared with tube-and-shell exchangers, are:

1. Only one-third to one-half the surface area for heat transfer is required so that savings in cost can be realized. The reduced surface area comes about because of the relatively high turbulence achieved in the corrugated flow paths between the plates, increased exchanger effectiveness through true countercurrent flow, and reduced fouling (one-tenth to one-fourth that of shell-and-tube exchangers) as a result of the turbulent fluid action and polished surface finish of the plates. The close spacing between adjacent plates is equivalent to the use of very-small-diameter tubes, with consequent high heat-transfer coefficients.

FIGURE 4.4
Large plate-and-frame heat exchangers. (*Courtesy Alfa-Laval and APV.*)

2. The reduced surface areas required permit substantial savings in the space required for plate-and-frame heat exchangers. Little clearance space around the exchanger is required to permit dismantling for cleaning, compared with the exchanger length required to permit the withdrawing of a tube bundle.
3. The improved exchanger effectiveness resulting from true countercurrent flow results in an increased temperature spread and allows a closer approach to ideality, a particularly significant feature in heat-recovery systems where the temperature potentials are limited.
4. The thermal rating of the plate-and-frame exchanger can be readily increased or decreased by varying the number of plates. This is a most attractive feature for systems likely to involve substantial future changes in load.
5. The increased effectiveness of the plate-and-frame heat exchanger increases the temperature rise of the cooling water, reducing the necessary flow and resulting in savings relative to piping, valves, pumps, and operating cost.
6. The internal void volume of a plate-and-frame heat exchanger is only 10 to 20 percent of that of the equivalent tube-and-shell exchanger. Less process medium is involved so that better process control and faster response times can be achieved.
7. The reduced surface requirement and volume of fluid contained combine to reduce the gross mass of the plate-and-frame exchanger to approximately one-sixth that of the equivalent shell-and-tube exchanger. Further savings result from lower shipping, installation, and foundation costs.

With such powerful advantages, it is hard to understand why the plate-and-frame exchanger is comparatively so little used. This likely arises from the characteristic conservatism of engineers, the widespread familiarity with tube-and-shell systems, and the large number of manufacturers of tube-and-shell exchangers. Furthermore, college engineering courses in heat transfer allocate only limited time to heat exchangers. Nearly all of this is concerned with tube-and-shell exchangers, and students perform some elementary design or rating calculations. Consequently, when these graduates find themselves at work in offices designing heat exchangers, they follow the procedures they have been taught. The major manufacturers of plate heat exchangers, APV and Alfa-Laval, have user-type computer programs that engineers may access through computer terminals and telephone connections for the purpose of in-office design at the early stages of a new project. It is hard to generalize about costs in today's unstable economic climate, except to say that there appears to be no significant difference between plate-and-frame and tube-and-shell exchangers for a given duty. Potential users would be well advised to obtain competitive quotations for both types of exchangers.

Disadvantages

The only significant disadvantage of plate-and-frame heat exchangers is the very large number of surfaces that must be sealed by gaskets. Elastomeric materials are normally used for most applications, but the maximum temperature and pressure at which these can be used are limited to about 2.7 MPa (400 lb/in^2) and 400 K (300°F). Compressed asbestos fiber gaskets may be used up to 600 K (600°F), but the pressure for which they are suitable is limited to about 1.8 MPa (240 lb/in^2).

Imperfect sealing of the gasket (particularly when reused, in direct contravention of manufacturers' recommendations) is quoted by users as the factor causing most difficulties with plate-and-frame heat exchangers.

Comparison with Shell-and-Tube Exchangers

A good technical review of plate-and-frame heat exchangers was presented by Marriott [1971] and includes the comparison of plate-and-frame and tube-and-shell heat exchangers reproduced in Table 4.1. He projects costs (in 1970 dollars) for the exchangers, in stainless steel and for a design pressure of 142 lb/in^2, of

TABLE 4.1
Comparison of Plate and Tubular Heat Exchangers for Typical Water/Water Duty.
(*After Marriott [1971].*)

Flow parameters		
	Hot side	Cold side
Flow, m^3/h	50	50
Temperature (in/out), °C	80/40	20/60

Results of the calculation		
	Plate	Tubular
Heat-transfer area, m^2	25	85
Overall heat-transfer coefficient (clean), kcal/(m^2·°C·h)	5200	1750
Fouling factor, m^2·°C·h/kcal	0.00006	0.0001
Overall heat-transfer coefficient (service), kcal/(m^2·°C·h)	3960	1500
Pass system (hot/cold)	1/1	8/baffled
Channels or tubes per pass	40	56
Plate or tube size	0.32 m^2	17/20 mm × 3 m
Calculated pressure drop, mwg	4	6
J, mwg/NTU	2	3
Pumping power, hp	1.1	1.65
Weight (empty), kg	615	2400
Weight (full), kg	720	3100
Overall size (including cleaning space), m	1.5 × 0.7 × 1.4	7.0 × 0.7 × 0.7
Floor area required, m^2	1	5

$2700 to $4000 for the plate-and-frame heat exchanger and $5500 to $6000 for the tube-and-shell (fixed-tube-sheet) exchanger. In mild steel the tubular unit was projected at about $3500, about the same as the plate-and-frame unit in stainless steel.

Marriott concluded that the plate-and-frame heat exchanger was well suited to general-purpose liquid/liquid duties in turbulent flow. He noted that a viscous fluid in laminar flow in a smooth tube was likely to be turbulent in a plate heat exchanger. Condensation at moderate pressures was economically attractive in plate-and-frame heat exchangers, but low-pressure, large-volume flows were better handled in conventional tube-and-shell crossflow condensers.

Marriott, of the Alfa-Laval Co., suggested the plate-and-frame heat exchanger was well suited to the common duty of cooling a process stream of fresh or treated noncorrosive, nonfouling water by means of a cheap, readily available, highly corrosive, brackish water supply. Titanium was normally used for construction, this being the "only material capable of withstanding indefinitely the action of brackish water from any source, coast, estuary, river, canal, lake or bore hole."

More recently, Fuller [1979], of the **APV** Company, has provided further confirmation of the reduced area and pressure-drop requirements of plate heat exchangers as compared with shell-and-tube exchangers. Table 4.2 shows area and pressure-drop requirements of plate and shell-and-tube exchangers for five different liquid/liquid heat-exchange duties. In all cases the area requirement for the plate heat exchanger is one-third to one-quarter that of the equivalent tube-and-shell exchanger.

Fuller also included the cost data reproduced in Fig. 4.5. These show the approximate cost in pounds sterling per square meter of heat-transfer area as a function of the exchanger size. Four characteristic curves are given, one for an all-stainless-steel plate heat exchanger and three for shell-and-tube heat exchangers in (1) all stainless steel, (2) stainless tubes and mild-steel shell, and (3) mild-steel tubes and shell. The curves are somewhat misleading unless it is recalled that the necessary area for the plate heat exchanger is very much less than that for the tube-and-shell exchanger. Superficial inspection of Fig. 4.5 shows that, area for area, the stainless-steel plate heat exchanger is substantially more expensive than the mild-steel tube-and-shell exchanger. However, on a *duty for duty* basis, the cost differential is greatly reduced and may even be reversed. As an example consider the large exchanger, case 2 of Table 4.2. The shell-and-tube exchanger requires 1125 m^2 at a cost (from Fig. 4.5) of £20/m^2, for a total cost of £22,500. The plate heat exchanger requires 337.5 m^2 of area with a cost of £70/m^2, for a total cost of £23,625. Although the cost per unit area is $3\frac{1}{2}$ times more for the plate exchanger, its required area is one-third that of the tube-and-shell exchanger. As a consequence, the cost difference is not significant. Furthermore, for the water/seawater duty specified for case 2, it is unlikely that a carbon-steel tube-and-shell unit would be selected because of high corrosion on the seawater side. Upgrading the quality of material on the seawater side to

TABLE 4.2

Comparison of Areas and Pressure Drops for Shell-and-Tube and Plate Heat Exchangers. (*After Fuller [1979].*)

Case	Flow rate, kg/h		Duty temperature, °C	Shell and tube			Plate exchanger		
	Fluid A	Fluid B		Area, m²	Pressure drop, kN/m²		Area, m²	Pressure drop, kN/m²	
					Fluid A	Fluid B		Fluid A	Fluid B
1	6,080 (hydrocarbon)	24,265 (water)	160 / 49 A 49 / 33.5 B	28	2.9	21	8.6	2	15
2	392,725 (water)	351,350 (seawater)	41.7 / 37.3 A 28 / 22 B	1125	131	131	337.5	75	60
3	26,820 (gasoline)	90,900 (water)	105 / 35 A 35 / 26 B	83.6	21.4	30	26.2	10	30
4	7,955 (solvent)	8,820 (water)	60 / 35 A 40 / 26 B	170	20	30	41.4	20	25
5	67,570 (desalter effluent)	214,775 (seawater)	105.5 / 41 A 38 / 20 B	139.5	75	55	35.3	30	70

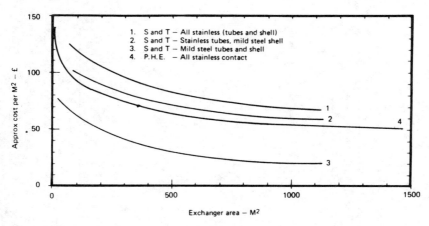

FIGURE 4.5
Relative costs of plate and shell- and- tube heat exchangers. (*After Fuller [1979].*)

reduce the corrosion (stainless steel in a mild-steel shell) increases the cost to about £60/m² for a total cost of £67,500, nearly three times the cost of the plate unit.

Comparative fouling studies of plate and shell-and-tube heat exchangers given the same cooling duty were carried out by Heat Transfer Research Inc. of Alhambra, California, an independent contract research organization. The fouling resistances measured for the plate exchanger were found to be one-quarter the TEMA recommended resistances for tube-and-shell exchangers. For the tube-and-shell exchanger, the fouling resistances measured were in fair agreement with TEMA recommendations. An account of these studies was given by Cooper et al. [1980] and follows Suitor [1976].

SPIRAL-PLATE HEAT EXCHANGERS

The spiral-plate heat exchanger was developed in Sweden by Rosenblad Patenter in the 1930s to recover waste energy from contaminated water effluents in pulp mills. Following World War II, its use was extended to other process applications, and it is now found in a wide variety of sizes and duties. Rosenblad Patenter was purchased by Alfa-Laval in 1965, and the production of spiral heat exchangers was continued. Similar units are also made by APV in association with the Japanese Nittoh Seisakusho Corp. Spiral heat exchangers are available as a general replacement for tube-and-shell exchangers in low- to moderate-pressure service and are particularly favored for difficult situations involving highly viscous liquids or dense slurries.

Construction

The elementary construction of a spiral heat exchanger is illustrated in Fig. 4.6. A pair of long strips are rolled around a split mandrel in scroll or spiral fashion.

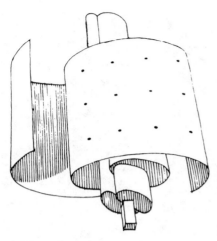

FIGURE 4.6
Elementary construction of a spiral heat exchanger. (*After Auth et al. [1978].*)

The strips are spaced a uniform distance apart by means of studs welded to the sheets, although for very viscous liquids and dense slurries the spacer studs are not included. One side of the strip is curved so as to contact the adjacent straight strip. This seam is welded so as to seal the spiral channel. Alternatively, in some cases, spacer bars are welded into each spiral channel at the sides, as illustrated in Fig. 4.7. Only one side of each channel is closed, and the closure is applied to alternate sides. The open sides of the two channels are closed by gasketed cover plates clipped to rigid flanges around the periphery of the cylindrical chamber, so that the spiral elements are enclosed as shown in the assembled unit in Fig. 4.8. This construction precludes the possiblity of leakage between the streams, yet allows ready accessibility to both channels for inspection, maintenance, or cleaning.

Both faces of the spiral element are ground flat during manufacture to

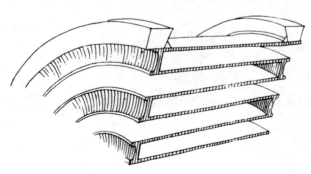

FIGURE 4.7
Sealing the spiral channels. (*After Auth et al. [1978].*)

FIGURE 4.8
Assembled spiral heat exchanger. (*Courtesy Alfa Laval.*)

facilitate sealing with the cover plates and thereby eliminate the possibility of fluid bypassing the spiral flow path. The end covers are machined slightly convex across the sealing face to provide some prestressing when the cover is bolted in place. This eliminates deflection or dishing of the end cover when internal pressure is applied. An inlet or outlet flange connection, the nozzle, is welded to the center of each cover. Cover gaskets are generally of asbestos, but various synthetic rubbers and Teflon are also used.

Spiral heat exchangers may be made of any metal that can be rolled or welded, including carbon steel, stainless steel, titanium, nickel, and the high-nickel alloys. Applications are limited to low- and medium-pressure devices up to 1.5 MPa (250 lb/in^2).

Operation

Flow paths in the spiral heat exchanger are illustrated in Fig. 4.9. Hot fluid enters at the center of the spiral element and flows to the outside. The cold fluid enters at the periphery and flows to the center. This true countercurrent flow is achieved with heat exchange taking place on both the upper and lower walls of the rectangular flow conduits.

In normal use for liquid/liquid applications, the axis of the exchanger is horizontal. Thus, any settled solids accumulate in the lower parts of the spiral channels and are reentrained when the flow rate increases, or the channel becomes partially blocked so the local flow velocity is high.

Many other flow arrangements are possible, two of which are shown in Figs. 4.10 and 4.11. For condensers and evaporators, it is convenient to set the exchanger with the axis vertical to facilitate separation of the liquid and vapor. Liquid drains and vapor vents are provided as required.

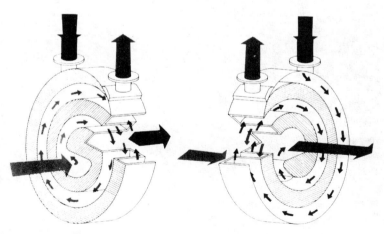

FIGURE 4.9
Flow paths in the spiral heat exchanger. (*After Auth et al. [1978].*)

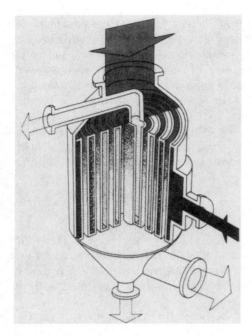

FIGURE 4.10
Vertical-axis spiral-flow/crossflow condenser. (*Courtesy American Heat Reclaiming.*)

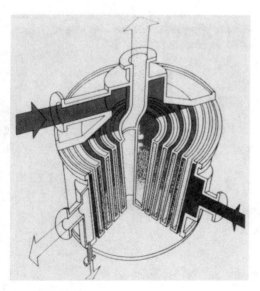

FIGURE 4.11
Combination spiral-flow/crossflow condenser. (*Courtesy American Heat Reclaiming.*)

The exchanger may be used in spiral-flow crossflow as shown in Fig. 4.10. This arrangement is particularly suited for applications where the fluid densities or mass flow rates of the two fluids are significantly different. It is possible to combine a high liquid velocity in the closed spiral element with a low pressure drop on the gas or vapor side in crossflow parallel to the axis.

The condenser shown in Fig. 4.11 is well suited for fluids made up of both condensable and noncondensable fractions. The coolant enters at the periphery and flows in the spiral passage to the center, where it leaves through a nozzle passing through the upper vapor dome. The vapor enters the vapor dome, which has a central cone section to allow the vapor to enter several open spiral channels in the center section. The liquid condensate falls to the bottom of the vertical channels and flows along the spiral to the liquid drain at the periphery. The noncondensable fraction flows through the upper section of the spiral channel to exit at the periphery through a separate nozzle and with a temperature close to the coolant inlet temperature.

For toxic fluids or those difficult to gasket and for liquid/gas exchangers, both spiral passages may be welded shut on one side and left open on the other side for closure with a single cover plate. Alternatively, the passages may be left open on both sides and sealed with end covers.

Advantages

Flow characteristics in the spiral conduits correspond to flow on the tube side of a shell-and-tube exchanger rather than the shell side. Therefore, with both fluids in tube-side flow, the spiral heat exchanger attains a higher overall heat-transfer coefficient. The rectangular flow sections can be dimensioned for optimal flow conditions, for there is no requirement for the flow section to be the same for both fluids or to even remain constant in passage from inlet to outlet.

The continuously curved flow section contributes to improved overall heat-transfer coefficients in the same way as curved pipe. Furthermore, at velocities resulting in laminar flow in straight-line tube flow, the continuously varying passage still realizes turbulent flow.

The true countercurrent flow increases the exchanger effectiveness. In combination with the higher overall heat-transfer coefficient, this permits a reduction of approximately 20 percent in the required heat-transfer area, with consequent savings in size, weight, and cost, including reduced transport, foundation, and installation charges.

The scrubbing effect of the fluids in both spiral passages inhibits deposition of sludge and other deposits. Fouling factors for spiral heat exchangers are commonly one-third the value of those recommended for conventional shell-and-tube exchangers. This makes the spiral exchanger preferred for fluids tending to sludge or those containing solids in suspension. Thermal insulation of the spiral exchanger is rarely required; the cool fluid enters at the periphery and flows in the outer passage.

Maintenance requirements tend to be low because the scrubbing action of the fluids inhibits the permanent deposition of scale and other solids. Furthermore, the side covers can be readily removed for inspection or mechanical cleaning. The single flow passage for each fluid facilitates cleaning by chemical means without removing the end covers.

The spiral arrangement permits an extremely compact unit. A transfer area of 30 m^2 can be contained in a spiral element 1 m in diameter and 1.5 m long. The only additional installation volume required is the servicing space to remove the covers.

The above was abstracted from a review paper about spiral heat exchangers by Auth and Loiacono [1978]. Technical information concerning the design of spiral heat exchangers was given by Minton [1970]. Applications of spiral heat exchangers were reviewed by Hargis et al. [1967].

PANEL-COIL HEAT EXCHANGERS

There is a family of plate-type heat exchangers with no accepted generic name for which the registered trade names Panelcoil (Dean Products Inc., New York) and Platecoil (Tranter Inc., Wichita Falls, Texas) are beautifully expressive. The basic construction of this family of exchangers is two embossed metal plates welded together, as illustrated in Fig. 4.12. The plates form a flat panel with a series of flow channels for the heat-transfer medium, coupled by relatively thick fins.

The double-embossed type has both plates embossed with identical matching flow channels. The single-embossed type has one flat plate and one embossed plate welded together, with a combined panel thickness approximately half that of the equivalent double-embossed panel. Of course, the section area of the flow channel in the single-embossed design is also half that of the double-embossed type.

There is little restriction on the patterns that can be embossed into the plate, so that flow channels of virtually any size, shape, and distribution can be readily obtained. The flow sections can be a single continuous coil or distributed in a more complicated fashion, as shown by the multizone Platecoil panel in Fig. 4.13. This is intended as a tank heater, with steam condensing in the "tubes" or flow channels. Note the difference between the large-diameter supply port and the smaller bottom draining condensate exhaust.

Panels can be fabricated in virtually any materials available in sheet form that can be welded or otherwise joined. Metals are used almost invariably. Carbon-steel plates are most economical where conditions permit, but many other combinations are routinely produced for corrosive fluids or other special service.

After the basic panel is fabricated, it can be rolled or pressed into a variety of different shapes or configurations. It can be simply mounted adjacent to a wall as part of the perimeter room heating system; it can be hung in a tank for

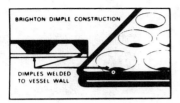

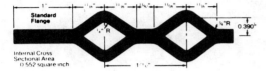

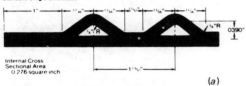

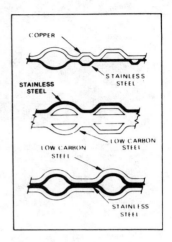

(a)

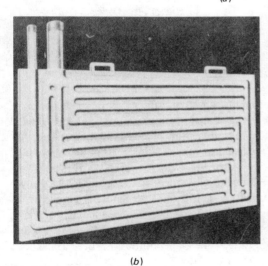

(b)

FIGURE 4.12
(a) Basic construction of a panel-coil heat exchanger. (b) Multizone Platecoil panel. (*Courtesy Tranter Inc.*)

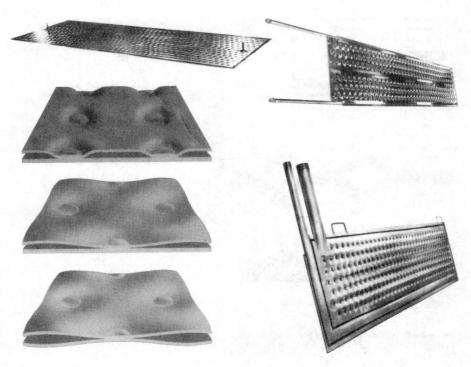

FIGURE 4.13
Forms and configurations of panel-coil heat exchangers. (*Courtesy Paul Mueller Co.*)

liquid heating; or it can be incorporated as part of the structure of a vessel or clamped on the inside or outside of a vessel. Multiple units can be ganged together in endless varieties for air or liquid heating. Panels can be made in flat pancake form or as cones with wide or shallow angle. Some of the many possible arrangements and combinations are shown in Fig. 4.14.

One interesting variant to the conventional form of construction is shown in Fig. 4.14. In this method of fabrication neither of the plates is embossed or preformed. The two flat plates are spot-welded together in a regular pattern, and then fluid at high pressure is applied between the plates, causing them to expand to provide interconnected cavities between the weld points. Another variant uses two flat plates, with flat disks separating the plates.

Panel-coil heat exchangers are remarkably versatile and low in cost and have been applied in a surprisingly wide range of industrial uses. They can be made in a variety of thicknesses to withstand any design pressure.

PLATE-FIN HEAT EXCHANGERS

An element of a plate-fin heat exchanger consists of a crimped or corrugated die-formed fin plate sandwiched between flat metal separator plates, as shown in

Figs. 4.15 and 4.16. Side bars are located along the extremities of the fin sections. The heat exchanger is formed as a stack of such elements, welded or dip-brazed to form a rigid matrix, as shown in Fig. 4.17.

Stacks of elements can be assembled and joined in virtually any size and shape. They are completed by welding on flow-distributing manifolds or headers, as shown by the large plate-fin heat exchanger in Fig. 4.17. Several such units may be eventually incorporated into very large heat exchangers, as shown in Fig. 4.18.

A very wide variety of fin plates has been used. Typical elements are shown in Fig. 4.19. In some cases a wavy corrugation is favored to promote turbulence and enhance the heat transfer. In other cases the fins are louvered or dimpled, again to enhance the heat transfer and promote turbulence. Measures such as these are effective in enhancing heat transfer but are invariably accompanied by increased fluid-frictional effects manifest in an increased pressure drop in the core. With low-density fluids, (i.e., gases) the extra work required to overcome the increased pressure drop often exceeds the increase in heat transfer accomplished. The exercise is therefore self-defeating, particularly as mechnical work is often worth up to 10 times its equivalent in heat transfer. Any proposed departure from the simplest corrugated fin form should therefore be examined critically.

There is no requirement that the alternate corrugated fin plates be of the

LNG VAPORIZER BANK ASSEMBLY

FIGURE 4.14
Forms and configurations of panel-coil heat exchangers. (*Courtesy Paul Mueller Co.*)

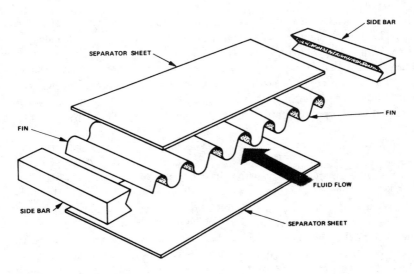

FIGURE 4.15
Construction details of a single-element plate-fin heat exchanger. (*After Lenfestey [1961].*)

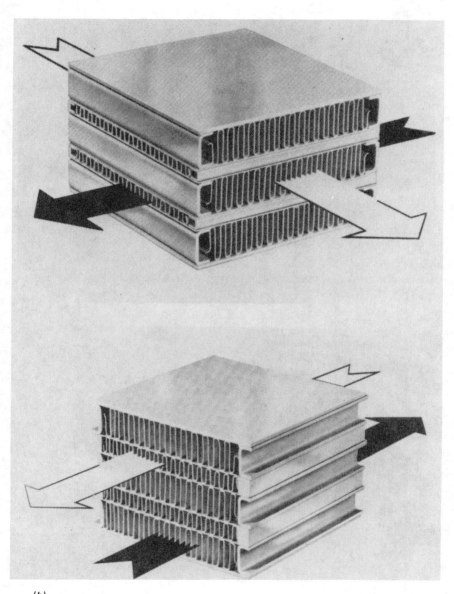

(b)
FIGURE 4.16
(a) Crossflow and (b) counterflow arrangements of plate-fin heat exchangers. (*After Lenfestey [1961].*)

FIGURE 4.17
Large plate-fin heat exchanger in the course of manufacture, with one flow header attached. (*After Lenfestey [1961].*)

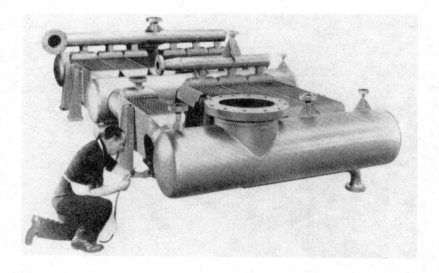

FIGURE 4.18
Large, complicated plate-fin heat exchanger in the course of manufacture. (*After Lenfestey [1961].*)

FIGURE 4.19
Elements of plate-fin heat exchangers.

same height, i.e., for uniform spacing of the separator plates. This is particularly advantageous in situations where there is a significant difference in the densities of the hot and cold fluid streams. Such differences occur when the two fluids are liquid on one side and gas on the other. Also, in gas-turbine recuperators, the hot exhaust products on one side are at atmospheric pressure whereas the air being preheated is at pressures of 4 to 10 atm. Similarly, in the recuperative counterflow plate-fin heat exchangers used in cryogenic systems, the return stream is at a much lower pressure (and hence density) than the high-pressure inlet stream en route to the expansion engine or valve. In such cases, the use of larger corrugation heights for the low-density stream allows the designer to approach a common Reynolds number for the flows of both streams. In this way, the constituent local heat-transfer coefficients can be made to correspond approximately, thereby optimizing the overall heat-transfer coefficient U (see Chap. 2).

Plate-fin heat exchangers are very widely used in cryogenic (very low temperature) systems and gas-liquefaction plants generally. Aluminum is commonly used throughout, with dip or furnace brazing as the preferred method of construction. Heat exchangers of extraordinarily compact construction can thereby be achieved at relatively moderate cost, small size, and low weight. Heat exchangers of high thermal capacity and effectiveness are required routinely in cryogenic applications so as to use the high input of compressor work most effectively.

Plate-fin heat exchangers are used extensively for closed-cycle gas turbines. In these applications, temperatures are very high and aluminum is not adequate; stainless-steel or nickel-alloy sheets are used instead, welded or brazed as required. The Garrett AiResearch Co. is noted for the complex and intricate stainless-steel sculptures of great beauty produced as recuperative heat exchangers for closed-cycle gas turbines and liquid-metal Rankine-cycle turbine systems for NASA and the U.S. Air Force.

Similar systems have been used for open-cycle gas turbines and heat-recovery exchangers in other fired systems. A difficulty preventing widepsread application is that the hot side soon becomes fouled by the products of combustion, and the small flow passages eventually become blocked. No really effective, low-cost, and convenient method of cleaning the very compact matrices of plate-fin heat exchangers has yet been developed. The relative cleanliness and purity of fluids in cryogenic and gas-liquefaction systems is one of the principal reasons why plate-fin heat exchangers have been able to assume a dominant position in such applications.

Plate-fin heat exchangers are sometimes used with a liquid/gas combination, but plate-coil, tube-fin, or tube-and-shell exchangers with liquid in the tubes are preferred in most cases.

A major development of interest and activity is foreseen (in the near future) for thermal recovery from relatively low temperature thermal-effluent streams. This will surely provide many opportunities for the application of plate-fin heat exchangers. Some will likely be of brazed-aluminum construction, but there appear also to be substantial fields of application for plate-fin exchangers of low-cost plastic, waxed paper, and aluminum foil with epoxied or other low-cost cemented construction. One form of low-cost aluminum heat exchanger for thermal-energy recovery systems is shown in Fig. 4.20. It consists of a stack of die-formed metal plates rolled at the seams of adjacent plates to form alternate flow channels with mutually perpendicular axes of flow. The assembled exchanger is shown in Fig. 4.21. It can be incorporated in the exhaust-energy recovery system shown in Fig. 4.22.

Some aspects of the use of plastics for heat exchangers have been examined by Koopman et al. [1979]. Earlier, Pescod [1974] summarized Australian research work on plate-fin heat exchangers of rigid polyvinyl chloride sheets intended for air-conditioning applications. The moderate pressure and temperature environment permitted the use of vacuum-formed plastic sheets so thin that the relatively low thermal conductivity of the plastic was not significant. A useful compilation of test and design data was included in the summary. For the design and comparative evaluation of plate-fin heat exchangers, the best compilation of heat-transfer and friction data is that given by Kays and London [1964].

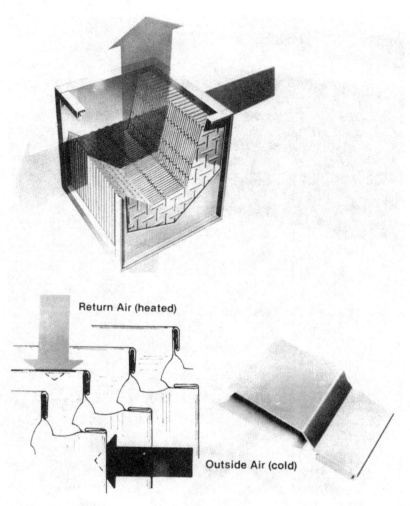

Return Air (heated)

Outside Air (cold)

FIGURE 4.20
Low-cost aluminum plate heat exchanger for energy recovery. (*Courtesy ACS, Elk Grove, IL.*)

SUMMARY

In this chapter we have briefly reviewed a number of different types of heat exchangers, all having the common feature that they do not include tubular flow channels.

Plate-and-frame heat exchangers are direct competitors of tube-and-shell exchangers for general service involving liquid/liquid transfer or condensing and evaporating. They offer very significant advantages over tube-and-shell exchangers and appear to be penetrating the refinery and process industries.

FIGURE 4.21
Assembled aluminum plate heat exchanger. (*Courtesy ACS, Elk Grove, IL.*)

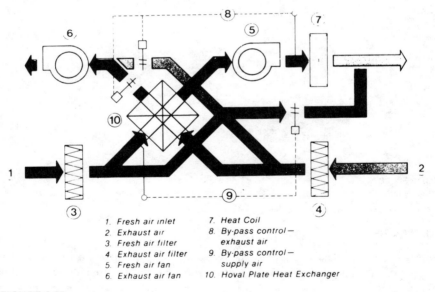

1. Fresh air inlet
2. Exhaust air
3. Fresh air filter
4. Exhaust air filter
5. Fresh air fan
6. Exhaust air fan
7. Heat Coil
8. By-pass control —
   exhaust air
9. By-pass control —
   supply air
10. Hoval Plate Heat Exchanger

FIGURE 4.22
Heat exchanger incorporated in exhaust-energy recovery system. (*Courtesy ACS, Elk Grove, IL.*)

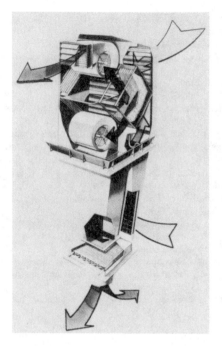

FIGURE 4.22 (*Continued*)
Heat exchanger incorporated in exhaust-energy recovery system. (*Courtesy ACS, Elk Grove, IL.*)

Spiral-plate heat exchangers are available as a general replacement for tube-and-shell exchangers and offer a number of significant advantages. They are particularly favored for heavily contaminated or highly viscous liquids and dense slurries.

Panel-coil heat exchangers can be readily fabricated from sheet metal or plastic material in an amazing variety of forms and are widely used over a broad spectrum of general service and process applications.

Plate-fin heat exchangers are made up of stacks of alternate die-formed and flat plates of metal, plastic, paper, and other materials. They are generally fabricated of dip-brazed aluminum plates for cryogenic service, and stainless-steel or alloy materials for gas-turbine exhaust or preheater service. The imminent development of energy recovery systems in buildings and process industries appears to offer very substantial applications for plate-fin exchangers in the future.

REFERENCES

Auth, W. J., and J. Loiacono: Plate and Spiral Plate Heat Exchangers, pp. 108–138 in "Practical Aspects of Heat Transfer," *Proc. 1976 Fall Lect. Ser., N.U. AIChE,* American Institute of Chemical Engineers, New York, 1978.

Cooper, A., J. W. Suitor, and J. D. Usher: Cooling Water Fouling in Plate Heat Exchangers, *Heat Transfer Eng.,* vol. 1, no. 3, pp. 50–55, 1980.

Dummett, G. A.: Invention or Innovation, Third Seligman Memorial Lecture, 1978.

Fuller, R.: Plate Heat Exchangers in Sea Water Cooling Systems, *Proc. I Mech. Eng. Conf.,* London, 1979.

Hargis, A. M., A. T. Beckman, and J. J. Loiacono: Applications of Spiral Plate Heat Exchangers, *Chem. Eng. Prog.,* vol. 63, no. 7, pp. 62–67, 1967.

Kays, W. M., and A. L. London: "Compact Heat Exchangers," 2d ed., McGraw-Hill, New York, 1964.

Koopman, R. N., D. Miller, and R. E. Heltz: A Look at Recovering Low Temperature Rejected Heat Using Plastic Heat Exchangers, *AIChE Symp. Ser.* 189, vol. 75, p. 304, 1979.

Lenfestey, A. G.: Low Temperature Heat Exchangers, in K. Mendelssohn (ed.), "Cryogenics," vol. 3, pp. 23–48, Heywood, London, 1961.

Lenfestey, A. G.: Compact Heat Exchanger Assemblies for Gas Separation Plants, *Proc. 2d Int. Cryog. Eng. Conf.,* pp. 47–49, Ilfiffe, Guildford, U.K., 1968.

Marriott, J.: Where, When and How to Use Plate Heat Exchangers, *Chem. Eng.,* April 5, 1971.

Minton, P. E.: Designing Spiral-Plate Heat Exchangers, *Chem. Eng.,* May 4, 1970.

Pescod, D.: Effects of Turbulence Promoters on the Performance of Plate Heat Exchangers, chap. 22 in N. Afgan and E. Schlunder (eds.), "Heat Exchangers: Design and Theory Sourcebook," Hemisphere, Washington, D.C., 1974.

Suitor, J. W.: PHE Fouling Study, HTRI rept. F-EX-1-8, Heat Transfer Research Inc., Alhambra, Calif., 1976.

5   Elements of thermal design

INTRODUCTION

The design of heat exchangers can be broadly classified into two separate but interrelated activities:

1. Thermal design
2. Mechanical design

Thermal design is concerned with the heat-transfer performance of the heat exchanger, to ensure that the heat exchanger is provided with adequate surface area to perform the required thermal duty. It includes an assessment of the deterioration in performance likely to occur in use with depositions and accretions on the heat-exchanger surfaces. Ways and means to minimize the effects of fouling are within the purview of the thermal designer.

Fluidic effects are also important. Heat transfer between moving fluids invariably involves fluid-friction effects arising out of the viscosity of the fluids. These viscous effects are manifest in a decrease in the fluid pressure from inlet to outlet. To overcome the pressure drop, fluid must be pumped or compressed. This requires an input of mechanical work from an electric motor or other form of drive. With compressible fluids—gases and vapors—the cost of producing a unit of mechanical work is usually several times greater than the worth of a unit of heat transferred. For this reason, the thermal designer is concerned with arranging the geometry and form of the heat exchanger to eliminate unnecessary frictional effects and minimize the pressure drop necessary to move fluids through the exchanger to accomplish the design heat-transfer duty.

The thermal designer seeks to accomplish these tasks against a backdrop of economic constraints—the costs of fabrication, transportation, installation, maintenance, and replacement, and the trade-off of reducing capital costs but increasing operating costs in the light of spiraling energy costs for electric motor drives for fans, pumps, etc. He must be alert to the temperature distribution of fluids in the heat exchanger, the propensity to corrosion of different materials and fluids, the possibility of polymerization of an organic fluid on an overheated surface, the precipitation of salts from brackish streams, and the deposition of particulates in low-velocity zones of solid-liquid slurries or heavily contaminated fluids.

The thermal designer must also be aware of environmental factors. Public concern over the discharge of thermal effluents, either to the air or to water, is important. The effects of unscheduled discharges (leaks) to the environment and the leakage of one fluid into another must be foreseen. Accidents with hazardous fluids must be anticipated, and measures to prevent or alleviate the consequences must be provided.

In similar fashion, occupational health and safety aspects are an urgent concern of the thermal designer, to eliminate the possibility of hazard to workers or members of the public, both in routine use and to alleviate the dangerous consequences of accidents. The increasing severity of legal liability and negligence settlements is an important factor.

The mechanical design of heat exchangers is concerned with making the unit strong enough to endure the design pressures and all foreseeable over-design pressures, with the seals and bearings that may be involved, with vibration effects, with corrosion, with methods of fabrication, with the arrangements for lifting, transporting, and installing the heat exchanger, and, finally, with the methods for cleaning, maintaining, and repairing the unit.

The elements of mechanical design are considered in Chap. 6; here we shall confine our attention to thermal design.

FLOW ARRANGEMENTS

Most heat exchangers function so that the hot and cold fluids, separated by a common wall, exchange heat through a combination of convective and conductive heat transfer.

The efficacy of the exchange process is very strongly affected by the way the flows are directed and arranged. There are three principal possibilities:

1. Both fluids have the same axis of flow. When the fluids move in the same direction, the flow is termed *parallel flow.* When they move in opposite directions, it is called *counterflow.*
2. The axes of flow of the fluids are mutually perpendicular. This is called *crossflow.*
3. The flows are partly along the same axis for both fluids and partly crossflow. This is termed *combined flow.*

Mixed and Unmixed Flow

It is also important to distinguish between *mixed* and *unmixed* flow. The distinction is illustrated in Fig. 5.1. In Fig. 5.1a, hot liquid flows in the tubes,

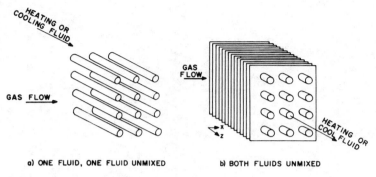

a) ONE FLUID, ONE FLUID UNMIXED   b) BOTH FLUIDS UNMIXED

FIGURE 5.1
Mixed and unmixed flow.

and cooling air passes over the tubes. The axes of flow of the liquid and the air are mutually perpendicular, so the flow can be described as crossflow. The liquid flowing in the tubes is said to be unmixed. Once fluid enters a tube, it is relatively unaffected by what happens in the other tubes.

In the case illustrated, the leading tubes will experience higher rates of heat transfer than tubes downstream in the airflow, for the cooling air is at the lowest (inlet) temperature when it passes over the leading tubes. Therefore, the temperature of liquid exiting the leading tubes will be less than that of liquid exiting the other tubes. If the tubes are all connected to a common exhaust header, the liquids will mix together to attain a common temperature.

The airflow is progressively heated in passing over the tubes and therefore approaches the downstream tubes at elevated temperature. Similarly the temperature of liquid in the tubes is not constant but decreases progressively from inlet to outlet. The temperature difference between the hot tube-side liquid and the cooling air is greatest at the tube inlet and least at the tube exit. This difference results in higher rates of heat transfer to the air near the tube inlet with a significantly elevated temperature increase in the air, as shown by the temperature profile in Fig. 5.2.

With bare tubes the flow is mixed, and the temperatures and flow conditions of the air at one location in the tube bank affect the temperature and flows in other parts. The tendency is to equilibrate the flow.

When the tubes are equipped with fins as shown in Fig. 5.1b, the airflow is unmixed. Air in the interstices between adjacent fins flows through the tube bank and is heated without reference to conditions arising elsewhere. The result is that the uneven temperature distribution illustrated in Fig. 5.2 is further enhanced. Furthermore, the heat transfer to the air is appreciably greater because of the increased heat-transfer area provided by the fins.

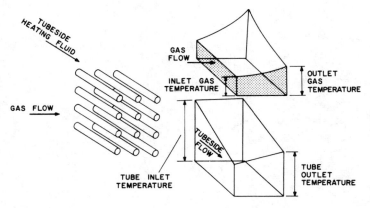

FIGURE 5.2
Temperature profile for crossflow heat exchanger.

TEMPERATURE DISTRIBUTION

The temperature distribution of the two or more fluids passing through a heat exchanger depends greatly on the flow arrangement. To see this, consider the simple double-pipe heat exchanger shown in Fig. 5.3. We shall call the fluid flowing in the tube the *tube-side* fluid and designate its temperatures with lowercase t's; we shall call the fluid in the annular space the *shell-side* fluid and designate its temperatures with capital T's. In both cases the subscript 1 refers to inlet conditions, and 2 to outlet conditions.

A surprising number of different flow regimes may arise with this apparently simple, even trivial, flow situation. Some of these are illustrated in Fig. 5.4.

The first possibility to consider is shown in Fig. 5.4*a*—a condensing evaporator where both fluids are changing phase. The hot fluid is condensing to liquid, and the cold liquid is evaporating to vapor. The physical process of boiling and condensing are complicated, but the temperature distribution is very uncomplicated; the temperatures simply remain constant throughout. The temperature potential for heat transfer is $\Delta T = t - T$ and is constant all over the system.

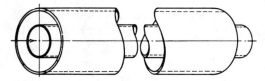

FIGURE 5.3
Double-pipe heat exchanger.

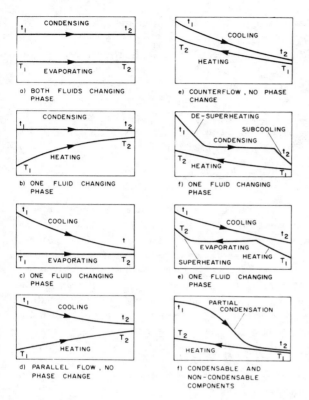

FIGURE 5.4
Fluid temperature distribution in a double-pipe heat exchanger.

Another possibility (Fig. 5.4*b*) arises when the hot fluid condenses (at constant temperature t) and heat is transferred to the cold fluid, thereby increasing its temperature in passage through the exchanger. The temperature potential for heat transfer now varies from $\Delta T_1 = t - T_1$ at the inlet to $\Delta T_2 = t - T_2$ at the exit.

The corresponding case in which one fluid only is changing phase arises when the cold liquid is evaporating (at constant temperature T), thereby cooling the hot fluid in its passage through the exchanger. This case is shown in Fig. 5.4*c*. As before, the temperature potential varies from $\Delta T_1 = t_1 - T$ at the inlet to $\Delta T_2 = t_2 - T$ at the exit.

The situation most commonly experienced is that in which neither fluid experiences a phase change, so the temperature varies continuously for both fluids from inlet to outlet. There are two cases to consider: parallel flow and counterflow.

The parallel-flow case is illustrated in Fig. 5.4*d*. In this case both fluids enter at the left of the figure and exit at the right. The cold fluid experiences a temperature increase, and the hot fluid a temperature decrease. The temperature potential varies from $t_1 - T_1$ at the inlet to $t_2 - T_2$ at the exit. The exit temperature T_2 of the cold fluid cannot exceed the exit temperature t_2 of the hot fluid; otherwise there would be a contravention of the second law of thermodynamics (heat will not flow from a cold body to a hot body). In the special case where the fluids are the same, say water, and where the mass flow rates are equal, the temperature changes $T_2 - T_1$ and $t_1 - t_2$ will be equal. The maximum temperature change either fluid could possibly have (when $t_2 = T_2$) is equal to half the total temperature difference at the inlet; i.e.,

$$(T_2 - T_1)_{\max} = (t_1 - t_2)_{\max} = 0.5 \, (t_1 - T_1)$$

The counterflow case is illustrated in Fig. 5.4e. Here the fluids flow in opposite directions. This is much better than the parallel-flow case. The exit temperature t_2 of the hot fluid can be lower than the exit temperature T_2 of the cold fluid, but there is no contravention of the second law. The temperature potential for heat transfer varies from $\Delta T = t_1 - T_2$ at one end of the exchanger to $\Delta T = t_2 - T_1$ at the other end. If the fluids are the same, the mass flow rates are equal, and the properties of the fluids do not change with temperaature, then the temperature difference $\Delta T = t - T$ remains constant through the exchanger. It is theoretically possible for the exit temperature of one fluid to approach the inlet temperature of the other, that is, $t_2 \rightarrow T_1$ and $T_2 \rightarrow t_1$. In theory, therefore, the temperature decrease of the hot fluid and the temperature increase of the cold fluid could be the same and equal to $t_1 - T_1$, twice the theoretical maximum of the parallel-flow case.

The significance of this counterflow advantage cannot be overemphasized. A comparison of apparently identical heat exchangers would show the thermal capacity of the counterflow exchanger to be *twice* the capacity of the parallel-flow exchanger. Heat exchangers should *always* be in counterflow and *never* in parallel flow when a choice is possible.

The condenser illustrated in Fig. 5.4f is perhaps more common than the condenser of Fig. 5.4b. In the former, the condensing vapor enters at a temperature greater than the saturation temperature. The vapor cools to the saturation condition in a process called *desuperheating;* then condensation occurs at constant temperature, and some further *subcooling* of the liquid takes place before the hot liquor leaves the exchanger. The cold liquid may flow in parallel or in the preferred counterflow direction as illustrated.

A corresponding situation, where the cold fluid enters as a liquid and is heated, evaporated, and then superheated is shown in Fig. 5.5g.

When the hot fluid consists of both condensable vapor and noncondensable gases, the temperature distribution is of the more complex form represented in a general way in Fig. 5.4h.

We have seen, in Fig. 5.4e, that with counterflow heat exchangers it is possible for the exit temperature T_2 of the cold fluid to be above the exit

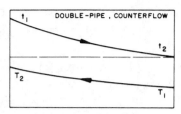

a) NO TEMPERATURE CROSS

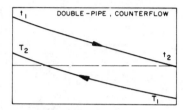

b) TEMPERATURE CROSS

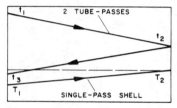

c) TEMPERATURE CROSS

FIGURE 5.5
Temperature distribution with and without a temperature cross.

temperature t_2 of the hot fluid. This is described as a *temperature cross* and is illustrated in Fig. 5.5. Three cases are shown:

1. The temperature distribution in a counterflow double-pipe heat exchanger without a temperature cross
2. The same with a temperature cross
3. The temperature distribution in a tube-and-shell heat exchanger with two tube passes and a single shell pass and having a temperature cross

The presence of a temperature cross is significant as indicative of a relatively small temperature potential between the fluids. This requires either a large area for heat transfer or relatively high velocity fluids to increase the heat-transfer coefficient with consequent high fluid-friction losses.

It is evident from the cases shown in Figs. 5.4 and 5.5 that there is rarely a uniform temperature difference between hot and cold fluids such as in Fig. 5.4a. The temperature difference varies from one end of the exchanger to the other and, in more complicated arrangements, at some intermediate point in the exchanger.

The minimum temperature difference between the fluids is called the *temperature approach*. Frank [1978] has given some useful rules of thumb about temperature differences and temperature approaches:

1. The temperature approach should normally be at least 5 to 7°C. For a refrigeration system, a closer approach (3 to 5°C) is sometimes justified.
2. In recuperative heat exchangers,* a temperature approach of at least 20°C should be maintained.
3. Cooling-water rise at low MTD† (< 40°C) should be less than 10°C. Cooling-water rise at high MTD (> 40°C) may be 10 to 20°C.
4. The cooling-water outlet temperature should be below 50°C.
5. The inlet coolant temperature should be maintained at more than 5°C above the process-fluid freezing point.
6. Excessive MTD should be avoided in heating and boiling systems (> 60°C).

MEAN TEMPERATURE DIFFERENCE

On grounds of simplicity and convenience, the use of the basic equation $\dot{q} = AU\,\Delta T$ to calculate the heat transfer in an exchanger is very attractive. However, as we have noted above, the temperature difference between the fluids varies widely and continuously from one end of the exchanger to the other. To overcome this difficulty, the concept of a mean temperature difference (MTD) is widely used.

Logarithmic Mean Temperature Difference

The customary MTD estimated is the logarithmic mean temperature difference (LMTD). This is defined as

$$\text{LMTD} = \frac{\text{maximum } \Delta T - \text{minimum } \Delta T}{\log\,(\text{maximum } \Delta T/\text{minimum } \Delta T)}$$

In most cases the maximum and minimum ΔT are taken to be the temperature differences between the fluids at the ends of the exchanger. Greater and lesser temperature differences may actually prevail in complicated exchangers, but these are usually unknown and relatively difficult to measure.

LMTD Correction Factors

When using the LMTD concept for heat exchangers other than the simple counterflow double-pipe or spiral heat exchanger, a correction factor Y must be used in the equation

---

*Assumed here to be exhaust-gas/air preheaters.
†Mean temperature difference.

$$\dot{q} = YUA\ \Delta T$$

The correction factor Y varies from 0 to 1. It allows for the complex combination of crossflow, parallel flow, and counterflow experienced in multiple-pass tube-and-shell heat exchangers and the general departures from ideality of other types of crossflow systems.

Correction factors are invariably presented graphically. Typical correction-factor charts are given in Fig. 5.6. These particular correction factors all relate to tube-and-shell exchangers. The upper diagram is concerned with one shell pass and two, four, or six tube passes. The middle diagram is concerned with two shell passes and four, eight, or twelve tube passes; the lower diagram with one shell pass and three, six, or nine tube passes.

The correction factor Y is determined from the appropriate chart using two other factors, X and Z, defined as follows:

$$X = \frac{t_2 - t_1}{T_1 - t_1} \qquad Y = \frac{T_1 - T_2}{t_2 - t_1}$$

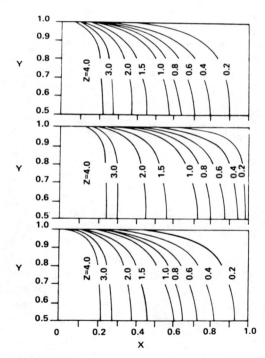

FIGURE 5.6

Chart of LMTD correction factor Y for tube-and-shell heat exchangers. Upper diagram is for one shell pass and two, four, and six tube passes. Middle diagram is for two shell passes and four, eight, and twelve tube passes. Lower diagram is for one shell pass and three, six, and nine tube passes. (*After Frank [1978].*)

where t_1 = tube-side temperature (inlet)

   t_2 = tube-side temperature (outlet)

   T_1 = shell-side temperature (inlet)

   T_2 = shell-side temperature (outlet)

For exchangers with large temperature differences between the inlet and outlet terminals, the correction factor Y is insignificant. As the terminal temperature difference becomes smaller, Y decreases. The effect becomes severe when there is a temperature cross (outlet temperature of hot fluid less than outlet temperature of cold fluid).

Frank [1978] recommends that tube-and-shell heat exchangers not be designed for use in applications where the Z factors approach a vertical slope. Under these conditions, small departures from the design point can result in a precipitous decline of the correction factor Y.

Correction-factor diagrams similar to Fig. 5.6 for other types of heat exchangers are widely used. Fraas et al. [1965] have given a comprehensive collection for most conventional types of exchangers.

Limitations of LMTD Concept

The LMTD concept and the associated correction factor Y should not be used for heat exchangers in which there is a phase change, boiling or condensing, of either fluid. Similarly, they should not be used for applications where the change in temperature of either or both fluids is so great that substantial changes in the physical properties of the fluid occur in transit through the exchanger.

Changes in the physical properties of fluids can be accounted for by application of the *caloric temperature* concept exemplified in the equation[*]

$$\dot{q} = A \, \frac{U_2 \, \Delta T_1 - U_1 \, \Delta T_2}{\log \, (U_2 \, \Delta T_1 / U_1 \, \Delta T_1)}$$

where ΔT_1 and ΔT_2 are the fluid temperature differences at the exchanger terminals, and U_1 and U_2 are the overall heat-transfer coefficients at the corresponding exchanger terminals.

The caloric temperature should not be used for cases where there are a close temperature approach (small ΔT) and relatively large temperature changes in transit through the exchanger. This is a situation characteristic of cryogenic heat exchangers. In these cases and others involving fluid phase change, a more complicated incremental analysis is necessary to realistically estimate the exchanger performance.

OVERALL HEAT–TRANSFER COEFFICIENT

At this point it is appropriate to recall (from Chap. 2) the composition of the overall heat-transfer coefficient U.

[*]From Frank [1978].

The process of energy transfer that occurs in a heat exchanger between two moving fluids separated by a metal wall is extremely complicated. It involves convective heat transfer from the hot fluid to the wall, conductive transfer through the wall, and further convective heat transfer from the wall to the cold fluid. The process is complicated by viscous effects in the fluids near the wall (the boundary layer) and by impediments to heat transfer arising from deposits, corrosion, or scale on one or both sides of the wall, referred to generally as fouling. Composite tubes of, say, stainless steel and carbon steel, for use with corrosive fluids inside the tubes may offer an additional impediment to heat transfer at the interface of the two metals.

All these various effects can be expressed in terms of a thermal resistance as discussed in Chap. 2 and then combined to obtain the overall thermal resistance R_{th}, where

$$\frac{1}{R_{th}} = \frac{1}{R_1 + R_2 + R_3 + R_4} = U$$

For most calculations it is more convenient to use the reciprocal value $(1/R_{th})$ than to use the total thermal resistance. This reciprocal value is given the name overall heat-transfer coefficient and designated by U. The value of U varies widely and has the units of watts per square meter per degree Celsius or Btu per hour per square foot per degree Fahrenheit.

The value of U can never exceed the numerical value of the minimum constituent of the thermal-resistance equation above. To illustrate this important point, let

$R_1 = 1/h_0$, where $h_0 =$ film heat-transfer coefficient on the shell side, say 8 W/(m$^2 \cdot$°C)

$R_2 = x/k_w$, where $x =$ tube thickness and $k_2 =$ thermal conductivity of tube material, say 0.008 m and 45 W/(m$\cdot$°C), respectively.

$R_3 = 1/h_1$, where $h_i =$ film heat-transfer coefficient inside the tube, say 2000 (W/(m$^2 \cdot$°C)

$R_4 = 1/f_0$, where $f_0 =$ fouling coefficient outside the tube, say 300 W/(m$^2 \cdot$°C)

$R_5 = 1/f_1$, where $f_1 =$ fouling coefficient inside the tube, say 1500 W/(m$^2 \cdot$°C)

Then,

$$U = \frac{1}{R_{th}} = \frac{1}{1/8 + 0.008/45 + 1/2000 + 1/300 + 1/1500}$$

$$= \frac{1}{0.125 + 0.00018 + 0.00050 + 0.00333 + 0.00067}$$

$$= \frac{1}{0.12968} = 7.71 \text{ W/(m}^2 \cdot \text{°C)}$$

In the above example, we neglected to include the effects of a difference between the outside and inside areas of the tubes.

The overall heat-transfer coefficient $U = 7.71$ W/(m²·°C) is less than any of its constituent parts. In this example one constituent, the shell-side heat-transfer coefficient h_0, was chosen to have the low value 8 W/(m²·°C). Because of this low value, the shell-side convective heat-transfer coefficient becomes the dominant thermal resistance.

The value of the overall heat-transfer coefficient U was only slightly below the value of the minimum constituent heat-transfer coefficient (7.71 compared with 8). This indicates that the thermal resistances of tube-wall conduction, tube-side convection, and fouling were trivial compared with the shell-side convective thermal resistance. Efforts to improve design would, in this case, focus on increasing the shell-side heat-transfer coefficient, perhaps by the use of fins or by increasing the velocity of the fluid, thereby increasing the Reynolds number and hence the Nusselt number (see Chap. 2).

It is clear, from the above, that the accuracy of the overall heat-transfer coefficient is limited by the reliability of the constituent parts. The tube-side and shell-side convective heat-transfer coefficients are primarily determined by physical properties of the fluids that are frequently not well defined and that experience considerable change in transit through the exchanger. The geometry of the exchanger exercises a considerable effect on the range of variation of the local fluid velocities, compared with those estimated on the basis of average bulk flow rates. The allowance for fouling both shell side and tube side is almost always chosen arbitrarily. Fouling is least when the exchanger is first put into service, and it increases progressively to eventually become the dominant thermal resistance, depending on the fluids and service conditions.

The overall heat-transfer coefficient U should therefore be regarded, at best, as an educated guess with a range of reliability no better than ±20 percent and frequently wider. As Frank [1978] remarks:

> It is mostly for these reasons that the over-enthusiastic application of sophisticated computational procedures is not justified. Little is accomplished by using a complex computer programme if the thermal conductivity of a fluid is not accurately known and if the fouling factors are arbitrarily applied . . . .

A chart, based on accumulated experience, for estimating values of U in preliminary calculations for a large number of different fluid combinations and process applications was given by Frank [1978]. The chart is reproduced as Fig. 2.6.

HEAT–EXCHANGER EFFECTIVENESS AND NTU

The LMTD approach to heat-exchanger design or analysis is most useful when the inlet and outlet temperatures are specified or can be easily determined. The log mean temperature difference can then be calculated from the terminal temperature differences of the fluids. By assuming an appropriate value of the overall heat-transfer coefficient, one can then proceed to determine the area of tubes required to perform a given thermal duty. Conversely, if the tube geometry is known, one can determine the thermal capacity of the exchanger and the mass flow rates of the fluids that can be handled.

The LMTD method is less useful when the temperature distribution of the fluids is not known but is to be determined from given exchanger geometry and fluid flow rates. The solution frequently involves an iterative procedure because of the logarithmic function in the LMTD. This situation is best handled by the other principal procedure of elementary thermal analysis, called the *NTU/ effectiveness method* and thought to have been pioneered by Prof. Kays and London of Stanford University. The method involves the use of two new concepts, the *heat-exchanger effectiveness* ϵ and a grouping of parameters called the *number of transfer units.*

Heat-exchanger effectiveness is defined simply as

$$\epsilon = \frac{\text{actual heat transfer}}{\text{maximum possible heat transfer}}$$

The actual heat transferred is the energy lost by the hot fluid or the energy gained by the cold fluid. To a first approximation, neglecting all extraneous energy losses, these are equal, so that

$$\dot{q} = \dot{m}_h c_h (T_{hi} - T_{ho}) = \dot{m} c_c (T_{co} - T_{ci})$$

where $\dot{q}$ = rate of heat transfer
  $\dot{m}$ = rate of fluid flow
  c = specific heat of fluid
  T = temperature
and the subscripts h and c refer to the hot and cold fluids, respectively, and the subscripts o and i to fluid inlet and outlet.

The product $\dot{m}c$ of the fluid mass flow rate and the specific heat is called the *capacity rate.* It is possible for the hot and cold fluids to have the same capacity rate. However, it is more likely that the capacity rates will *not* be equal and can then be identified as the minimum capacity rate C_{min} and maximum capacity rate C_{max}.

If the heat lost by the hot fluid is equal to the heat gained by the cold fluid, then

$$\dot{q} = C_{max} (\Delta T)_{min} = C_{min} (\Delta T)_{max}$$

In other words, the fluid with the minimum capacity rate will experience a *greater* temperature change than the fluid with the maximum capacity rate. This must be true given the equality of heat loss from the hot fluid and heat gain by the cold fluid.

Now it is possible to estimate the maximum heat transfer that could possibly take place in the heat exchanger. This would occur when one of the fluids experienced the maximum possible temperature swing, $T_{hi} - T_{ci}$. According to the above, this could only occur for the fluid with the minimum capacity rate. The maximum possible heat transfer is then

$$\dot{q}_{max} = C_{min}(T_{hi} - T_{ci})$$

and therefore the effectiveness is

$$\epsilon = \frac{\dot{m}_h c_h (T_{hi} - T_{ho})}{C_{min}(T_{hi} - T_{ci})} = \frac{m_c C_c (T_{co} - T_{ci})}{C_{min}(T_{hi} - T_{ci})}$$

In a further development of the theory, the combination UA/C_{min} is identified as a useful grouping of parameters and called the NTU, for "number of transfer units." U is the familiar overall heat-transfer coefficient, and A the appropriate area for heat transfer.

The numerical value of the NTU is a handy indicator of the level of sophistication of the heat exchanger, the degree to which the materials involved have been used effectively. The higher the NTU value of an exchanger, the more sophisticated the design. Run-of-the-mill exchangers have NTU values in the range 2 to 6. Advanced designs of ultracompact cryogenic heat exchangers have been fabricated with NTUs as high as 140 [Cowans, 1974].

The theoretical analysis of heat exchangers using NTU/effectiveness concepts and working from first principles is fairly complicated. Fortunately, the theory for most common types of exchangers is well established, and charts for most useful ranges of applications were presented by Kays and London [1963]. Specimen charts are given in Figs. 5.7 and 5.8. The chart in Fig. 5.7 relates to counterflow double-pipe or counterflow single-pass tube-and-shell heat exchangers. It shows the exchanger effectiveness ϵ as a function of the number of transfer units (NTU) for several different ratios C_{min}/C_{max} in the range 0 to 1. This chart shows clearly that the exchanger effectiveness increases as the number of NTU increases. The effectiveness also increases as the capacity ratio decreases.

A similar chart is given in Fig. 5.8. This relates to a tube-and-shell heat exchanger with one shell pass and two tube passes. Comparison of the two charts reveals in dramatic fashion the loss in exchanger effectiveness arising from a departure from true counterflow operation with two tube passes, one in counterflow and the other in parallel flow.

A comprehensive collection of charts of NTU/ϵ for most conventional exchangers was given by Kays and London [1963] and has been reproduced in many other texts.

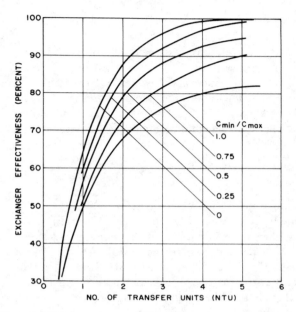

FIGURE 5.7
Chart of effectiveness versus NTU for counterflow, double-pipe heat exchanger. (*After Kays and London [1964].*)

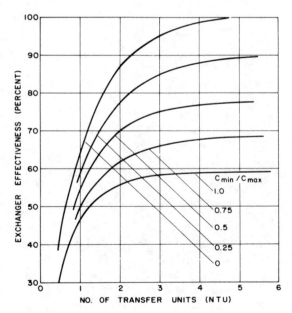

FIGURE 5.8
Chart of effectiveness versus NTU for shell-and-tube heat exchanger; one shell pass, two tube passes. (*After Kays and London [1964].*)

ILLUSTRATIVE EXAMPLE

To demonstrate the LMTD and NTU/effectiveness methods of analysis, we consider a tube-and-shell heat exchanger with one shell pass and two tube passes. The exchanger is used to cool oil (in the tubes) with water (in the shell). The flow rate of the oil is 170 kg/min. The inlet and outlet temperatures are 110 and 75°C, respectively, and the specific heat of the oil is 1.9 kJ/(kg·°C). Cooling water is available at 20°C. Following the guidelines of Frank [1978] given above, we shall

1. Keep the outlet water temperature below 50°C.
2. Select a cooling-water rise in the range 10 to 20°C.

Let us, therefore, assume a cooling-water outlet temperature of 40°C.

To estimate a value for the overall heat-transfer coefficient, we consider the composite chart in Fig. 2.6. The thermal fluid coefficient, the horizontal scale on Fig. 2.6, includes a sector for water ranging from 180 to 240 Btu/(h·ft^2·°F) for heat transfer from the shell side. The mean value is 210 Btu/(h·ft^2·°F). The upper scale for the process-fluid heat-transfer coefficient contains a sector for oils ranging from 50 to 100 Btu/(h·ft^2·°F). The mean value is 75 Btu/(h·ft^2·°F). The combined overall value may be determined by drawing a line through the two points, 75 on the upper scale and 210 on the lower scale, and reading an overall value on the middle scale. Alternatively, the overall value may be calculated as

$$\frac{1}{U} = \frac{1}{210} + \frac{1}{75}$$

or

$$U = 55.3 \ \text{Btu/(h·ft}^2\text{·°F)}$$

Since 1 Btu/(h·ft^2·°F) is equal to 6 W/(m^2·K) approximately,

$$U = 330 \ \text{W/(m}^2\text{·K)}$$

The rate of heat lost by the oil is

$$\dot{q} = \dot{m}_o c_o (\Delta T_o) = 170 \times 1900 \ (110 - 75) = \frac{11.3 \ \text{MJ/min}}{60 \ \text{s/min}} = 188 \ \text{kW}$$

This is also the rate of heat gain by the water, i.e.,

$$\dot{q} = \dot{m}_w c_w (\Delta T_w)$$

so that

$$\dot{m}_w = \frac{\dot{q}}{c_w \, \Delta T_w} = \frac{11.3 \times 10^6}{4180 \, (40 - 20)}$$

where the specific heat of water is $c_w = 4.18$ kJ/(kg·°C). Therefore, the required water flow rate is $\dot{m}_w = 135$ kg/min.

All the fluid temperatures are known, so the LMTD may be calculated. On the shell (water) side, the inlet temperature is 20°C, and the outlet temperature is 40°C. On the tube (oil) side, the inlet temperature is 110°C, and the outlet temperature is 75°C. Now

$$\text{LMTD} = \frac{\Delta T_1 - \Delta T_2}{\ln \, (\Delta T_1 / \Delta T_2)}$$

Calculating the temperature difference at each end for counterflow operation, we have

$$\Delta T_1 = 110 - 40 = 70°\text{C}$$

$$\Delta T_2 = 75 - 20 = 55°\text{C}$$

Then

$$\text{LMTD} = \frac{70 - 55}{\ln \, (70/55)} = 62°\text{C}$$

The correction factor Y to be applied because we have a tube-and-shell exchanger rather than a double-pipe counterflow heat exchanger may be determined by reference to Fig. 5.6. The upper diagram refers to an exchanger with one shell pass and two tube passes. The diagram factors X and Z may be determined as

$$X = \frac{t_2 - t_1}{T_1 - t_1} = \frac{75 - 110}{20 - 110} = 0.39$$

$$Z = \frac{T_1 - T_2}{t_2 - t_1} = \frac{20 - 40}{75 - 110} = 0.57$$

In the above, t and T refer to tube-side and shell-side temperatures, and subscripts 1 and 2 to inlet and outlet conditions, respectively. From the upper diagram of Fig. 5.6, when $X = 0.39$ and $Z = 0.57$, the value of the correction factor Y is 0.98.

Since

$$\dot{q} = YUA \, \Delta T_m$$

where

$$\Delta T_m = \text{LMTD}$$

we have

$$A = \frac{\dot{q}}{U \, \Delta T_m \; Y} = \frac{1.88 \times 10^5 \; \text{W}}{330 \; \text{W/(m}^2 \cdot \text{K)} \times 62°\text{C} \times 0.98} = 9.38 \; \text{m}^2$$

The total area of the tubes for heat transfer is 9.38 m^2.

The tube area is calculated from the equation

$$A = \pi d l n$$

where d = outside diameter of tubes
  l = length of tubes
  n = number of tubes

Clearly there are many different possible combinations of tube diameter d, tube length l, and number of tubes to produce the required area. On the one hand we may have a short, fat exchanger with many tubes of small diameter, and on the other a very long, thin exchanger with a few large-diameter tubes.

The combination selected depends on a number of factors, including the requirements or constraints of the installation, the relative cost of manufacturing or purchasing different configurations, and the permissible or available pressure drop. Frank [1978] recommends that for preliminary work the tube-side velocities be selected in the range 3 to 6 ft/s, and the tube-side velocities for water be in the range 5 to 8 ft. The final decision as to exchanger geometry will probably be based on technoeconomic considerations including a balance of high fluid velocities and increased heat transfer against the increased pumping costs.

It is tempting to increase the surface area beyond that calculated as necessary, so as to provide a margin of reserve to allow for subsequent fouling effects. The additional area can be provided by increasing the number of tubes or extending the length of the tubes. In nearly all cases the best approach to providing a safety margin is to increase the length of the tubes, although, of course, this will likely increase the pressure drop of both the tube-side and shell-side fluids. Increasing the number of tubes to allow for deteriorioration due to fouling will likely prove to be a self-fulfilling prophecy. The increased number of tubes will result in decreased fluid velocity and thermal performance, with a consequent increased tendency for fouling.

The above work illustrates the use of the **LMTD** method in the preliminary selection of the proper size of heat exchanger for an application for which the fluid temperatures are known or can readily be determined. Now let us consider the situation in which an exchanger is already in place or has been selected. It is of interest to establish what change in the operation will occur as a result of some change in the operating conditions.

Let us assume that the water flow rate in the above example is reduced to half the value determined to be necessary to cool the oil. The reduced water flow rate will, of course, reduce the amount of heat transferred in the exchanger. Consequently, the oil will experience a reduced temperature decrease and will exit at a higher temperature. It is possible to speculate that the reduced flow of water will experience a greater increase in temperature and exit at a higher temperature. These changes in the oil and water exit temperatures are unknown, and it is not possible to determine the LMTD directly as before.

One possibility would be to assume a value for the rate of heat transferred $\dot{q}$ and then, from the equation

$$\dot{q} = \dot{m}_o c_o \, \Delta T_o = \dot{m}_w c_w \, \Delta T_w$$

calculate the temperature increases and decreases of the water and oil, ΔT_w and ΔT_o. The LMTD and the LMTD correction factor could then be determined. Finally, the rate of heat transfer could be calculated from the equation

$$\dot{q} = UAY \, (\text{LMTD})$$

If the rate of heat transfer were not the same as that assumed initially, it would be necessary to start again with a revised guess, probably the mean of the original and final values. This can become a tedious business, particularly in situations involving an ensemble of heat exchangers where some flows are split or otherwise diverted.

The alternative approach, using the NTU/effectiveness method, is preferred. For our example we assume the oil flow rate and *inlet* temperature remain at 170 kg/min and 110°C, respectively. Similarly, we assume that the inlet water temperature remains at 20°C and that the mass flow rate declines to 67 kg/min, half the original rate.

To use the NTU/effectiveness method, we first determine the capacity rates for both fluids, the product of the mass flow rate and the specific heat:

For oil:

$$\dot{m}_o c_o = C_o = \frac{170}{60} \, 1900 = 5383 \text{ W/°C}$$

For water:

$$\dot{m}_w c_w = C_w = \frac{67}{60} \, 4180 = 4667 \text{ W/°C}$$

The water is the minimum-capacity-rate fluid and will therefore experience a greater temperature change (increase) than the oil (decrease). The capacity ratio is

$$\frac{C_{min}}{C_{max}} = \frac{4667}{5383} = 0.867$$

The number of transfer units may now be determined:

$$\text{NTU} = \frac{UA}{C_{min}} = \frac{330 \times 9.2}{4667} = 0.650$$

The exchanger effectiveness may be determined by reference to the NTU/effectiveness chart given in Fig. 5.8, with NTU = 0.65 and $C_{min}/C_{max} = 0.87$. Under these conditions, the exchanger has a very low effectiveness ϵ of approximately 40 percent. This means the actual temperature swing of the minimum-rate fluid, water, is only 40 percent of the theoretical maximum value. In an ideal exchanger the water exit temperature could, in principle, increase from the inlet temperature of 20°C to the oil inlet temperature of 110°C. Therefore, the maximum possible temperature swing is

$$\Delta T_{max} = 110 - 20 = 90°C$$

If the exchanger effectiveness is only 40 percent, the actual temperature swing of the water is $\Delta T_{act} = 0.4 \times 90 = 36°C$. The exit temperature of the water is therefore $T = 20 + 36 = 56°C$. The rate of heat transfer in the exchanger with reduced water flow is

$$\dot{q} = \dot{m}_w c_w \, \Delta T_w = 4667 \times 36 = 168 \text{ kW}$$

Heat is lost by the oil at the same rate, so

$$\Delta T_o = \frac{\dot{q}}{\dot{m}_o c_o} = \frac{168 \times 10^5}{5383} = 31.2°C$$

The oil exit temperature is $110 - 31 = 79°C$ with the reduced water flow. The water flow rate was reduced by 50 percent from 135 to 67 kg/min; yet the thermal operation of the heat exchanger declined from 188 to 168 kW, a loss of only $(188 - 168)/188$ or 11 percent.

Finally, to check the calculation, let us rework the revised case by the LMTD method. The oil inlet and exit temperatures are

$$t_1 = 110°C \qquad t_2 = 79°C$$

The water inlet and exit temperatures are

$$T_1 = 20°C \qquad T_2 = 56°C$$

Then

$$\Delta T_1 = 110 - 56 = 54°C$$

$$\Delta T_2 = 79 - 20 = 59°C$$

and so

$$\text{LMTD} = \frac{54 - 59}{\ln (54/59)} = 56.8°C$$

$$X = \frac{t_2 - t_1}{T_1 - t_1} = \frac{79 - 110}{20 - 110} = 0.344$$

$$Z = \frac{T_1 - T_2}{t_2 - t_1} = \frac{20 - 56}{79 - 110} = 1.16$$

Hence, from the upper diagram of Fig. 5.6, $Y = 0.95$ and therefore

$$Q = UAY \, (\text{LMTD}) = 330 \times 9.38 \times 0.95 \times 56.8 = 168 \text{ kW}$$

This is in agreement with the value estimated by the NTU/effectiveness method.

FLUID-FRICTION EFFECTS IN HEAT EXCHANGERS

The thermal design of heat exchangers is directed to the provision of adequate surface area for heat transfer so the exchanger can handle the thermal duty for which it is required. Fluid-friction effects in the heat exchanger are equally important. They determine the pressure drop of the fluids flowing in the system and, consequently, the pumping or fan-work input necessary to maintain the flow. Provision of pumps or fans to generate the input work adds to the capital cost and is a major part of the operating cost of the exchanger. Savings in exchanger capital cost achieved by the purchase of a compact unit with high fluid velocities may soon be lost as increased operating costs. The final design and selection of a unit will, therefore, be influenced just as much by effective use of the permissible pressure drop and the cost of fan or pump horsepower as the temperature distribution and provision of adequate area for heat transfer.

The overall pressure drop in the exchanger is the cumulative total of a number of components. The primary loss generally occurs in fluid flow through tubes, in crossflow over the tube bank (which may or may not be finned) or, as in shell-and-tube heat exchangers, the combined crossflow and axial flow characterized by repeated passes through the tube bundle in baffled shell-side flow. Secondary losses occur as a result of the sudden contractions and expansions as fluid enters and leaves the exchanger through the inlet and outlet nozzles or enters and leaves the tube bundles. Reversals in the flow direction in multiple-pass exchangers also increase the pressure drop. The use of roughened surfaces increases the heat transfer but does so at the expense of increased pressure drop. Extended surface fins or tubes exposed to low-density airflows may be corrugated, curved, or louvered to enhance the heat transfer. Frequently, the cost in terms of increased fluid friction requires an input of mechanical work greater than the realized benefit of increased heat transfer.

The magnitude of the pressure drop for various flow situations in tubes, over tube banks, or in shell-side flow is generally estimated by an equation of the form

$$\Delta p = \frac{cf\rho u^2 LN}{D}$$

where Δp = pressure drop
   c = constant, characteristic of the flow situation
   f = fraction factor, characteristic of the flow situation
   ρ = fluid density
   u = fluid velocity
   L = length of tube
   N = number of tube passes
   D = diameter of tubes or some other characteristic dimension

The exact form of the equation depends on the flow situation. For flow over tube banks, the number of rows of tubes, the depth of the bank, and the arrangement of the tubes (whether they are in line or staggered) are obviously important. In shell-side flow, the type, number, spacing, and clearance of the baffles affect the magnitude of the pressure drop. Another important factor with many fluids having a strongly temperature-dependent viscosity is the ratio of the bulk fluid viscosity to that of fluid adjacent to the wall. This ratio of the viscosities is found to influence the pressure drop according to the 0.14 power and is sometimes included in the denominator of the pressure-drop equation as $\phi = (\mu_f/\mu_w)^{0.14}$.

The most important parameter affecting pressure drop is the fluid velocity. The pressure drop increases as the square of the velocity. A change in the fluid velocity is therefore more significant than a change in the other factors. Doubling the fluid velocity increases the pressure drop by a factor of *four*. Doubling any other factor in the numerator simply doubles the pressure drop; the relationship is linear. The velocity u of the fluid, this important parameter, depends on the mass rate of flow G, the density ρ of the fluid, and the cross-sectional area A of the flow, since $G = \rho A U,$ according to the well-known continuity equation. Sometimes the pressure-drop equation is expressed in terms of the mass rate of flow G rather than the velocity u and so must also be included as the squared term G^2.

The cross-sectional area A of the flow in the above equation depends on the diameter of the tubes. Thus, in computing the tube-side pressure drop, the total flow area is

$$A_T = \frac{\pi}{4} D_i^2 n$$

where A_T = total flow cross-sectional area
   D_i = tube internal diameter
   n = total number of tubes included in pass
For crossflow heat exchangers, the minimum cross-sectional area of flow (where the maximum velocity occurs) is in the "pinch" space between two adjacent tubes. The cross-sectional area for flow is

$$A_c = (S - D_o)Ln$$

where A_c = flow section area
   S = pitch, centerline distance between the tubes
   D_o = external diameter of tubes
   L = length of tubes
   n = number of transverse rows of tubes
   The friction factor f included in the equation for the pressure drop Δp is empirical and has been determined experimentally for great numbers of different possible tube diameters and arrangements, for smooth tubes and rough tubes, for different baffle and shell arrangements, and so on. Frank [1978] provided a good summary of these experimental data in chart form. A more extensive collection was given by Fraas and Ozisik [1965] and by Kays and London [1964].

The tube-side pressure drop and heat-transfer coefficient may be deter-mined much more accurately and reliably than the crossflow or shell-side pressure drops and heat-transfer coefficients. This is because the flow geometry is so well controlled and precisely determined. The nominal bore of the tubes is maintained to close limits, the number of tubes is accurately known, and the flow passage is relatively long, without abrupt changes in the flow section or direction.

Low-velocity, viscous liquids are more likely to become turbulent in shell-side flow than in tube-side flow. When a design study indicates that the liquid will be laminar in tube-side flow, the effect of changing to shell-side flow should be determined. Turbulent flow on the shell side may be an influential factor in determining the flow location. If the study shows the flow to be so slow or the fluid so viscous that it is still laminar on the shell side, it is probably better to revert to tube-side flow, for the fluid behavior can then be more accurately predicted.

SUMMARY

In this chapter, we have discussed some of the elements concerned with the thermal design of heat exchangers. The chapter is not intended as a comprehensive review of thermal design procedures or as a guide to anyone contemplating the design of a heat exchanger. We have included virtually nothing concerned with a large and important class of exchangers in which the fluid experiences a change of phase—condensers, evaporators, boilers, and reboilers. The paper by Frank [1978] that has been referred to many times contains an equally useful distillation of guidelines and preliminary design recommendations for phase-change systems. Taborek [1977] gives a brief overview of heat-exchanger design techniques that is highly recommended for those wishing to carry the subject further. An earlier [Taborek, 1974] critical review of design methods is more comprehensive and detailed.

For budding designers of heat exchangers, the books by Fraas and Ozisik [1965] and by Kays and London [1964] are highly recommended. The large book by Afgan and Schlunder [1974] contains a compilation of articles by many experts on all aspects of heat transfer. The ultimate reference, the Heat Exchanger Design Handbook, edited by Schlunder, has been in preparation for over 6 years and is now in press.

REFERENCES

Afgan, N. H., and E. U. Schlunder: "Heat Exchangers: Design and Theory Sourcebook," Hemisphere, Washington, D.C., 1974.

Cowans, K.: A Countercurrent Heat Exchanger that Compensates Automatically for Maldistribution of Flow in Parallel Channels, pp. 473–444 in K. Timmerhaus (ed.), "Advances in Cryogenic Engineering," vol. 19, Plenum, 1974.

Fraas, A. P., and M. N. Ozisik: "Heat Exchanger Design," Wiley, New York, 1965.

Kays, W. M., and A. L. London: "Compact Heat Exchangers," 2d ed., McGraw-Hill, New York, 1964.

Schlunder, E. U.: "Heat Exchanger Design Handbook," Hemisphere, Washington, D.C., 1981.

Taborek, J.: Design Methods for Heat Transfer Equipment—A Critical Survey of the State of the Art, pp. 45–74 in N. Afgan and E. U. Schlunder (eds.), "Heat Exchangers: Design and Theory Sourcebook," Hemisphere, Washington, D.C., 1974.

Taborek, J.: Evolution of Heat Exchanger Design Techniques, D. Q. Kern Memorial Lecture, *17th Nat. Heat Transfer Conf.*, Salt Lake City, 1977 (reprinted in *Heat Transfer Eng.*, vol. 1, no. 1, pp. 15–29, 1979).

6

Elements of mechanical design

INTRODUCTION

Nearly all heat exchangers serve the double duty of transferring thermal energy from one fluid to another and acting as a containment vessel for fluids under pressure. Most exchangers consist of two such pressure vessels that are intricately combined.

The separate but related functions of the heat exchanger are recognized by performing the design in two stages: thermal design and mechanical design. Thermal design is concerned with the provision of adequate surface area and flow guidance to ensure that the exchanger is capable of the required thermal duty with an acceptable fluid pumping power. Mechanical design is concerned with ensuring the integrity and durability of the exchanger as a pressure vessel.

These two activities, thermal design and mechanical design, are sometimes carried out simultaneously by the same designer. More frequently, however, specialized groups are concerned only with one or the other activity. This has the advantage that designers can develop a deeper understanding of the fundamentals and an awareness of recent developments in their specialty. It also has the danger that the mechanical and thermal designs will each proceed independently of the other, and it can lead to unnecessary constraints, expense, or delays as a result of a breakdown in communication between the two groups.

Some elements of thermal design have been surveyed in Chap. 5. Here we take a look at some of the concerns of those responsible for the mechanical design of heat exchangers.

SCOPE OF MECHANICAL DESIGN

The mechanical-design process embraces a very wide range of technical interest and activity, all directed to ensuring safe operation of the exchanger unit for the required life at minimum total cost of original purchase and subsequent operating costs, including inspection, maintenance, and repair.

The primary objective is, of course, to provide enclosures that are sufficiently strong to completely contain the fluids flowing through the system and, internally, to prevent leakage of one fluid to another. This requires detailed calculation and specification of the required thicknesses for the walls of shells, end covers, tubes, tube sheets, flanges, and nozzles. The materials to be used for construction must be selected not only from the aspect of strength to contain the design pressures, but also for their compatibility with the fluids to be used and with the other exchanger construction materials. Some fluids are corrosive to some metals and not to others. Combinations of different metals, in the presence of fluids acting as an electrolyte, function as electrochemical cells with consequent rapid decomposition. Some combinations of metals in mutual contact degrade rapidly at the joint interface by fretting corrosion when subject to vibration. Over time the physical properties of the materials may degrade, owing to changes in the metallurgical stucture arising from metal fatigue or interaction with trace corrosive agents in the fluids.

In addition to the normal operating conditions the design must be adequate to cope with transitory but hazardous situations. These may arise in normal service during start-up, shutdown, or unscheduled interruptions in fluid flow rates. The possibility of pressure surges in the fluid lines may exist. In many locations the possibility of an earthquake cannot be ignored. Provision must be made for adequate and safe lifting points; otherwise very high stresses in localized areas may arise during fabrication, transportation, and installation.

Vibrations arising from a number of sources are a matter of concern, for a coincidence of the forcing frequency with the natural frequency of the tubes can lead to disastrous damage to the exchanger.

The prospect of future cleaning operations to remove accumulated sludge and scale must be anticipated with the provision of means to dismantle, remove, clean, repair, replace, and reseal the exchanger assembly. Means to identify and then remove and replace defective tubes may be included in the design. Alternatively, the design may be sufficiently flexible to allow defective tubes to be plugged and thereby eliminated from the fluid circuit.

An important aspect of mechanical design is concerned with gasket-sealing the high-pressure stationary joints and flanges of the bonnet covers and inlet and outlet nozzle connections. Another area of importance to sealing is the design of packing boxes used for packed floating-tube-sheet or floating-head designs. Mechanical design is also concerned with the detailed design and spacing of baffles and the tube support plates to minimize the possibility of undue tube motion or fluid bypass flow.

CODES AND STANDARDS

Mechanical design is much more closely regulated by codes and standards than thermal design. This is because failure of the exchanger as a pressure vessel could be extremely hazardous to personal safety. Failure of the exchanger to fulfill its thermal duty is not usually hazardous so much as an embarrassment to the maker and the user.

Throughout the world there is a bewildering array of standards and codes for pressure vessels and heat exchangers. One hopes that the Pressure Vessel Standards of the International Standards Organization will eventually be adopted and become the uniform standards for all regulatory authorities. This aspiration is likely to remain unfulfilled for a considerable time to come, so until then it will be necessary to continue with the existing plethora of national standards.

In North America the regulatory authorities for the various federal departments, states or provinces, and cities have generally adopted the ASME code.* This is a voluminous compendium of material in which the design rules for various types of pressure vessels and components are given. Allowable stresses are specified as a function of alloy, metal temperatures, and the nature of the applied stress. Section VIII is generally referred to as most applicable to heat exchangers. The ASME code is the subject of continuous scrutiny and improvement by committees of ASME members and the principal regulatory authorities.

In Britain and many countries of the Commonwealth, various British Standards Institute (BSI) standards form the basis of the pressure-vessel regulations of the national regulatory bodies. Many other countries, in Europe and elsewhere, also have their own pressure-vessel standards that appear to have much in common with the ASME and BSI standards. At the same time, there are sufficient idiosyncratic differences to make it imperative to consult the local standards for offshore assignments.

In addition to the pressure-vessel standards that are specifically directed to safety, various other standards for heat exchangers are directed more to economy, operational features, and maintenance. The best-known of these supplementary standards is undoubtedly the TEMA standards.† These are specific to tube-and-shell heat exchangers and contain a wealth of data and information relating to both mechanical and thermal design. The standards were assembled by committees of the association and have been developed through various editions to represent a distillation of the experience of all manufacturers. Various other compilations of a similar nature have been assembled by other special-interest groups and published to facilitate uniformity and good practice. The principal compilations are listed in Chap. 10.

Yokell [1978] has given an entertaining and detailed review of the mechanical design and fabrication of tube-and-shell heat exchangers, with a good discussion of the use and interpretation of the various codes and standards.

PROPERTIES OF MATERIALS

Materials used for heat exchangers are selected primarily because of their:

---

*The Boiler and Pressure Vessel Code of the American Society of Mechanical Engineers.
†Standards of the Tubular Exchangers Manufacturers Association, New York (sixth edition, 1978).

1. Strength properties at the appropriate temperature levels
2. Ability to resist corrosion, chemical degradation, or erosion in the environment and in contact with fluids during exchanger operation
3. Cost
4. Availability
5. Lack of tendency to fouling and scale formation
6. Ease of cleaning

Strength Properties

The property most significant from the aspect of stress analysis is the stress-strain
curve. A typical stress-strain curve for a metal is shown in Fig. 6.1. The curve
shows the relationship between the stress imposed and the strain experienced
when a specimen of the metal is stretched, in tension, in a testing machine.
Initially (section *AB*) the specimen behaves elastically like a piece of rubber and
stretches by an amount proportional to the stress imposed. If the stress is
reduced, the specimen returns to its original length and shows no evidence of
having been subjected to strain. This part of the curve is called the *elastic region.*

Section *BC* of the curve is called the *plastic region.* When the applied stress
exceeds that at *B,* the material experiences an increased rate of strain per unit of
stress and adopts a *permanent set.* Now when the load is removed the material
does not return to its original length but follows a new path such as the line *DE.*

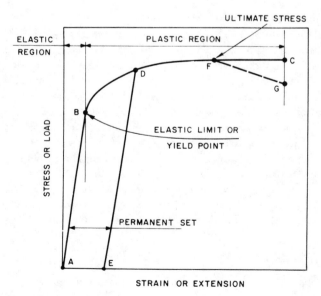

FIGURE 6.1
Typical stress-strain curve for a metal.

When the load is completely removed, the specimen remains permanently extended by an amount AE.

If the load is not removed as at point D but is increased still further, the plastic strain continues until eventually the specimen breaks.

Some materials are very ductile (have a high strain). Sometimes they experience a very large extension near the point of eventual fracture. This large local strain results in a marked reduction in the surface area in that region and, as a result, the stress (load per cross-sectional area) actually decreases (section FG).

Point B, the transition between the elastic and plastic regions, is called the *elastic limit* or *yield point,* and the stress at B is called the *yield stress.* The stress at C (or the maximum stress F) is called the *ultimate stress.*

Heat exchangers are invariably designed for the material to remain in the elastic region, section AB of the curve in Fig. 6.1. Experience has shown the possible deviations are substantial. It is therefore customary to use a factor of safety in design, so the maximum calculated stress is considerably less than the limiting yield stress or ultimate stress. Typically the factor of safety (or uncertainty!) is 2 for the yield stress or 4 for the ultimate stress, but design to a reference yield stress is preferred. The use of a factor of safety allows for some variation in the properties of the material, uncertainties in the loading conditions and stress-concentration factors, a margin of safety for unexpected loadings, and poor workmanship.

Sometimes the transition between the elastic and plastic regions is difficult to determine, so there is uncertainty about the location of the yield point. In these cases it is customary to use the *proof stress* as an alternative. This is illustrated in Fig. 6.2. A line is drawn on the stress-strain curve parallel to the straight part of the curve but displaced from it by a given strain—usually 0.1 percent of the length of the test specimen (5 or 20 cm). The stress at the point P where this line crosses the stress-strain curve beyond the yield point B is called the *proof stress.* To avoid the possibility of confusion it is customary to define the given strain and the specimen gauge length, so that one refers to the "0.1-percent proof stress on a 2-in gauge length" or whatever other reference strain has been used.

Effect of Temperature

The general effect of temperature on the tensile properties of metals is that the yield stress and the ultimate stress both decrease with an increase in temperature. A typical strength characteristic for a steel alloy as a function of temperature is shown in Fig. 6.3. The yield stress tends to be related to temperature in a more or less straight-line relationship; there is a progressive decrease in strength with increasing temperature. The ultimate stress tends to remain relatively unaffected by temperature changes at lower temperatures and to decrease sharply at higher temperatures.

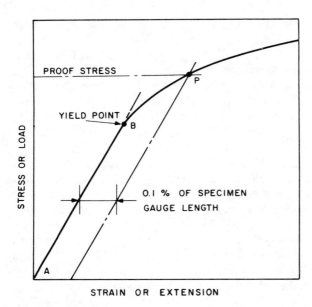

FIGURE 6.2
Proof stress and yield stress.

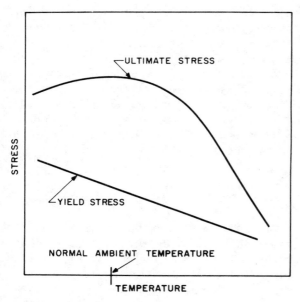

FIGURE 6.3
Effect of temperature on the short-term yield and ultimate stresses of a metal.

The characteristic decrease in strength with an increase in temperature prevails with all metals likely to be found in heat exchangers. The range of possible variation, however, is so very wide and diverse that it is not possible to establish rule-of-thumb guidelines of any substance. Tables of data and charts for the specific materials of interest must be consulted.

Creep

When a material is loaded in any way, it deforms, virtually instantaneously, by an amount proportional to the load. The deformation may be permanent if the induced stresses exceeded the yield stress. When the load is sustained, some further deformation occurs but at a low and progressively diminishing rate. This additional deformation is called *creep.*

The effect is negligible at low temperature and stress levels but becomes increasingly significant as the temperature increases. Figure 6.4 shows the effect of temperature on the creep strength of an alloy steel. The upper curves show the conventional ultimate stress to cause fracture and the yield stress. Of the three curves below these, the upper one shows the stress that will result in rupture in 1000 h. The lower curves show the stress that will result in a 1 percent creep in 10,000 and 100,000 h.

Fatigue

The stresses induced in a structural element will fluctuate with time as the loading changes. The change may arise in any number of ways: fluctuating wind loads, emptying and filling of a pressure vessel or tank, heating and cooling of a system, rotation of a loaded shaft, dynamic vibration, and so on. Experience has shown that materials subject to periodically fluctuating loads fail at stress levels significantly below the ultimate stress. This is called *metal fatigue.* The range of the stress variation, the number of stress reversals, the temperature of operation, and the heat treatment, cold working, or fabrication technique employed are all important factors. For a given metal in a specified condition, the relationship between the range of stress variation and the number of reversals is usually presented as shown in Fig. 6.5, a diagram called the *S-N curve* for some specified temperature. The number of cycles N is usually plotted on a logarithmic scale. The *S-N* curve is a useful indicator of permissible stress as a function of life. Combinations of stress level and number of stress reversals below the curve ABC in Fig. 6.5 are safe and permissible in design. Combinations above the curve, such as point D, are not permissible. They could never be realized in practice. At the given range of stress variation f, the element would fail at the low number of stress reversals N_1. Alternatively, to achieve the desired life N_2, the range of stress variation would have to be reduced to f_2.

Fatigue failure starts as a small crack at some point of local stress concentration—a sudden change in cross section, a hole, or an occlusion in the

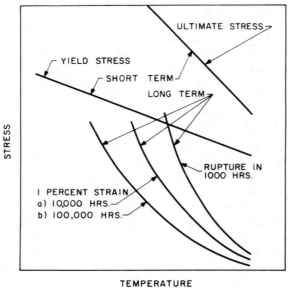

FIGURE 6.4
Typical creep-strength characteristics for steel alloy.

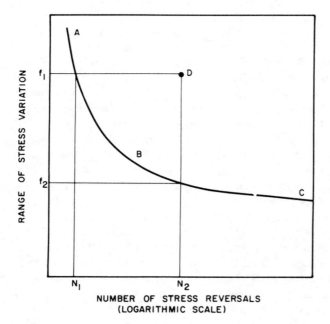

FIGURE 6.5
Typical fatigue-strength characteristics (*S-N* curve) for a metal.

metal. The crack then spreads progressively across the element until eventually the stress in the reduced cross section exceeds the ultimate stress and the element is sufficiently weakened for sudden, catastrophic failure to occur.

Stresses can arise from the imposition of external mechanical loads such as pressure forces, weight of pipeline or tank contents, wind or snow loads, etc., or from internal strains arising from differential thermal expansion.

In heat exchangers, fatigue failure by thermal strain cycling is more common than failure by pressure strain cycling or vibration. Thermal stresses differ from pressure stresses in that they are relieved by a small amount of plastic strain. This is very effective in ductile materials. However, only a few repetitions of strain beyond the yield point into the plastic region are necessary to initiate the formation of thermal fatigue cracks. This is a form of accelerated fatigue where only a few dozens or hundreds of repetitions will induce failure, instead of the customary several hundred thousands. For this reason, in the design of heat exchangers every consideration must be given to avoiding the onset of high stress levels due to temperature effects.

Impact Strength

The *toughness* of a material as measured by the Izod or Charpy impact strength is an important criterion of the ability to resist sudden or shock loadings and of the resistance of the material to crack propagation. The impact strength is determined by measuring the energy absorbed to fracture a specimen of the material. This is accomplished by holding the specimen rigidly in a vise in the path of a heavy weight swinging like a pendulum from a given height and, hence, having a known potential energy. A notch of specific dimension is usually, but not always, cut into the specimen, and the impact strength is frequently referred to as the "notch toughness" of a material.

This property is of most importance in low-temperature engineering. Materials that are ductile at normal temperatures and above become brittle at low temperatures and snap off like a carrot when given a light blow that they normally would sustain without incident. This ductile-brittle transformation is accompanied by a dramatic change in the Izod or Charpy impact strength, as shown in Fig. 6.6. The temperature at which the decrease in impact strength occurs is called the *transition temperature*. Rubber is perhaps the most familiar example of an elastic ductile material that becomes hard and brittle at low temperatures. Low-carbon mild steel is a common structural material that has a transition temperature of about $-20°C$.

Fortunately, not all materials experience the ductile-brittle transformation. Many metals (nickel, copper, and aluminum, for example) exhibit no sudden decrease in their impact strength down to the lowest temperatures. Rather, their impact strength improves or declines only slightly as the temperature decreases. This difference in behavior arises from the molecular structure of the metal. Most metals can be characterized as having either a body-centered or a face-centered

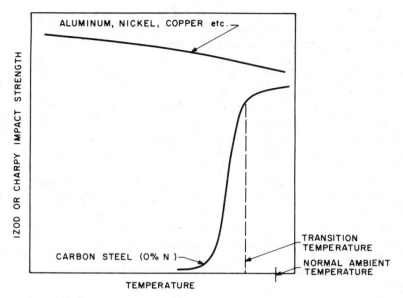

FIGURE 6.6
Effect of temperature on the fracture toughness of metals.

cubic lattice arrangement. Metals with a face-centered cubic lattice experience the ductile-brittle transformation; those with a body-centered cubic lattice do not.

The addition of nickel to carbon steel delays the onset of the ductile-brittle transformation, as shown in Fig. 6.7. This is why $3\frac{1}{2}$-percent nickel alloys are used for propane and liquid petroleum gas (LPG) construction, and 9-percent nickel alloys for liquid natural gas (LNG) construction. Stainless steels, aluminum, and copper have high impact strength at low temperatures and are widely used for cryogenic engineering equipment.

Impact strength is an excellent criterion of the tendency to crack propagation in a material. Once initiated in a local zone of high stress concentration, a crack will readily propagate through a material of low impact strength. The end or tip of the crack itself is a source of stress concentration, so the crack will continue to propagate until catastrophic failure occurs or until it penetrates a zone of higher impact strength.

Heat Treatment

The mechanical properties of a metal are primarily dependent on the crystal structure of the metal. Heating, rapid quenching, or slow cooling can cause changes in the metallurgical structure.

Carbon steels have a characteristic transformation temperature above which hard iron carbides dissolve uniformly in an *austenitic* phase. Quenching of the heated metal freezes the carbides, forming a hard, brittle structure called *martensite.* Slow cooling allows dissociation of the iron carbides from the *ferrite* phase, resulting in a soft ductile steel.

Tempering is the process of reheating a quenched martensitic steel to a temperature below the transformation temperature, followed by moderate cooling. This improves the ductility and toughness of the steel. The higher the tempering temperature, the greater the reduction in hardness.

Annealing is the process of heating to a temperature greater than the transformation temperature, followed by slow cooling. This results in soft, ductile metal of the lowest possible hardness.

Normalizing is the process of heating to a temperature 50 or 60°C above the transformation temperature, followed by rapid air-blast cooling (a moderate rate of cooling compared to water quenching). This results in a fine-grain steel that is softer than quenched steel but harder than annealed.

Cold Working and Forming

The mechanical properties of metals can be dramatically changed by the cold working or forming it experiences in the various manufacturing processes of rolling, pressing, and bending. *Cold forming* is the process of causing a metal to yield at ambient temperature by mechanical means. The external fiber elongation is best limited to less than 5 percent. *Cold working* is the term used to describe

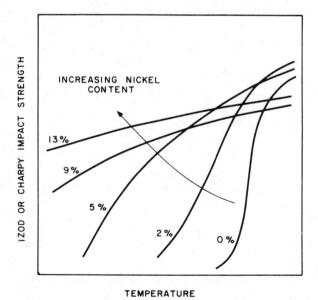

FIGURE 6.7
Effect of the addition of nickel on the fracture toughness of carbon steel.

hardening of the metal following cold forming. The hardening arises from the buildup of residual stresses in the metal and may be eliminated by annealing or heating to a temperature below the transformation temperature, followed by slow cooling. Sometimes a hard surface on the metal is induced by shot-peening, a process that causes cold working of the metal close to the surface. Some metals, particularly certain aluminum alloys, have markedly different strength properties in the fully annealed and the cold-rolled (cold-worked) conditions. If the design is conditional on the metal retaining its cold-rolled strength properties, great care must be exercised during welding to minimize the stress-relief effects that ensue from the heating process.

Weldments

Welded joints or other weldments can result in substantial changes in the mechanical properties of the metals in the welding zone. The effect is particularly pronounced for metals that are heat-treated or cold-worked to establish a particular metallurgical structure with desired strength characteristics. In the heat-affected zone around the weld, the metal may become annealed or its crystal structure may be modified, resulting in a substantial change in the metallurgical properties. In the weld itself, the weld metal will almost certainly be different from the base metal and consequently will have its own character-istic strength properties—frequently superior to those of the base metal.

Preheating of the welded joints is frequently adopted for high-carbon and for low-chrome–molybdenum steel alloys. Preheating of the joint to be welded minimizes the rapid flow of heat away from the molten joint metal and thereby minimizes the tendency for the formation of hard, brittle martensite, subject to cracking.

Another treatment sometimes used in welding processes is *stress relieving*. After the weld is completed, the area of the joint is heated to a temperature just below the transformation temperature and allowed to cool relatively slowly, so as to relieve the residual stresses arising from the welding process.

Thermal Properties

The thermal properties of interest for heat exchangers include the:

1. Coefficient of thermal expansion
2. Specific heat
3. Thermal conductivity

By far the most important of these is the coefficient of thermal expansion. This is the amount by which a material changes in length (linear coefficient) or in volume (volumetric coefficient) per unit length or per unit volume for each 1-degree change in temperature. The magnitude of the coefficient varies widely from a maximum of 2×10^{-5} m/(m·°C) for aluminum to half this value for carbon steel and to relatively negligible values for some ceramics and glasses.

The coefficient of thermal expansion is important in heat exchangers because of the high stress levels induced if the elements are mechanically constrained and not free to move to accommodate the dimensional changes arising from temperature variations. Consider, for example, a type BEM tube-and-shell heat exchanger having fixed tube sheets, as shown in Fig. 3.8. The hot fluid would likely pass through the tubes, so the tube-metal temperature would be appreciably greater than the shell-metal temperature. Therefore the tubes would tend to increase in length more than the shell. With a fixed-tube-sheet design the tubes would be loaded in compression, since they are inhibited from expanding by the shell. Similarly, the shell would be loaded in tension by the effort of the tubes to expand. Stresses arising in this way are relieved by including a flexible expansion joint in the shell casing, as shown in Fig. 3.8, or by modifying the fixed-tube-sheet design to allow one end of the tube bundle to float, as shown in Fig. 3.7.

STRESS CONSIDERATIONS

Stress considerations are relatively unimportant in many heat exchangers operating at moderate temperatures and pressures. They become increasingly significant at higher temperatures and pressures and ultimately dominate design and material selection. Fraas and Ozisik [1965] suggest that the effects of static weight and pressure loads can be

> ... predicted with roughly the same confidence as the heat transfer and pressure drop characteristics (that is, with a probable error of 20 to 50 percent depending on the complexity of the system) and the problems involved are roughly equivalent in difficulty to those in the fluid flow and heat transfer fields. The life of a structure under repeated, severe, thermal cycling conditions is much more difficult to predict analytically—so much that the uncertainty in the life to failure may be as much as a factor of 10.

Pressure Effects

The fluid pressures to be contained within the exchanger will influence and, at high pressures, largely dictate the thickness and the type or quality of metal used to construct the exchanger. The fluid pressure levels affect not only the strength requirements of the exterior envelope but also the thickness of the walls separating the two fluids.

In some cases, for example, the plate-and-frame heat exchanger shown in Fig. 4.2, the principal loads arising from the fluid pressure can be carried entirely by compression bolts and robust end covers that play no part in the heat-transfer process. In other designs, say, tube-and-shell or panel-coil heat exchangers, the same elements may serve double duty as heat-exchange surfaces and as primary structural members.

Shape

The optimal shape to contain a given volume of fluid at a specified pressure is a sphere. A spherical vessel will have a thinner wall and so weigh less than any other shape.

For a relatively low pressure, thin-wall vessel, the wall thickness necessary to contain the pressure may be determined from the equation

$$\sigma = \frac{\text{load}}{\text{area}} = \frac{p(\pi/4)D^2}{\pi D t} = \frac{pD}{4t}$$

where σ = stress due to pressure
p = internal pressure
D = diameter of sphere
t = wall thickness

When the design stress σ is some fraction x of the yield stress σ_y, then

$$t = \frac{PD}{4x\sigma_y}$$

Spherical vessels are relatively expensive to fabricate and, in many cases, less convenient than cylinders, the next best shape from the point of view of wall thickness and weight. This is one reason why tubes and cylindrical shells are so widely used. The other reasons are that it is easy and low in cost to drill a circular hole for a tube, to turn a circular tube-sheet disk on a lathe, or to roll a flat sheet into a cylindrical shell.

For a low-pressure, thin-wall, cylindrical vessel, the wall thickness necessary to contain the pressure can be determined from the equation

$$\sigma = \frac{\text{load}}{\text{area}} = \frac{pD}{2t}$$

Note that, for a cylinder having a given internal pressure and diameter, the wall thickness t is *twice* that necessary for a sphere.

Cylinder Ends

The ends of cylinders are subject to severe bending stresses and high stress-concentration factors, requiring the use of thick, heavy flanges or end covers to eliminate high local stresses. Some of the many variations for cylinder ends are shown in Fig. 6.8. There is no single optimal solution and the method adopted depends on the situation.

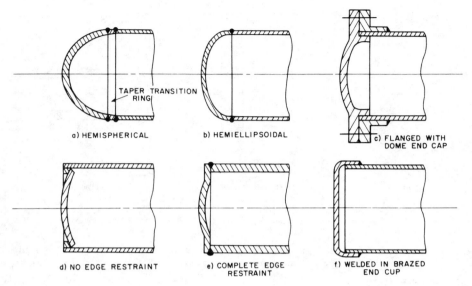

FIGURE 6.8
Design variations for cylinder end closures.

Penetrations

In most pressure vessels provision must be made for the fluid inlet and outlet connections and, in large vessels, access for inspection, cleaning, and maintenance. Local stress concentrations of up to three or four times the average value arise in the vicinity of circular holes in a vessel subject to internal fluid pressure. The stress-concentration factor is a function of the radius of the hole and increases with decreasing hole size. The effect of this local stress concentration can be overcome by increasing the thickness of the metal around the penetration by as much as three times. Typical methods of reinforcement are shown in Fig. 6.9.

Many pressure-vessel designs feature a hemispherical or ellipsoidal shell or bonnet, attached, by welding or bolted flange joint, to a cylindrical section. Where the design permits, it is good practice to concentrate the penetrations through the shell in the end bonnet. The bonnet can then be made sufficiently thick to obviate the need for local reinforcement in either bonnet covers or shell body.

Flanges

The use of a flanged joint is attractive because it facilitates subsequent disassembly of the exchanger shell for maintenance, repair, and cleaning. However, use of the flange inevitably sets up additional bending stresses in the shell, which must be compensated for by increasing the thickness of the shell

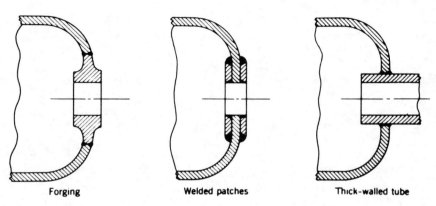

Forging	Welded patches	Thick-walled tube

FIGURE 6.9
Methods of reinforcing pressure-vessel penetrations. (*After Fraas and Ozisik [1965].*)

near the wall. This and the robust nature of the flange itself can contribute appreciably to the overall weight of an otherwise thin-walled, lightweight, low-pressure vessel.

Details of typical flange joints are shown in Fig. 6.10. The flange attached to the pressure vessel is shown as a "welding-neck" flange on the left and as a "slip-on" flange on the right. The bottom flange is a "blind flange" used to cover the end of the vessel.

Flange joints are not practical for large diameters because of the large amount of mass that they add. Furthermore, their use in very high pressure

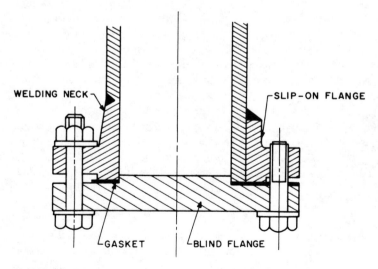

FIGURE 6.10
Flange joints.

systems is to be discouraged, except where there is no alternative, because of the large increase in mass and the possibility of leaks.

Manholes

Access to large vessels may be allowed by providing a manhole. It is obviously advantageous to design the manhole so the cover plate is located inside the shell, as shown in Fig. 6.11. The internal pressure then acts on the cover plate to assist in the sealing process. A lightweight locking system is necessary to locate and hold the cover in place when there is no internal pressure.

Circular manholes are not customary. With internal sealing, a circular cover could not be passed through the hole after the vessel was sealed. It would need to be included before the vessel was finally welded up, and there would be no way of replacing it. To overcome this difficulty, manholes are usually elliptical in shape. The cover and cover gasket can then be passed through the shell, provided the minor axis of the cover is smaller than the major axis of the manhole.

Enclosure Design

The above considerations of the stress due to pressure all combine to encourage design of the exchanger enclosure as a long, thin, cylindrical shape with hemispherical ends, minimal flange joints, and concentration of the enclosure penetrations, as shown in Fig. 3.14—a tube-and-shell exchanger with a two-pass shell, a two-pass "hairpin" tube bundle, and a single flanged joint.

Gaskets

In heat exchangers gaskets are used for static seals at nozzle flanges and bonnet covers, to prevent the escape of fluids from the exchanger or the intermixing of the streams in the exchanger.

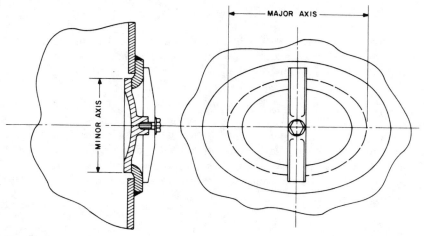

FIGURE 6.11
Manhole design showing internal seal and cover-plate preload support beam.

There are various types of gaskets in general use. Asbestos gaskets of one sort or another are quite common, but concern about the health hazards associated with asbestos may increasingly inhibit future applications. For non-hazardous, nonflammable service, asbestos-based sheet material is used, from which the gaskets are easily cut or stamped. For higher pressure applications, metal-jacketed, asbestos-filled gaskets provide a more rigid element. A variety of metals are used, depending on the fluids involved. It is essential the gasket metal be in the annealed state to facilitate sealing. Spiral-wound, asbestos-filled metal gaskets are used for high-pressure service. Frequently a solid metal stop ring is included to prevent overcompression.

Solid soft-metal gaskets are used for high-pressure applications. These may be flat metal gaskets of lead, copper, or a copper alloy. For very low temperature systems, indium or lead wire in male and female flange grooves is widely favored. For hydrogen service at high pressure, solid metal rings of oval or octagonal section are used in conjunction with special flange-face ring grooves.

Elastomeric gaskets are made from a variety of synthetic rubbers. These gaskets are designed to seal the flange surfaces with only minor compression. They are customarily found in the form of O rings of circular cross section, seated in grooves cut in one flange face and sealing against the flat face of the adjacent flange. Sometimes PTFE Teflon plastic and a variety of other plastic materials are used in O-ring or solid flat gaskets.

In multiple-pass tube-and-shell heat exchangers, rib gaskets are used to separate the various flow streams, to inhibit bypassing or fluid stream intermixing.

Gaskets should, of course, never be used after being compressed between two flanged surfaces, even if the system was never put into service. This is a particularly difficult practice to enforce. Frequently gaskets appear to be "as new," and concerned employees feel they are effecting economies by reusing what appear to be perfectly good elements. The savings achieved in nine cases can frequently be overwhelmed by the cost of the increased maintenance, lost production, environmental regulations broken, and health hazards resulting from fluid leakage with the tenth reused gasket.

Thermal Stresses

Thermal stresses arise as a consequence of the increase in the length or volume of a material with an increase in temperature. If the dimensional changes are unrestrained, there is no thermal stress. If the dimensional changes are restrained, the stresses induced can become sufficiently high to exceed the yield point and so cause plastic strain and permanent deformation.

To see this, consider a rod of length L, heated uniformly so the temperature increases by ΔT. Now the rod material will have a characteristic property, the coefficient of linear thermal expansion α, the amount by which a unit length of the material will increase per unit temperature change.

If the rod is unrestrained, the length of the rod increases, as a result of the temperature change, by the amount

$$\Delta L = \alpha L \ \Delta T$$

Now consider the alternative situation, where the ends of the rod are held so rigidly that no increase in length is possible. This corresponds to the situation where a rod of length $L + \Delta L$ at some uniform temperature is *compressed* by ΔL to length L. This would be accomplished by exerting a compression force on the rod. If the stress induced is less than the yield stress, it can be calculated from the relationship

$$E = \frac{\text{stress}}{\text{strain}}$$

where E is called the *modulus of elasticity*. Then the stress σ is

$$\sigma = E \times \text{strain} = E \ \frac{\Delta L}{L + \Delta L}$$

and, approximately,

$$\sigma = \alpha E \ \Delta T$$

For carbon steel, the modulus of elasticity is $E = 30 \times 10^6$ lb/in^2, and the coefficient of thermal expansion is 6.5×10^{-6} in/(in·°F). Therefore the thermal stress induced in a completely restrained mild-steel structure when heated through a temperature change of $100°F$ is approximately

$$\sigma_t = 30 \times 10^6 \times 6.5 \times 10^{-6} \times 100 = 20{,}000 \ \text{lb/in}^2$$

The yield stress of mild steel is 39,000 lb/in^2. A temperature increase of only $200°F$ would therefore cause a thermal stress equal to the yield stress in the material. There may also be some pressure stress involved in the loading. Depending on the nature (tensile or compressive) of the pressure and thermal stress, the combined stresses may add or substract.

Aluminum has a modulus of elasticity of 1×10^7 lb/in^2 and a coefficient of thermal expansion of 1.3×10^{-6} in/(in·°F). The thermal stress induced in a completely restrained aluminum structure when heated by $100°F$ is

$$\sigma_t = 10 \times 10^6 \times 13 \times 10^{-6} \times 100 = 13{,}000 \ \text{lb/in}^2$$

The yield stress of aluminum ranges from 13,000 lb/in^2 for commercially pure metal to a maximum of 40,000 lb/in^2 for wrought (cold-rolled) aluminum alloy.

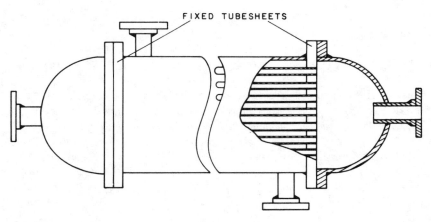

FIGURE 6.12
Thermal-stress problems in a single-tube-pass, single-shell-pass heat exchanger. With hot fluid in the tubes, differential thermal expansion will cause the tubes to be in compression and the shell in tension.

Of course, it is rare that structures are fully restrained. Nevertheless, the above example serves to illustrate the point that thermal-stress considerations are by no means trivial.

Thermal stresses can arise in two different ways. Different materials will likely have different coefficients of thermal expansion. Therefore, a structure composed of different materials will likely have some thermal stresses induced even when heated in such a way that the whole structure changes temperature uniformly. Conversely, a structure may be made of a single material or of materials carefully selected to have the same coefficient of thermal expansion. This structure too will likely experience thermal stress if different components operate at different temperatures. Finally, severe thermal stresses may arise in a structure during periods when the temperatures are changing. Large parts having high mass will experience a slower rate of temperature change than lightweight structural parts, even though the same temperature eventually prevails throughout.

Contractions due to cooling can result in thermal-stress problems just as severe as those caused by expansion due to heating. Stress problems due to contraction are of concern in cryogenic systems.

Heat exchangers are particularly susceptible to thermal-stress problems, for their *raison d'etre* involves substantial differences in temperature among the various parts of the exchanger. An obvious source of thermal-stress problems is illustrated by the tube-and-shell heat exchanger shown in Fig. 6.12. This is a single-tube-pass, single-shell-pass exchanger with fixed tube sheets. The hot fluid will pass through the tubes, and the cold fluid through the shell. The mean temperature of the tubes will therefore be significantly above the mean temperature of the shell. This will be reflected in different thermal expansion of the tubes and of the shell; as a result of the fixed-tube-sheet design, the tubes will be subject to compressive stress and the shell will be subject to tensile stress. In practice, these thermal stresses are eliminated in fixed-tube-sheet designs by including a flexible expansion joint in the shell, as illustrated in Fig. 3.8. To avoid the use of the relatively complicated, expensive, and nonreplaceable expansion joint, it is necessary to abandon the fixed-tube-sheet design in favor of a floating-head or U-tube-bundle arrangement. In some cases the hairpin U tubes cannot be used if a straight-through single-pass arrangement is required. Use of the floating-head exchanger is always possible, but it tends to be expensive and introduces the possibility of leakage between streams or to the environment.

Some exchanger applications may appear to have no need, under normal operating conditions, for a shell expansion joint in the fixed-tube-sheet design. Difficulties will likely arise, however, from the more severe conditions that may obtain during transient operation, start-up, shutdown, or some process upset.

Sometimes, to avoid any possiblity of interleakage and mixing of the two fluids, it is necessary to resort to a double-tube-sheet design. Two possibilities for this are shown in Fig. 6.13. When the temperatures of the hot and cold tube sheets are significantly different, the differential radial displacements of the tube sheets may impose appreciable bending and shear stresses on the tubes in the gap between the tube sheets, as illustrated in Fig. 6.14. The stresses thus imposed on the tubes can be minimized by increasing the gap between the tube sheets. Extending the gap does not affect the heat transfer in any way, but it does, of course, increase the pressure drop, add the cost of extra metal, and increase the space required for installation and shipping.

Thermal-stress problems often arise as a result of unequal tube temperatures. Even in single-tube-pass fixed-tube-sheet designs, there are likely to be significant differences in the temperatures of the various tubes. This comes about because of different rates of fluid flow through the tubes (maldistribution of flow) or, on the shell side, over the tubes, as a result of fouling accretions, misplacement of baffles, etc.

In multiple-pass tube arrangements, even more severe thermal gradients will appear. In the "ribbon" flow-tube layout, the temperature difference between adjacent banks of tubes is usually 30°C (often taken as a design rule of thumb for the limiting average metal temperature difference between adjacent tube banks). In the "quadrant" flow-tube layout, the hottest and coldest tubes are adjacent, with a temperature difference often as high as 90°C. In the "mixed" flow arrangement, temperature differences of 60°C commonly occur between adjacent tube banks. The ribbon flow will likely induce less severe thermal stresses than any of the alternative arrangements.

Tube Sheets

From the above consideration of pressure and thermal stress, it is evident that the tube sheets of a tube-and-shell heat exchanger fulfill a vital and demanding

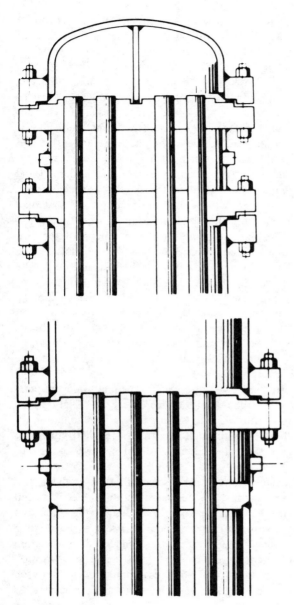

FIGURE 6.13
Double-tube-sheet designs. (*After Yokell [1978].*)

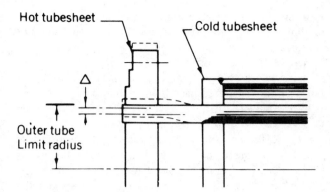

FIGURE 6.14
Shear stress on tubes in double-tube-sheet designs arising from differential thermal expansion of tube sheets. (*After Yokell [1978].*)

role. They act first as a diaphragm, separating the hot and cold fluids in the bonnet and shell of the exchanger. They sustain the differences in pressure between the tube-side and shell-side fluids and the temperature difference between the fluids. The tube sheet is the primary support of the tubes in both the radial and longitudinal directions. Holes are drilled in the tube sheet for the tubes, and the tubes are attached to it by a variety of methods: welding, brazing, or, more usually, by expanding the tube into the hole by rolling or swaging. The material of the tube sheets between the tube holes, called the *ligaments,* provides the strength to maintain the integrity and stability of the tube sheet and the tube bundle.

Tube sheets and tube-sheet design are not presently covered by the ASME code, since they are not vital to the safe containment of the pressurized fluids; however, extensive coverage of tube-sheet design is included in the TEMA standards (see Chap. 10). There are many who advocate extension of the ASME code to embrace tube sheets, and this development is expected in the near future.

Miscellaneous Stresses

In addition to the pressure and thermal stresses considered above, a number of other stress considerations may be important. Superimposed loads may be created by stacking exchangers one upon the other, or by mounting some equipment or insulation on the surface. The connecting piping may impose axial, bending, or torsional forces on the inlet and outlet nozzles.

Large, heavy exchangers filled with dense fluids may be inadequately supported, so that high stresses are induced at the support saddle locations. Sometimes the weight of the exchanger and its contents are the principal stress loads.

Wind loads are important in external locations, especially for a long, vertical exchanger free-standing on its base or suspended on columns. For horizontal units in northern climates, snow loads may be significant.

In many areas of the world the possibility of earthquakes must be considered, and techniques must be adopted to minimize the hazards consequent in their occurrence. Special attention to this aspect is important where noxious or dangerous fluids are to be contained. All nuclear containment vessels should be earthquakeproof.

Significant stresses can arise in the course of fabrication, transportation, and installation. Provision should be made for reinforcement around the lifting points, and instructions prepared and carefully followed for loading, unloading, and installation procedures. Abuse should be anticipated and specifically countered with warning notices stenciled on the vessels. Similar procedures should be followed in routine and major maintenance work, to reduce or eliminate the prospect of damage arising from local overload conditions.

Impact loads may arise as the result of rapidly fluctuating pressures, thermal shock, or water hammer. The provision of fluid damping systems in the flow conduits may be necessary to avoid these.

VIBRATIONS IN HEAT EXCHANGERS

There are various sources of vibrations in heat exchangers. When the frequency of vibration coincides with the resonant natural frequency of some element in the heat exchanger, serious damage can occur as a consequence of the high amplitudes attained.

Concern about vibration appears to be manifest only in regard to tubular exchangers. Plate-and-frame, plate-fin, panel-coil, and spiral heat exchangers appear to be less susceptible to vibration problems. This is due to the general rigidity of construction, with very many points of contact between the elements. The resonant natural frequency is high, and substantial damping is incorporated. Among tubular exchangers the use of extended surfaces in air-cooled tube-and-fin exchangers so stiffens the tube that vibration problems rarely arise except in the very largest coils, and these can be rectified fairly easily. Vibration problems are therefore limited in the main to tube-and-shell exchangers, and even here the problems are normally not serious except in very large exchangers so highly rated that the fluid velocities are unusually high.

Despite these exemptions and qualifications, vibration problems remain a serious cause of heat-exchanger failure, particularly with regard to the tubes. Tube vibration may result in failure by metal fatigue, by fretting corrosion of the tubes in the tube sheet, or by impact and contact rubbing of the tubes with one another or with the baffles.

The bond between the tube and the tube sheet, maintained by stored compressive stresses induced by rolling during fabrication, may be lost via vibration-induced creep relief. Sometimes the tube-sheet ligaments crack as a result of the dynamic loads induced by vibration of the tubes.

In flow-induced vibrations, the energy required to sustain vibration must be extracted from one or the other of the fluids, most commonly on the shell side. This is reflected by an increase in the shell-side fluid pressure drop.

Excessive vibration imposes additional bending stresses on the tube, so stress corrosion may be intensified. Finally, intolerable noise levels have been known to result from aeolian propagation and "organ piping," with large-amplitude motion in the longitudinal interstices between the tubes.

Resonant Frequency

Every structure of elastic components has a natural frequency of vibration or oscillation. Common examples are a cork bobbing in water or a mass suspended by a spring as shown in Fig. 6.15. When the mass is pulled down (so that the spring is extended) and then released, the mass will oscillate at a certain frequency, say f_n, which depends on the magnitude of the mass and the spring characteristics. This natural frequency of the spring-mass system is

$$f_n = \frac{1}{2\pi}\sqrt{\frac{k}{M}}$$

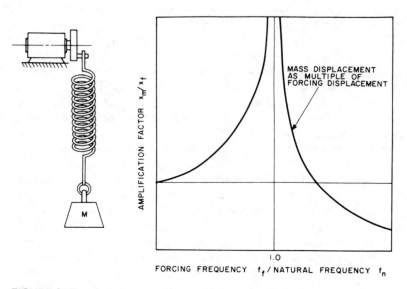

FIGURE 6.15
Undamped spring-mass system and frequency-response characteristic.

where f = natural frequency of spring-mass system
  k = spring stiffness (force to cause a unit displacement)
  M = applied mass (includes one-third the mass of the spring)

In the apparatus shown in Fig. 6.15, the upper end of the spring is attached to a pin on a small wheel driven by a variable-speed motor. The upper end of the spring therefore oscillates at the same frequency as the motor. This is called the *forcing frequency* f_f. The *amplitude* (extent of movement) in the vertical direction is equal to the pitch-circle diameter of the pin movement, say x_f.

Motion of the upper end of the spring causes the lower end (and the attached mass) to move at the same frequency and with an amplitude x_m. At low speeds the mass has exactly the same amplitude as the upper end of the spring so the ratio $x_m/x_f = 1$. This ratio is called the *amplification factor*. As the forcing frequency increases, the amplitude of vibration of the mass increases, even though the amplitude of the spring at the upper end remains constant. Now the amplification factor $x_m/x_f > 1$, as shown in the graph of amplification factor versus frequency ratio in Fig. 6.15.

Damping

When the forcing frequency f_f coincides with the natural frequency f_n, the amplitude of oscillation of the mass becomes very high. The mass dances wildly on the end of the spring. In theory the amplitude of the motion is infinite, but the drag forces due to air around the mass and hysteresis effect in the spring combine to prevent the development of infinite amplitude. However, they may not be sufficient to prevent catastrophic damage to the system.

This natural *damping* effect may be enhanced by the addition of some form of damping device to minimize the amplitude. This is shown in Fig. 6.16 as a *dashpot* connecting the mass to a fixed location or ground. The dashpot shown is simply a piston sliding in a cylinder. Gas or oil in the cylinder passes from one side of the piston to the other as the mass moves and so exerts a moderating influence on the movement of the mass.

The response characteristics of the damped system are also shown in Fig. 6.16, for different degrees of damping. When there is no damping (no oil in the dashpot), the response characteristic is exactly the same as that of the undamped system of Fig. 6.15. Increased damping progressively moderates the amplitude of vibration, particularly at the natural frequency. In heavily damped systems the amplification factor x_m/x_f is always less than 1, except at the very lowest frequencies.

The system described above is the simplest possible example of mechanical vibration, but it is a useful demonstration of resonant operation and damping. Practical systems are generally much more complicated. There may be several masses involved with various modes of vibration, and consequently with several different characteristic resonant frequencies, called higher-order *harmonics* of the fundamental natural frequency.

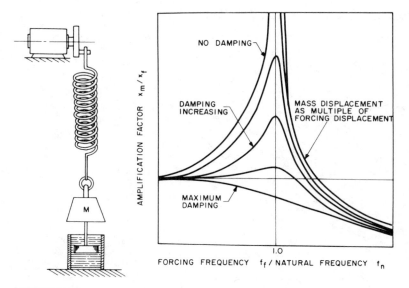

FIGURE 6.16
Damped spring-mass system and frequency response characteristics.

In many systems the damping effects occur somewhat fortuitously and are difficult to predict with any precision. Damping absorbs the excitation energy and limits the maximum amplification factor.

In general there are two principal damping components: structural damping and fluid damping. Structural damping arises principally from hysteresis effects on the structural elements of the system and on their relative motion. It depends mainly on the material properties of the structural elements, on the way the elements are joined or connected, and the degree of lubrication between the moving elements.

Fluid damping depends principally on the density and viscosity of the fluids surrounding the vibrating system. The relative significance of fluid damping decreases as the amplitude of vibration increases, whereas the converse is true for structural damping (see Fig. 6.17).

It is clear from the above that the best way to avoid resonant vibration problems is to operate so that the forcing frequency is less than 75 or 80 percent of the natural frequency of the system. If the forcing frequency is fixed by the conditions of operation and is dangerously close to the resonant frequency of the system, the alternatives to minimize vibration problems are:

1. Increase the damping.
2. Increase the natural frequency. This may be done by changing the elastic characteristics (the spring forces) of the system or by decreasing the mass of the system.

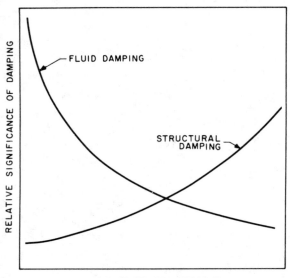

FIGURE 6.17
Relative significance of structural and fluid damping as a function of vibration amplitude.

Tube Vibrations

A single tube mounted between fixed tube sheets is shown in Fig. 6.18a. To simplify matters we can neglect any restraints offered by the tube sheet and assume the tube to be simply mounted on knife-edge supports as shown in Fig. 6.18b. The tube can be excited in a number of ways, for example, by vibrating the supports or by passing a pulsating fluid in crossflow over the tube or through the tube. The tube will vibrate as a consequence of the excitation, and it can do so in a number of ways, as shown in Fig. 6.18. The fundamental mode is shown in Fig. 6.18c. The tube oscillates within the envelope shown in much the same way as a violin string moves when plucked. The natural frequency of the vibration is calculated from the equation

$$f_n = \frac{C}{L^2} \sqrt{\frac{EI}{M}}$$

where f_n = natural frequency of vibration
  C = constant
  L = span
  E = modulus of elasticity of tube material
  I = second moment of area (moment of inertia) of tube cross section
  M = mass of tube and contents per unit length

At higher frequencies the tube may oscillate as shown in Fig. 6.18*d* (the first harmonic) or as shown in Fig. 6.18*e* (the second harmonic). In these modes of vibration the tube behaves as a two- or three-span beam with *nodes N* that do not move as the tube vibrates.

Another possibility is that a higher-order harmonic vibration may be superimposed upon the fundamental mode, with the result shown in Fig. 6.18*f*. Motion at the fundamental frequency always dominates because the energy level required to excite the fundamental mode is less than that needed to excite the higher-order harmonics. The complex vibration may therefore be separated into a large-amplitude component at the fundamental frequency and small-amplitude components at higher-order harmonic frequencies.

Baffles are provided in tubular exchangers almost without exception. The principal function of the baffles is to guide the shell-side flow through a

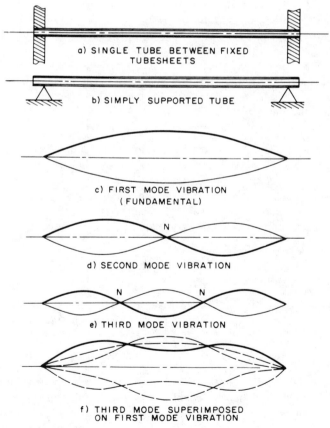

a) SINGLE TUBE BETWEEN FIXED
TUBESHEETS

b) SIMPLY SUPPORTED TUBE

c) FIRST MODE VIBRATION
(FUNDAMENTAL)

d) SECOND MODE VIBRATION

e) THIRD MODE VIBRATION

f) THIRD MODE SUPERIMPOSED
ON FIRST MODE VIBRATION

FIGURE 6.18
Single-tube vibration modes.

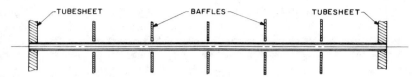

FIGURE 6.19
Exchanger-tube with multiple baffle supports.

circuitous path, part crossflow and part axial flow, over the tubes. They provide the important secondary function of supporting the tubes and restraining tube motion.

The tubes in heat exchangers do not therefore comprise the simple single-span system shown in Fig. 6.18, but the more complicated multispan continuous-beam system shown in Fig. 6.19. The natural frequency of a continuous beam is calculated with an equation similar to that given above, but the value of the constant C varies. The natural frequency depends on the number of spans, the distribution of the beam length among the spans (uniform or nonuniform span), the end-span support conditions (clamped or simply support-ed), the axial loading, the flexural rigidity, and the weight per unit length. Given these complexities, the determination of natural frequencies can become quite complicated. Formulas and guidelines are given in the TEMA standards for straight tubes with equal and unequal multiple spans and for U tubes.

An important point made by Borthman et al. [1978] is that

> ... for continuous beams of five or more component spans the first four harmonics of the fundamental natural frequency can fall within a narrow band of 40 cycles/sec. Under fluctuating dynamic loading the span function moments undergo fluctuation. The influence of fluctuating junction moments is to broaden the range of frequencies over which the beam responds to vibratory excitation at its highest dynamic gain.

Excitation of Tube Vibration

There are numerous mechanisms for exciting general vibrations in heat exchang-ers, with tube vibrations as the principal concern. Some of these mechanisms are briefly considered below.

Mechanically Coupled Excitation

The vibration may be mechanically transmitted through the structure from an external source. This is called mechanically coupled excitation, and it may occur in shipboard or other compact machinery installations where the exchanger is mounted on a common support structure with reciprocating or rotary pumps or compressors. The vibration may also be transmitted through the piping and result in high fluctuating loadings on the exchanger nozzles.

Hydraulically Coupled Excitation

Hydraulically coupled excitation vibrations may be transmitted as regular or intermittent pressure pulses in the fluids circulating in the exchanger. These pressure fluctuations may arise from the normal operation of reciprocating machinery or valves external to the exchanger.

Flow-Induced Vibrations

The mechanism of greatest concern in most cases is flow-induced vibration, almost invariably caused by shell-side flow. Four basic mechanisms for shell-flow-induced vibrations are recognized:

1. Vortex shedding
2. Turbulent buffeting
3. Fluctuating axial drag
4. Fluid-elastic swirling

Vortex shedding. When a fluid flows in crossflow over a tube, the downstream wake is very highly disturbed and there is some reverse flow, as indicated in Fig. 2.13. Above a critical Reynolds number the phenomenon of vortex shedding is initiated. Contrarotating vortices develop in the immediate downstream wake, first on one side of the tube and then on the other. The vortices grow and are then "shed" from the tube at regular intervals so that a state of regularly disordered flow prevails downstream of the tube, as shown by Fig. 6.20. The train of vortices are often referred to as Karman vortex streets.

This alternate shedding of the vortices is accompanied by changes in the pressure distribution in the fluid surrounding the tube. This leads, consequently, to the establishment of a harmonic force transverse to the direction of the flow, acting on the tube and providing the excitation force to induce vibration of the tube. The frequency of vortex shedding (and hence the frequency of the excitation force) is determined from the Strouhal number Sh, defined as

$$\text{Sh} = \frac{f_s D}{u}$$

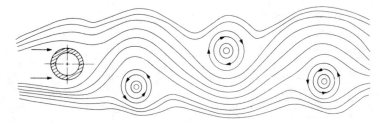

FIGURE 6.20
Crossflow vortex shedding: Karman vortex street.

where f_s = vortex-shedding frequency
       u = fluid velocity
       D = tube diameter

For a single tube in crossflow, Sh is a function of the Reynolds number and has a value that is approximately constant and equal to 0.2 in the range $200 < \text{Re} < 100{,}000$. It is interesting to note that in this range the frequency of vortex shedding is independent of both the fluid density and the viscosity and depends only on the fluid velocity and inversely on the tube diameter. The constant value of 0.2 for the Strouhal number implies that the vortex-shedding frequency is one-fifth that deduced by dividing the stream velocity by the tube diameter. Therefore, the vortex spacing in the downstream wake might be assumed to be five times the tube diameter. Photographs of Karman streets in experimental flow studies show that the vortex pitch is actually about three times the tube diameter. This difference arises because the actual time-averaged wake velocity is about half the free-stream velocity.

The vortex-shedding phenomenon is well understood for flow over a single tube. The situation is considerably more complicated with multiple-tube arrays and depends greatly on the geometry and arrangement of the tube bank. In these cases it is customary to base the Strouhal number on the *spacing* between the tubes rather than on the tube diameter, and values in the range $0.5 < \text{Sh} < 1.2$ may be found in the literature. Two types of flow oscillations that have been recognized with in-line tube banks are shown in Fig. 6.21. Crossflow through

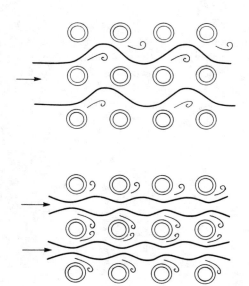

FIGURE 6.21
Vortex shedding and flow oscillations with in-line tube banks.

staggered tube arrays can also induce flow oscillations leading to tube vibrations, although the probability of serious difficulty appears to be substantially reduced with staggered, irregular arrays.

Crossflow patterns are strongly influenced by the presence of fins on the tubes. Furthermore, the fins, particularly longitudinal fins, stiffen the tube and increase the natural resonant frequency of individual tubes and of the tube bank. The effect on vibration of adding fins to otherwise bare tubes is difficult to predict with great precision.

Turbulent buffeting. Turbulent flow over a tube exerts fluctuating forces over a broad frequency spectrum; the forces arise from the fluid-pressure fluctuations. From a confused and apparently random spectrum, the tube will select, and become excited in vibration by, forces at frequencies corresponding to its fundamental and higher harmonic natural frequencies.

Brothman et al. [1978] recognize four distinct components of turbulent buffeting:

1. The vortex-trail turbulence from upstream tubes
2. The random "white-noise" broad-spectrum pressure fluctuations characteristic of shell flow
3. Diffused vortices from upstream portions of the bundle
4. The "slit-cavity-slit" (compression-expansion-compression) flow patterns arising from flow in a tube bundle through successive confined areas between the tubes (slits) and downstream interstices between tube rows (cavities)

In all these components the pressure and, hence, force fluctuations on the tube prevail over a wide frequency spectrum. Furthermore, they occur in both axial and transverse crossflow.

Tube response to turbulent buffeting depends principally on the natural frequency and the structural and fluid damping of the tube and its environs, specifically the tube diameter, shell-flow Reynolds number, crossflow velocity, and fluid density. At Reynolds numbers greater than 100,000, the energy associated with broad-spectrum turbulent buffeting becomes increasingly significant compared with the energies associated with the periodic vortex-shedding mechanism. At a Reynolds number of 200,000, periodic vortex shedding has become imperceptible against the hurly-burly of turbulent flow.

The broad-band distribution of energy in turbulent flow poses particular problems to the mechanical designer, for there are no characteristic discrete exciting frequencies to be avoided as with vortex shedding. The only design options to eliminate or reduce tube vibration are to drastically reduce the fluid velocity or increase the number of baffles and the structural damping.

Fluctuating axial drag. In a tube-and-shell heat exchanger, the shell-side fluid flow is a combination of transverse and axial flow. The axial-flow component generates an axial drag on the tubes just as the crossflow generates a transverse drag. The axial drag on the tubes fluctuates with variations in the shell-side axial flow. This superposes dynamic buckling loads on the compressive loading of the tube existing under operating conditions. Brothman et al. [1978] assert that

> For a tube which is under a substantial compressive static load the tolerable or critical axial velocity may be lowered to a value such that superposed dynamic drag loadings can cause buckling mode vibration.

The exact nature of buckling-mode vibrations is unclear, but they go on to indicate that

> ... the critical axial flow velocity is determined by the flexural rigidity of the tube, the length and diameter of the tube, the transverse loading of the span by other excitations and the static column loading of the span. The deflection response is controlled by the hydraulic radius of the flow channel and the square of the ratio of the axial flow and critical axial flow velocities.

Fluid-elastic whirling. A long, thin cable hanging as a catenary between end supports is excited into a vibrational orbital motion when the crossflow velocity exceeds a critical value. These fluid-elastic vibrations are self-exciting and arise from the variation in lift force as the fluid impingement angle changes on the moving cable.

The same fluid-elastic mechanism sometimes gives rise to troublesome vibrations in heat-exchanger tube bundles. The crossflow velocity excites a coupled orbital whirling motion of two or more tubes in a row, as illustrated in Fig. 6.22. Whirling is sustained by the harmonically varying lift and drag displacement forces created by the harmonically varying slit widths. The motion becomes progressively more violent above a critical crossflow velocity and can become catastrophically destructive. The critical crossflow velocity depends principally on the natural frequency of the coupled tubes, the system density, and the density of the crossflow fluid.

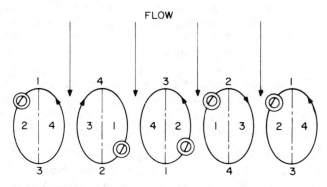

FIGURE 6.22
Coupled fluid-elastic whirling.

Acoustic Oscillations

Acoustic oscillations occur occasionally with a crossflow of gas or vapor in tube-and-fin or tube-and-shell exchangers. The cavities and slits between the tubes transverse to the flow correspond to closed-end organ pipes. Flow oscillations similar to those illustrated in Fig. 6.21 occur in these spaces, and acoustic oscillation results when the frequency of these transverse flow variations coincides with the acoustic frequency or harmonic of the column between the tubes. The acoustic oscillations are propagated at sonic velocities from standing waves in the fluid column. There are various modes of oscillation, at frequencies ranging from the fundamental through successive harmonics.

Acoustic oscillations are not dangerous in themselves so much as unbearably noisy. However, they may excite large-amplitude vibration of the tubes or duct walls if their frequency coincides with the resonant frequency or a harmonic of these elements. They may also synchronize the vortex-shedding characteristics of the tubes, thereby increasing the magnitude of the vortex-shedding flow instabilities.

The frequency of resonant oscillations in the tube interstices depends on the length of the column between baffles, the fluid molecular weight, the temperature, and the specific-heat ratio ($\gamma = C_p/C_v$). A simple approach to resolving the problem of acoustic oscillations is to interpose supplementary acoustic baffles to shorten the length of the organ pipes and so "detune" the system.

Resolution of Vibration Problems

The elementary discussion above has illustrated some of the complexities of vibration problems that may occur in heat exchangers. In an interesting historical survey, Gardner [1974] has pointed out that

> Whatever vibration problems may have existed in the infancy of the heat exchanger industry had been essentially eliminated by restraints imposed on the maximum unsupported tube spans and tube-to-baffle clearances. These were codified by the TEMA standards at a time when shell diameters of exchangers rarely exceeded 40 in. (1000 mm). The rules adopted were basically empirical, and no velocity restraints were imposed. Nevertheless, the rules worked and tube vibration problems were very seldom encountered over a period of 20 or more years.
>
> In the early 1950's, for the first time, the nuclear power industry demanded the transfer of vast flows of heat, not through water tube boilers as in the past, but through heat exchangers. As a result, in order to reduce the number of shells required, the diameters of the heat exchangers designed increased by a factor of 2 or more,

i.e. to 80 to 100 in. (2000 to 2500 mm). The designers, as often as not, were architect-engineer employees without previous experience or intuitive sense of propriety in design. The trouble began in the late 1950's as these items of equipment finally came into operation at specified flowrates, whereupon a disturbing incidence of tube failures by vibration became evident.

The resolution of heat-exchanger vibration problems follows two distinct paths. The first and obviously preferred method is to anticipate problems at the design stage. Remedial measures can be readily applied on the drawing board at very low cost. Unfortunately, it is difficult to foresee all the vibration problems that might arise, and so solutions are often required to problems that become manifest during the actual operation of an exchanger.

At the design stage it is rarely possible to anticipate the problems of mechanically or hydraulically coupled excitation emanating from external sources. These problems can usually be resolved by measures external to the heat exchanger: the addition of mass to the structure or surge chambers to the fluid circuit, the use of flexible mounts or couplings, and minor changes in the speed of operation of pumps or reciprocating equipment.

Flow-induced vibrations can be resolved by changing the resonant frequency of the tubes or the fluid flow velocity. We saw earlier that the natural frequency of the tube is $f_n = (C/L^2) \sqrt{EI/m}$, where L is the span, the distance between tube supports or the baffle spacing. Now the frequency of vortex shedding is $f_s = \mathrm{Sh}\, u/D$, where Sh is the Strouhal number; when $f_s/f_n < 1$, there can be no large amplification factors, as shown in Fig. 6.15 or 6.16.

In a tube-and-shell heat exchanger the crossflow velocity u varies inversely as the baffle spacing L. Therefore, provided the Strouhal number remains constant as u and, hence, Re change, the forcing frequency f_s must also vary inversely as the baffle spacing. Since the natural frequency f_n of the tube varies inversely as the baffle spacing squared, then to a first approximation the ratio f_s/f_n varies directly with the baffle spacing L. A decrease in the baffle spacing decreases the ratio f_s/f_n to a value more remote from the critical value $f_s/f_n = 1$.

At the same time, a decrease in the baffle spacing increases the crossflow velocity, and it will be recalled that the frictional pressure drop is a function of the *square* of the velocity. The shell-side pressure drop is therefore increased as the baffle spacing is decreased, and this alone may provide a limit to the alleviation of vibration possible by this method.

Another approach sometimes adopted is to use a nonuniform baffle spacing so that adjacent tube spans are of different lengths and consequently have different resonant frequencies. This spreads the risk of some flow-induced vibration over a wider range of flow velocities but reduces the likely amplitudes of vibration for any given span. The adjacent spans increase the structural-damping characteristics to prevent the buildup of dangerous amplitudes of vibration.

Increased structural damping may also be gained by reducing tube clearance in the baffles and by increasing the thickness of the baffle tube support in a variety of ways.

Novel designs of baffles or supplementary tube supports are sometimes effective in eliminating troublesome vibrations. Small and Young [1979] described a tube support baffle constructed of a series of rods having a diameter equal to the clearance between tubes, inserted between tube rows in both the horizontal and vertical directions, and fastened to a support ring. Various arrangements of the basic concept have been explored, as shown in Fig. 6.23. Although hailed by Small and Young as a new development, the rod baffle was described nearly half a century earlier by Sieder [1935] in a patent now long expired.

The alternative approach of reducing the fluid crossflow velocity requires no structural alterations to the exchanger. Reduction in the fluid crossflow

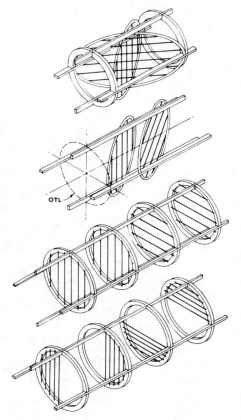

FIGURE 6.23
Rod baffle arrangements. (*After Small and Young [1979].*)

velocity eliminates the development of resonant vibrations and is the principal method of resolving the problems of flow-induced vibrations in an existing heat exchanger. Unfortunately, a reduction in the crossflow velocity will almost invariably degrade the thermal performance of the exchanger and may be unacceptable in a complicated process system. As partial compensation, a reduction in pressure drop also follows from the reduced crossflow velocity.

At the design stage the crossflow velocity may be reduced by changing the baffle design and arrangements. One of the many possiblities is shown in Fig. 6.24. Fig. 6.24a shows a single-pass shell with conventional segmental baffles, with the fluid in crossflow across the whole diameter of the shell. In Fig. 6.24b the baffles are of the double-segmental type, resulting in a split, parallel shell-side flow with the crossflow velocity reduced to approximately half that of Fig. 6.24a. A nonflow condition, created across the horizontal diametral plane, has the same result as the inclusion of an actual partition. With double-segmental baffles the pressure drop in the shell is markedly reduced, for not only is the fluid velocity reduced but the actual number of tube layers traversed by each stream is reduced.

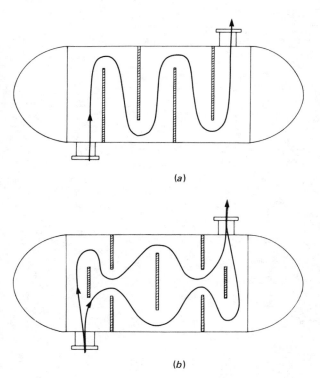

(a)

(b)

FIGURE 6.24
Baffle designs affecting crossflow fluid velocity: (a) single shell-side stream, (b) split, parallel shell-side streams.

Reference to Figs. 6.15 and 6.16 may suggest that operation with ratios of forcing frequency to natural frequency (f_s/f_n) *greater* than unity would resolve tube vibration problems equally as well as operation with $f_s/f_n < 1$. It would permit the use of very high fluid velocities and consequent high thermal performance. This is a logical and valid conclusion that is applicable to the exchanger once it is "up to speed." Difficulties would, however, arise in starting and· shutdown, as the exchanger passed through the critical vibration regions of the fundamental and harmonic modes. With a lightly damped system the starting transients might be severe enough to preclude attainment of the stable operating conditions.

Further difficulties might follow from an unplanned reduction in the mass rate of flow of the shell-side fluid. The reduction might be sufficient to put the f_s/f_n ratio close to one of the critical resonant values. In principle, the high-ratio system has much to commend it, but for the above reasons it is customary to design exchangers for an f_s/f_n ratio that is always less than unity.

CORROSION

All materials corrode to some extent in all environments, but there is great variety in the rates of decay and in the dominant mechanisms of corrosion. The useful lives of nearly all engineering systems are limited principally by corrosion of one sort or another. Skill in selecting the appropriate material to ensure adequate life at the lowest possible cost is therefore an important function of mechanical design.

A review of the mechanisms of corrosion could be included here but appears to be more appropriate to Chap. 8, the discussion of heat-exchanger degradation and failure.

SUMMARY

In this chapter we have reviewed some aspects of the mechanical design of heat exchangers. The strength properties of materials are important from the aspect of the exchanger's function to contain the fluids under pressure and to prevent leakage between the fluids or to the environment. Stresses arising from internal pressure, thermal effects, and miscellaneous loadings have a major influence on the shapes, thicknesses, and metals used for exchanger components.

The vibration characteristics of an exchanger are an important factor affecting the integrity of the mechanical design. Various sources of vibrations in heat exchangers were reviewed, and procedures for resolving these problems were discussed.

REFERENCES

Brothman, A., W. R. Rowley, A. Devore, G. B. Hollar, and A. Horowitz: Heat Exchanger Vibrations Problems, pp. 76–85 in "Practical Aspects of Heat Transfer," American Institute of Chemical Engineers, New York, 1978.

Fraas, A. P., and M. N. Ozisik: "Heat Exchanger Design," Wiley, New York, 1965.

Gardner, K. A.: Anticipation of Problems in the Design of Heat Transfer Equipment, chap. 2 in N. Afgan and E. U. Schlunder (eds.), "Heat Exchangers: Design and Theory Sourcebook," Hemisphere, Washington, D.C., 1974.

Small, W. M., and R. K. Young: The Rodbaffle Heat Exchanger, *Heat Transfer Eng.*, vol. 1, no. 2, pp. 21–27, 1979.

Sieder, E. H.: Heat Exchanger, U.S. Patent 2,018,037, October 2, 1935.

Yokell, S.: Design and Fabrication of Heat Exchangers, pp. 139–167 in "Practical Aspects of Heat Transfer," American Institute of Chemical Engineers, New York, 1978.

7

High-
and low-
temperature
systems

INTRODUCTION

Heat exchangers are used over a very broad spectrum of temperatures and pressures. The greatest number of applications are those in which the temperature is not significantly above or below the ambient temperature; this range is classified for convenience as the middle range of temperature. Middle-range temperature applications pose the least demanding technical regimen. In many cases conditions are sufficiently relaxed for the design and selection of heat exchangers based on factors such as economics, production methods, or aesthetics not related to operating temperature.

For both high-range and low-range temperatures there are fewer options, and design becomes progressively more dependent on temperature effects. Some of the more important considerations affected by temperature are:

1. The mechanical properties of the structural materials
2. The reaction rate or corrosion kinetics of the structural-material–fluid combination
3. The transport properties (thermal conductivity, viscosity, specific heat) of the fluids
4. Thermally induced stresses or displacements of the structure
5. The relative significance of radiation heat transfer
6. The efficiency of seals, packing, etc.

The high, middle, and low temperature ranges mentioned above are not clearly distinguished by accepted and customary engineering usage. A low-temperature coil in air-conditioning applications may operate *above* ambient temperatures. Similarly, the high-temperature process exchanger for alcohol distillation or milk pasteurization would be considered as a low-temperature system by those concerned with steam generation for power production. Any definition of temperature range is therefore arbitrary. For the purpose of our discussion, we shall define the temperature ranges as follows:

1. Low-range temperatures: $0 < T < 300$ K, further subdivided into the:
   a. Cryogenic temperature range: $0 < T < 120$ K
   b. Noncryogenic temperature range: $120 < T < 300$ K
2. Middle-range temperatures: $300 < T < 600$ K
3. High-range temperatures: > 600 K

HIGH-TEMPERATURE SYSTEMS

Steam Boilers

Power Generation

High-temperature heat exchangers are principally used for (1) generating high-pressure steam for power production and (2) energy recovery from high-temperature thermal effluents.

Steam raising for power production is vital to our technology-based society because of its dependence on low-cost, reliable electric power. In the main, power is generated in large base-load electric power stations where high-pressure steam is expanded in turbines to drive electric generators. A small fraction of the total power is produced by gas-turbine or diesel engines driving power generators, usually in small, remote, or temporary installations.

Energy to generate steam is derived mainly from the combustion of fossil fuel—coal, natural gas, or oil. Nuclear energy provides a small but increasingly significant fraction of the total steam-generating capability.

With the incessant increase in demand for power (10 to 11 percent per year, or a doubling every 7 years) and progressive improvements in materials and manufacturing techniques, base-load electric power stations have increased in size to become the largest and most complex technological creations of man; the cathedrals of the twentieth century. Typically a station is comprised of two principal parts, located in separate buildings. One contains the steam generators or boilers. The other contains the steam-turbine power generators and condensers. Another large ancillary structure contains the condenser cooling-water/atmospheric-air heat exchanger. Multiexpansion turbines having a capacity of 2000 MW on a single shaft have been built. Boilers capable of generating 5 Gkg/h of steam (10,000,000 lb/h) have been constructed, with pressures and temperatures ranging up to 38 MPa (5500 lb/in^2) and 900 K (1750°F).

High-capacity boilers are large, complicated systems of heat exchangers. Figure 7.1 is a cross section of a large coal-fired boiler for base-load power stations. An impression of its size can be gained from the human figure at the bottom, near the slag-tap furnace.

Preheated water at high pressure is supplied to the vertical water tubes surrounding the combustion space. The water is heated in the tubes by radiant

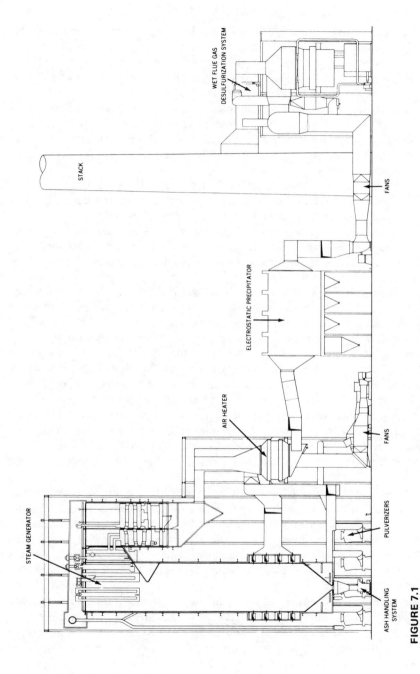

FIGURE 7.1
Typical large, coal-fired steam boiler for base-load electric power station. (*Courtesy Babcock and Wilcox, New Orleans.*)

and convective heat transfer from the products of combustion of the coal. The pressurized water passes to the upper steam drum and separates in liquid and vapor. The vapor is led from the steam drum through a series of heat exchangers—the *primary superheater,* the *secondary superheater,* operating at a higher temperature, and finally the *attemperator,* where the reheat steam temperature is adjusted before delivery to the turbines.

Steam expands in the turbines to an exceptionally low pressure [less than 4 KPa (29 in Hg vacuum)] and is then condensed. The condensate is pressurized by the feedwater pump and returned to the boiler through a series of heat exchangers—up to four stages of regenerative feed heater and the economizer feedwater heater shown in Fig. 7.1.

Combustion air enters the system near the base of the stack in Fig. 7.1 and passes through a forced-draft fan. A steam-heated coil is located downstream of the fan to raise the air temperature above the dew point of the air so as to evaporate any moisture or fog in the incoming air. The air then passes through an exhaust-gas/air preheater. This is a very large gas-to-gas tube-and-fin or plate-fin recuperative heat exchanger or, alternatively, a packed-bed Ljungstrom regenerative exchanger.

The heated air then mixes with a fraction of the products of combustion recirculating in the system. This is done to moderate the temperature levels attained in combustion. The mixture passes to the combustion space and combusts with coal that has been crushed or pulverized to specific sizes.

The high-temperature exhaust products then begin a tortuous passage through the boiler en route to the stack. Heat is transferred, at first primarily by radiation, to the membrane water-tube arrangement surrounding the combustion space. Further heat transfer, primarily by convection from the combustion products, takes place in a sequence of heat exchangers—the secondary super-heater, the primary superheater, the feedwater economizer, and finally the exhaust-air heater. Final cleanup of the products of combustion is accomplished in the electrostatic precipitator before the products of combustion are discharged to the stack.

This brief description is sufficient to show that the high-capacity power boiler is actually a complicated sequence of gas-to-liquid, gas-to-vapor, liquid-to-liquid, and gas-to-gas heat exchangers. On the liquid or vapor side the pressures and temperatures are very high, close to the maximum possible with present-day materials. On the gas side the pressure is low, a little above atmospheric, but the temperatures are exceptionally high, particularly in the combustion zone and immediately downstream.

Heat-transfer coefficients for liquid in the water tubes are very high; they are somewhat lower, but still high because of the density, for steam in the superheater tubes. On the gas side, heat transfer to the tube walls is primarily by radiation in the combustion zone, with a very large temperature difference between the gas and the wall. Further downstream, at lower gas temperatures, the primary superheater and economizer tubes are finned to compensate for the low gas-side convective heat-transfer coefficient. Pin fins, short lengths of metal bar, are customarily welded to the outside of the economizer tube to provide the extended finned surface.

Nuclear Systems

Nuclear steam power generation follows a somewhat different pattern. A line diagram for a nuclear power station is shown in Fig. 7.2. In this type, a pressurized water reactor, steam for expansion in the turbine is generated in a large, high-temperature, high-pressure heat exchanger known appropriately as the steam generator. The steam leaving the steam generator follows a path similar to that in the fossil-fuel boiler plant.

Heat for the steam generator is generated by the radioactive decay of uranium-based fuel in the nuclear reactor, shown in Fig. 7.3. The heat is conveyed from the reactor to the steam generator by a fluid loop, which may be pressurized heavy water, liquid metal, or a gas—carbon dioxide, helium, etc. A possible arrangement for a pressurized water reactor and associated steam generator is shown in Fig. 7.4.

The operation of a nuclear reactor imposes stringent requirements on the metals used to contain the fuel elements. They cannot be selected primarily for their high-temperature strength properties as in fossil-fueled boilers. As a consequence, the steam temperature and pressure levels of nuclear systems are much below those commonly found in contemporary fossil-fuel stations. Since the maximum temperature is the primary parameter affecting thermal efficiency, the result is that nuclear power plants are less efficient than fossil fuel plants by a substantial margin: 30 percent compared with 40 percent.

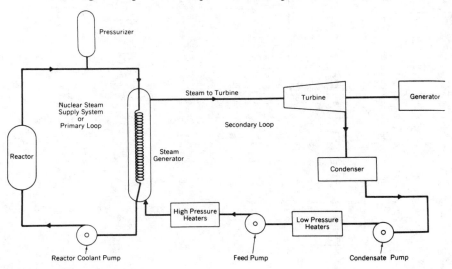

FIGURE 7.2
Line diagram for nuclear steam power generating station.

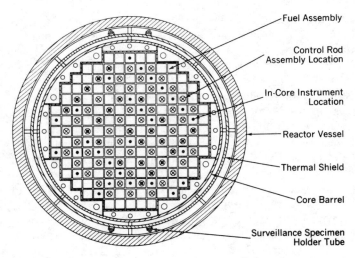

FIGURE 7.3
General arrangement of nuclear reactor. (*Courtesy Babcock and Wilcox, New Orleans.*)

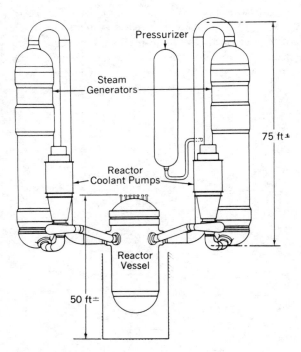

FIGURE 7.4
Arrangement of pressurized-water nuclear reactor and steam generator. (*Courtesy Babcock and Wilcox, New Orleans.*)

The thermal efficiency is the fraction of the supplied heat that is converted to useful work or electric power. Efficiencies of 30 or 40 percent may appear low, but they are, in fact, among the highest thermal efficiencies achieved by practical energy-conversion systems. An efficiency of 30 percent means that 70 percent of the heat supplied must be rejected as exhaust or waste heat. In base-load electric power stations this is an enormous quantity of energy; for every unit of electric power generated, two and one-half units are rejected as low-temperature heat from the station. Most of the waste heat passes, in the condenser, from low-pressure steam to the condenser cooling water. Thence it is dissipated either to a river, a lake, or estuarial waters or, by means of a cooling tower, to the atmosphere. A small fraction of the exhaust heat from power stations is presently utilized for domestic and industrial heating in a process known as district heating. As the energy crisis deepens, more district heating schemes will be adopted. The vast quantities of low-grade energy dissipated from power plants, particularly nuclear plants, is the reason for siting them on rivers and coasts—the main sources of cooling water. The prime sites are already occupied, and atmospheric air cooling is being used to an increasing extent.

Industrial Steam Plants

In addition to huge fossil or nuclear steam plants for power generation, there are many steam plants of modest capacity in oil refineries, chemical plants, and other industrial and commercial enterprises. They provide steam and hot water for a variety of process heating applications.

A typical industrial *integral-furnace* boiler installation is shown in Figs. 7.5 and 7.6. Although not so large and extensive as a base-load power boiler, this type of steam generator is still a substantial and complicated ensemble of equipment. It is available in a range of sizes capable of generating up to 0.5 Gkg/h (1,100,000 lb/h) of steam at pressures and temperatures up to 10 MPa (1500 lb/in^2) and 800 K (950°F).

These boilers are used in refineries, chemical process plants, and industrial concerns to provide steam. They can operate on a variety of solid, liquid, and gaseous fuels, including waste products such as sawmill waste wood and various refinery or chemical-plant waste liquids and gases. They are primarily intended to provide steam at intermediate or low pressures for process heating, but it is common for the steam to be generated at higher pressures and temperatures than required for this purpose. The steam is expanded in simple turbines to the required pressure, and the work obtained can be used to drive pumps or fans but in most cases is used to generate electric power. The additional cost of generating steam at the higher temperature and pressure is small, so the power obtained comes as a virtually free bonus at nearly 100 percent thermal efficiency.

Many different types of boilers are used, in a wide variety of sizes. They can be classified generally as either *fire-tube* boilers or *water-tube* boilers, depending on whether the tubes contain the combustion products or the water. Another important distinction, that between *heating* boilers and *process*

FIGURE 7.5
Industrial integral furnace boiler. (*Courtesy Babcock and Wilcox, New Orleans.*)

FIGURE 7.6
Industrial integral furnace-boiler installation. (*Courtesy Babcock and Wilcox, New Orleans.*)

boilers, is made on the basis of operating pressure and temperature. Heating boilers are limited in their operating pressure and temperature to

1. For steam: 15 lb/in² gauge or 30 lb/in² absolute (208 kPa)
2. For hot water: 160 lb/in² gauge or 175 lb/in² absolute (1.2 MPa) and 250°F (394 K) at or near the boiler outlet

In North America, heating boilers are fabricated in accordance with Section IV of the ASME Boiler and Pressure Vessel Code.

Process steam boilers include all steam boilers operating at pressures in excess of 15 lb/in² gauge or 30 lb/in² absolute (208 kPa). They are fabricated in accordance with Section I of the ASME Boiler and Pressure Vessel Code. Common design pressures are 150, 200, 250, and 300 lb/in² (100, 140, 175, and 200 kPa). Boilers used for power production operate at very much higher pressures and temperatures. They are called power boilers and are also fabricated under Section I of the ASME code.

Steam and water heating are much favored in large industrial plants for many heating, evaporating, concentrating, and cooking processes. With steam heating the temperatures can be precisely controlled by relatively simple pressure regulators. Also, the complicated firing equipment can be concentrated into one large, well-equipped facility under the supervision of a responsible plant engineer. This is very advantageous as compared with a multiplicity of independently fired systems, all complete with automatic control systems and operational idio-syncrasies, even with the advent of the microprocessor and a quantum jump in the sophistication of automatic instrumentation.

Marine Boilers

Steam boilers are always used on ships. Even when the main propulsion power plant is a diesel engine, boilers provide steam for auxiliary power and heating service. When the propulsion power requirements exceed 20,000 hp, steam-turbine units become competitive with diesel engines. Above 35,000 hp, steam turbines are used exclusively.

Marine boilers for propulsion power are generally of the two types shown in Figs. 7.7 and 7.8, the two-drum integral-furnace boiler and the single-drum sectional-header boiler. The two-drum type ranges in steam capacity to 200 Mkg/h (400,000 lb/h) at pressures to 8.3 MPa (1200 lb/in²) and temperatures to 810 K (100°F). The sectional-header type ranges in capacity to 70 Mkg/h (150,000 lb/h) at pressures up to 4 MPa (600 lb/in²) and temperatures to 760 K (910°F).

Waste-Heat Boilers

Very large quantities of heat are available in the exhaust-gas streams from gas turbines, large diesel engines, steel-making furnaces, cement kilns, and chemical or refinery process plants. Many industrial, commercial, municipal, and agricultural operations produce waste products that can be incinerated in properly

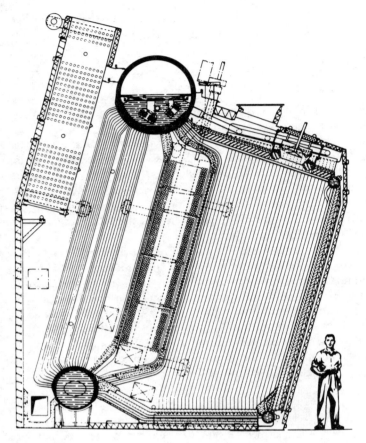

FIGURE 7.7
Top-fired, two-drum, oil-fueled integral furnace–marine-boiler with horizontal superheater.
(*Courtesy Babcock and Wilcox, New Orleans.*)

designed furnaces as the preferred means of disposal. On incineration they produce large quantities of heat. To an increasing degree, as the cost of fuel increases, it becomes worthwhile to utilize the heat available in these thermal effluents to generate steam or hot water for process-heating purposes or power production.

Waste-heat boilers are available in a wide range of sizes and types, with special equipment as required to accommodate highly dust-laden gases (from cement kilns) or the combustion of unusual fuels (such as bagasse, the matted cellulose fiber and fine particles produced as the refuse from milling sugar cane).

Figure 7.9 shows a large waste-heat boiler designed to utilize the thermal energy in dust-laden or heavily contaminated gas streams from a cement kiln.

Figure 7.10 is a waste-heat boiler of smaller size suitable for coupling to a gas turbine or other relatively clean, high-temperature exhaust stream.

Sources of Additional Information

Steam generation and the heat exchangers associated with it are an important part of modern technology. Steam boilers are among the largest, most expensive

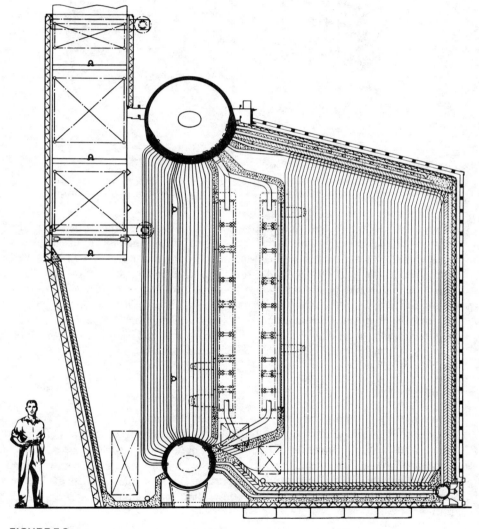

FIGURE 7.8
Oil-fired, sectional-header, merchant-type marine boiler. (*Courtesy Babcock and Wilcox, New Orleans.*)

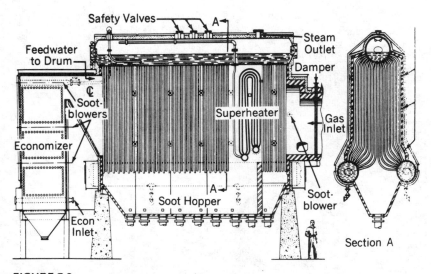

FIGURE 7.9
Three-drum, bent-tube waste-heat boiler fitted with lance ports and soot blowers, designed to operate with heavily dust-laden gases. (*Courtesy Babcock and Wilcox, New Orleans.*)

creations of man and the most challenging technically, for they operate at the limit of manufacturing and metallurgical capability. The above discussion is adequate only as the most superficial introduction to the topic.

One of the best books on the subject is "Steam—Its Generation and Use." This excellent book, published by the Babcock and Wilcox Co., New Orleans, is maintained up to date in successive editions (38th edition in 1972). Technical information about steam boilers may also be obtained from the American Boiler Manufacturers Association, Arlington, Virginia, and, of course, from the individual members of the association and other manufacturers.

Fired Heaters

Steam heating is used extensively for industrial-heating purposes. There are also numerous applications for directly fired heaters for heating air and other gases, vapors, and light and heavy liquids, and for evaporators.

Fired heaters are used where steam supplies are not available or where process fluids must be heated to temperatures greater than normal. They are sometimes used in general-purpose heating applications in preference to steam heating.

One form of directly fired heater is shown in Fig. 7.11. It consists basically of a double-pipe heat exchanger with longitudinal fins on the exterior of the internal pipe. Combustion of the fuel-air mixture takes place in the center-tube, and heat is transferred by radiation and convection through the center tube to

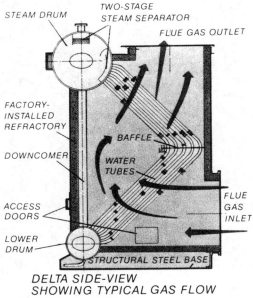

STEAM DRUM

TWO-STAGE
STEAM SEPARATOR

FLUE GAS OUTLET

FACTORY-
INSTALLED
REFRACTORY

BAFFLE

DOWNCOMER

WATER
TUBES

FLUE
GAS
INLET

ACCESS
DOORS

LOWER
DRUM

STRUCTURAL STEEL BASE

DELTA SIDE-VIEW
SHOWING TYPICAL GAS FLOW

FIGURE 7.10
Typical waste-heat boiler for gas turbine or other high-temperature exhaust stream. (*Courtesy Deltrac Corp., Minneapolis.*)

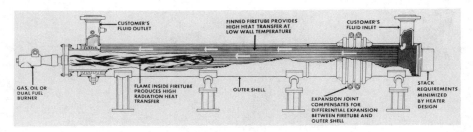

FIGURE 7.11
Directly fired heater. (*Courtesy Brown Fintube, Tulsa.*)

the process fluid in the annulus. The fuel can be gaseous or liquid, and a very wide range of combustibles can be handled.

A variation of the arrangement shown in Fig. 7.11 has an externally finned fire tube arranged as a hairpin U tube but without the outer tube. This is a fired immersion heater for warming the contents of tanks of asphalt or bunker C oil prior to withdrawal.

Manufacturers claim that the advantages of directly heated systems include a quick heat-up to operating temperatures and an equally rapid shutdown. This fast response arises out of the minimal use of metal and refractory lining, with no large thermal mass to be heated or cooled. The heater adjusts quickly to changes in load, for there is little process fluid in the annular space; thus, changes in flow condition and temperature can be rapidly detected, and adjustments made to fuel and air supplies. The use of longitudinal fins on the outside of the fire tube increases the area for heat transfer up to 10 times the bare-tube area. This assures good heat transfer to the process fluid and avoids local hot spots in the process stream and low wall temperatures.

Fired heaters have the advantage of being easy to install, since they are compact and lightweight. They can be used in vertical, horizontal, or inclined locations. They are available in a range of heat capacities up to 360 kW (1.2×10^6 Btu/h), and multiple units are installed for greater capacities. This modular approach allows units to be turned off at part load and allows maintenance or repair to be carried on without complete shutdown. Standard heaters are available for pressures and temperatures of the process fluid up to

920 K (1200°F) and 2 MPa (300 lb/in²). Special heaters may be had for process fluid conditions up to 1360 K (2000°F) and 7 MPa (1000 lb/in²).

Indirectly fired air heaters are commonly used for domestic, commercial, and industrial space heating, with gas or oil as the fuel. One well-known type of domestic, gas-fired, forced-air furnace is shown in Fig. 7.12, along with a detail of the unusual "flattened-tube" heat exchanger. The fuel gas is burned in the combustion zone at the bottom of the tube, and the products of combustion pass through the tubes. Air is heated in passage over and between the tubes en route to the space to be heated. The whole process of combustion and air heating occurs at near ambient pressures, with the equipment fabricated from light sheet metal. The economy gained by mass production and customer eye appeal are more important for this type of product than for large, high-pressure industrial or process heat exchangers whose design is largely dictated by the ASME code or other regulatory requirements.

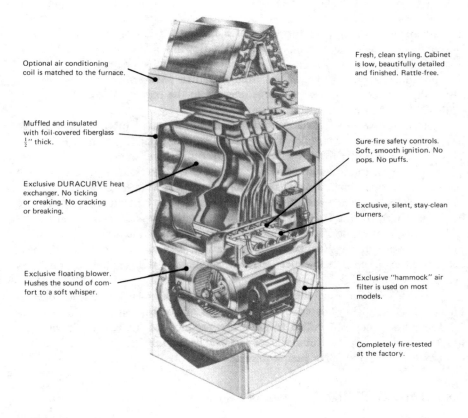

Optional air conditioning coil is matched to the furnace.

Muffled and insulated with foil-covered fiberglass $\frac{1}{2}$" thick.

Exclusive DURACURVE heat exchanger. No ticking or creaking. No cracking or breaking.

Exclusive floating blower. Hushes the sound of comfort to a soft whisper.

Fresh, clean styling. Cabinet is low, beautifully detailed and finished. Rattle-free.

Sure-fire safety controls. Soft, smooth ignition. No pops. No puffs.

Exclusive, silent, stay-clean burners.

Exclusive "hammock" air filter is used on most models.

Completely fire-tested at the factory.

FIGURE 7.12
Domestic gas-fired forced-air furnace with detail of heat exchanger. (*Courtesy Lennox Industries, Dallas.*)

Exhaust/Air Preheaters

High-temperature heat exchangers are used in exhaust-heat recovery systems. Figure 7.10 shows a typical industrial waste-heat boiler, and Figure 7.11 includes an exhaust-gas/air preheater transferring heat from the combustion products to the incoming combustion air. There are many similar recovery systems in which heat is recovered from high-temperature exhaust streams and used to heat incoming air en route to the combustors, to reduce fuel requirements.

Exhaust-gas/air preheaters may be of the recuperative or regenerative type. Recuperators are heat exchangers with separate flow paths for the two fluids. Regenerators are simply porous solids through which the hot and cold fluids flow alternately.

Regenerative heat exchangers are considered in more detail in Chap. 11, but their principal use in thermal recovery systems warrants inclusion of a brief reference here. One form of regenerative heat exchanger, called the Ljunström thermal wheel, is shown in Fig. 7.13. It consists of a porous matrix of finely divided material arranged in the form of a flat disk. The disk rotates and passes alternately through the hot and cold fluids flowing in ducts connected to the casing of the thermal wheel. An element of the wheel passing through the hot-fluid duct is heated by the hot fluid and then passes to the cold-fluid duct. There the heat stored in the matrix is released to the cold fluid, and the element cools to the cold-fluid temperature.

Recuperative heat exchangers can be various forms of tubular or plate-type heat exchangers. One example of a recuperative exchanger is shown in Fig. 7.14. This is an incinerator for the elimination by direct-flame oxidation of effluents contaminated with organic solvents, phenols, aldehydes, oil mists, phthalic anhydride, sulfides, mercaptans, rendering odors, and sewage gases. The process gas and vapor effluent enter the incinerator and pass through the recuperative heat exchanger on the way to the fuel/air plenum chamber. Gas or oil fuel is added to raise the calorific value of the contaminated effluent above the critical value necessary to ensure complete oxidation. Use of the recuperative heat exchanger to preheat the inlet effluent minimizes the amount of fuel gas necessary.

Another example of thermal recovery is the slab reheat furnace shown in Fig. 7.15. Steel slabs, heated by the products of combustion, pass through the furnace before they are rolled to size. The products of combustion leaving the furnace flow through a tubular recuperative heat exchanger that preheats the combustion air to minimize fuel requirements.

Some Aspects of High-Temperature Service

The loss of strength by all materials as the temperature increases is the most important determinant of high-temperature service. The metallurgical properties of the various steel alloys and other structural materials are all well documented

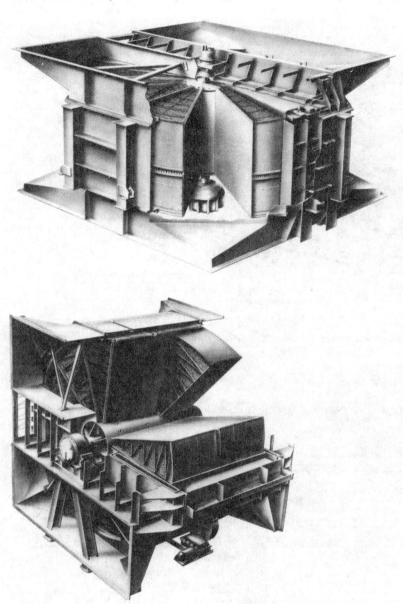

FIGURE 7.13
Ljungström thermal-wheel regenerative heat exchanger. (*Courtesy C. E. Air Preheater Co., Wellsville, N.Y.*)

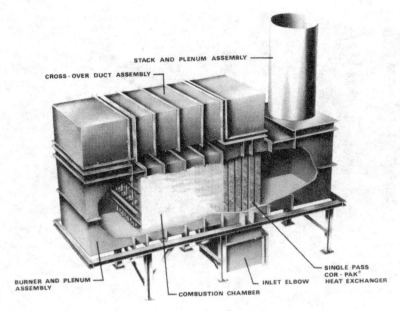

STACK AND PLENUM ASSEMBLY

CROSS-OVER DUCT ASSEMBLY

SINGLE PASS
COR-PAK®
HEAT EXCHANGER

BURNER AND PLENUM
ASSEMBLY

INLET ELBOW

COMBUSTION CHAMBER

FIGURE 7.14
Direct-flame fume incinerator. (*Courtesy C. E. Air Preheater, Wellsville, N.Y.*)

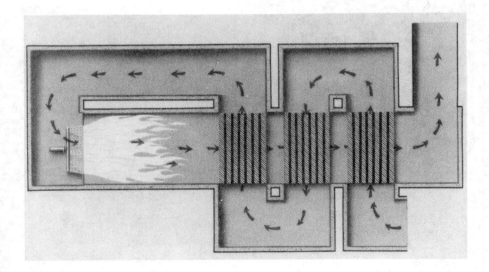

FIGURE 7.15
Steel rolling-mill slab reheat furnace with recuperative combustion-air preheater. (*Courtesy C. E. Air Preheater, Wellsville, N.Y.*)

in handbooks. Limiting temperatures of operation for different materials are carefully detailed in the applicable sections of the ASME Boiler and Pressure Vessel Code. There is such wide diversity in the metals used for high-temperature systems as to make it impossible to provide brief guidelines. It is sufficient to warn readers that the selection of materials for use at high temperatures is best left to specialists in the field. Materials that are excellent in moderate-temperature service may experience crystallographic modification at high temperatures, profoundly affecting their strength and durability. This is particularly true for materials that are heat-treated or cold-worked in the course of manufacture.

Another important effect of high-temperature operation is the enhancement of rates of corrosion and fouling deposition. Organic materials polymerize to introduce a mechanism of fouling not experienced at low temperatures: large, complex molecules break up, and carbonaceous deposits are formed. This occurs principally at local "hot spots," where fluid velocities may be low as a result of flow maldistribution or improper fluid management. The resulting deposition aggravates the situation by posing additional thermal resistance to heat transfer, with consequent further increase in the local temperature. The provision of extra heat-transfer surface in a heat exchanger can be the basis for a self-fulfilling prophecy of a high rate of fouling.

High-temperature heat is an important economic resource, worthy, to an increasing extent, of conservation to reduce the fuel input requirement. Extraneous heat leaks to the environment are to be minimized as far as possible. This is generally accomplished by thermal insulation, which is asbestos-based in most high-temperature systems. It is thermodynamically advantageous to design systems with the minimum possible temperature differences between surfaces or fluid streams. The counterflow principle should be used wherever possible. The highest temperature zones should be shrouded by lower-temperature zones so that the minimum-temperature surface interfaces with the environment.

Thermal loss occurs by conduction, convection, and radiation. Conduction losses can be minimized by the use of long, thin conduction paths. Convection losses are minimized by exposing a low-temperature surface to the air. A layer of asbestos or other insulation applied to a hot plate will dramatically reduce the heat lost by convection. A significant reduction in convective heat loss can be gained by minimizing air movement around the hot surface. This can be achieved by the use of baffles or an enclosure to eliminate the possibility of cross winds.

Radiation heat transfer can be minimized by providing a low-emissivity finish to the hot surface, i.e., a polished, reflecting metal surface. This is not practical in many cases, but the same effect can be achieved by shrouding the hot surface with a polished, reflecting metal foil. Aluminum is perfectly satisfactory, provided the temperature is not so high as to cause discoloration. Multiple layers of metal foil, crinkled or crimped to provide some separation, are more effective than a single layer, although little is gained by using more than four layers. Frequently such a radiation shield will also function effectively as a "spoiler" of natural convection streams.

There are clear and undoubted advantages in the use of exhaust-gas heat exchangers to recover thermal energy from exhaust streams. However, the overexuberant application of this principle may result in severe corrosion problems. These will arise when acid-related liquids or compounds in the combustion products cool below their dew points or deposition temperatures before discharge to the atmosphere. The combination of these liquids or compounds with water formed in the combustion of the hydrogen component will form highly corrosive agents in the exhaust system. Sulfur in the fuel resulting in sulfuric acid formation is very common, and other equally hazardous reaction products are often found. The practice of *not cooling* the combustion products below their dew point simply increases the burden on the environment. Acid rainfall, emanating from power plants burning high-sulphur coal in the United States and Canada and from the nickel plant at Sudbury, Ontario, has caused public concern about the future of recreational areas and lakes. The solution of these difficulties has so far eluded the politicians, even though it is simply a matter of treating the fuel or the exhaust to remove sulphur and its compounds. Our failure to resolve this problem will be deplored by generations yet to come, who must endure the consequences.

LOW-TEMPERATURE SYSTEMS

Heat exchangers are crucial components of low-temperature systems. Temperatures of application range all the way from near zero to 300 K, the ambient temperature, but are mainly concentrated in the *cryogenic* temperature range. Etymologically, the word cryogenic is derived from the Greek *kryos*, meaning frost or icy cold, but it is now reserved for temperatures below 120 K.

At temperatures below this limit, the so-called "permanent" gases liquefy. The liquefaction temperatures of some gases (at a pressure of 1 atm) are given in Table 7.1.

One of the principal activities in cryogenic engineering is the liquefaction and separation of gases for industrial purposes. Air is liquefied and separated into constituent fractions, approximately 21 percent oxygen, 78 percent nitrogen, and 1 percent argon, along with carbon dioxide, water vapor and miscellaneous trace gases.

Oxygen is used industrially for steel-making, gas welding, and industrial chemical processing. It has extensive medical life-support applications. Nitrogen is the inert gas or cryogenic liquid commonly used for low-temperature refrigeration applications, the flash freezing of frozen food products, preservation of bull semen, metallurgical processing, and low-temperature-properties research. Argon is an inert gas used extensively as the reducing shroud in the tungsten-inert-gas (TIG) aluminum welding process.

Another activity important to the development of cryogenic technology

TABLE 7.1

Liquefaction Temperatures of Certain Gases

Gas	Liquefaction temperature, K
Methane	112
Oxygen	90
Argon	87
Fluorine	85
Carbon monoxide	82
Air	79
Nitrogen	77
Neon	27
Hydrogen	20
Helium	4

was the recovery of helium from natural gas. Helium is a rare element with unique properties. It was unavailable in sizable amounts until, in the 1920s, the U.S. Bureau of Mines instituted a program to recover helium gas from selected natural-gas supplies. The use of helium has increased dramatically as a result of the conservation program. Vast amounts have now been accumulated in underground storage chambers, and the conservation program was abandoned in the mid-1970s; this was a controversial measure that excited intense opposition from the cryogenic community.

A large and increasing sector of cryogenic engineering is concerned with the liquefaction of natural gas. Natural gas is a complex mixture of many components, principally methane, with sizable proportions of propane, butane and other hydrocarbons, nitrogen, carbon dioxide, carbon monoxide, and hydrogen sulfide. It is produced in the recovery of crude oil and for many years was simply "flared off" at the well site. Its value as a fuel gas was recognized early in the United States, and a network of transmission and distribution pipelines was established to market the gas for industrial, commercial, and domestic heating. Much later, similar systems were established in Canada, Europe, and elsewhere. However, even today, in the Middle East, far too much of the gas produced as a by-product of oil production is flared at the well site to no useful purpose—a criminally wasteful measure. A small but increasing fraction of the gas is cooled to low temperatures, liquefied, loaded on specially constructed, highly insulated marine tankers, and carried to the United States, Japan, or Europe, where it is revaporized and used for heating purposes or as chemical feedstock for fertilizer and plastics production. The liquid natural gas industry is already the biggest international cryogenic engineering activity and is in a state of continuing rapid development.

Superconductivity is another area of cryogenic engineering, presently quite small but with every indication of developing into a major field. Superconductivity is the name given to the phenomenon whereby certain metals and alloys suddenly lose their electrical resistance in the course of a temperature reduction of a degree or so. The effect was first observed at about the turn of the century by Kammerlingh Onnes, a renowned Dutch physicist, at the University of Leiden, using mercury in liquid helium at 4 K. Since then many other developments have occurred to bring us to the threshold of the "coming age of superconductivity" referred to by people of vision. Compounds have been developed that exhibit the phenomenon of superconductivity at temperatures as high as 18 K, and the upper limit progressively increases with further research attention.

Future superconducting systems will revolutionize electronic computers (through the use of Josephson tunnel diodes), ship propulsion, magnetohydrodynamic power generation, the magnetic levitation of high-speed railway trains, the long-distance transmission of electric power, and the use of incredibly sensitive magnetometers for all manner of applications.

Infrared detectors have response characteristics that are functions of temperature. Highly sensitive systems have been developed that require cooling to cryogenic temperature levels for their operation. They are used in military systems for night-vision equipment and for heat-seeking missile-guidance purposes. There are civilian applications for such equipment in security equipment, resource development, and energy conservation.

Cryogenic Heat Exchangers

Heat exchangers play a vital role in all the equipment mentioned above. The principles on which cryogenic heat exchangers operate are exactly those prevailing at higher temperatures, but the emphasis is different in many aspects.

Most systems for generating refrigeration at cryogenic temperatures do so by the expansion of gas from a high pressure to a low pressure. One type of cryogenic refrigeration system that is used very extensively is the Claude system, shown in Fig. 7.16 as an equipment line diagram and as a temperature-entropy diagram. The Claude cycle operates as follows:

Gas at condition 1 enters the compressor and is compressed to some state 2. The process may be a single stage of compression, but more likely is divided into several stages, with intercooling between stages of compression to reduce the compressor work. The gas is then cooled to ambient temperature 2 in an aftercooler. The intercoolers and aftercoolers are likely to be water-cooled tube-and-shell type heat exchangers with gas in the tubes and water in the shell.

The cooled, compressed gas then enters a sequence of counterflow recuperative heat exchangers and is cooled to progressively lower temperatures by the flow of low-temperature, low-pressure vapor returning to ambient temperature through the heat exchangers. Following passage through the first recuperator, the high-pressure stream, cooled to state 3, is split into two streams. One stream continues on through the second recuperator and is cooled to state 4. The other

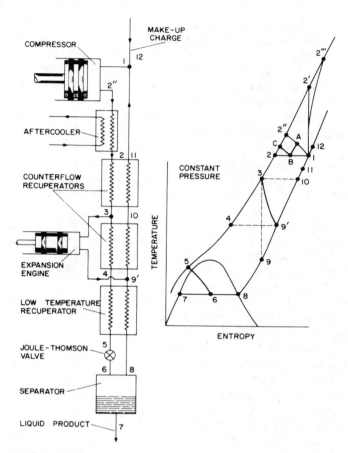

FIGURE 7.16
Claude-cycle cryogenic refrigeration system.

stream is expanded in an expansion engine to a low pressure. The expansion process is isentropic or near-isentropic, so the exhaust temperature from the engine at 9 or 9' is substantially lower than the inlet temperature 3. The expansion-engine exhaust stream mixes with the returning low-pressure vapor stream and passes through the intermediate-stage recuperator.

Expansion of compressed gas in an expansion engine thus provides refrigeration at some intermediate temperature to cool the remainder of the high-pressure stream en route to even lower temperatures. This process of bleeding off and expanding part of the high-pressure stream to cool the remainder may be extended to as many as five stages at progressively lower temperatures. The expansion engines may be turbines in larger systems, or piston engines in smaller systems.

Eventually the cooled stream, of reduced mass flow, passes through a final stage of recuperative counterflow heat exchange and attains the minimum temperature of the high-pressure fluid 5. It is then expanded to low pressure in a Joule-Thomson valve at constant enthalpy to state 6. If the final condition lies within the liquid/vapor envelope as shown in Fig. 7.16, the resulting fluid will actually consist of different fluids: saturated liquid at condition 7 and saturated vapor at condition 8. If the system is a liquefier, the two fluids are separated into the useful liquid product and the return stream of low-pressure vapor passing through the recuperators back to the compressor.

The compressor of the Claude system will be a multistage reciprocating compressor if the pressure is high. Screw compressors are being used to an increasing extent because of their reduced maintenance requirement, but they are capable of only moderate pressure ratios as compared with reciprocating machines.

Practical systems are considerably more complicated than that described above. They may have several stages of expansion, dual pressure systems, and associated Claude-cycle liquefiers to provide a precooling stage, perhaps a liquid-nitrogen precooler for a hydrogen or helium liquefier. Nevertheless, the simplified system adequately demonstrates the key significance of heat exchangers in cryogenic systems. The principle of counterflow operation is used extensively, and every effort is made to extract the last measure of recuperative action from the system.

The prevention of heat leaks to the cold parts by conduction, convection, and radiation is important. It is customary to consolidate the cold components in a "cold box" with adequate thermal insulation to isolate the cold components from ambient temperatures. So far as possible, the cold box is a vertical structure with components of similar temperature arranged at the same level, the coldest at the bottom. Each unit is individually insulated in a "muff" of foil thermal insulation, and the interior of the cold box is filled with a granular thermal insulating material such as perlite or the proprietary Min-K insulant. A typical cold box for a gas-liquefaction plant is shown in Fig. 7.17a.

Heat exchangers of many and varied types are used in cryogenic systems. Plate-fin heat exchangers of the type shown in Figs. 4.17 and 4.18 are much favored. Stacks of die-formed aluminum plates are dip- or furnace-brazed with headers and manifolds welded or brazed on to complete the assembly. The substantial difference in density between the high-pressure and low-pressure fluids can be accommodated by different depths of the fin plate to provide a compensating change in velocity and so approximate similar Reynolds numbers for the two flows. The fluids used in cryogenic systems are clean, so that contamination and blockage of the multiple flow passages is not of the same concern as in combustion systems.

Tubular heat exchangers are widely used. Figures 7.18 and 7.19 show typical large tubular heat exchangers for a liquid-natural-gas plant. Very high pressure fluids are contained in the large number of relatively fine-bore tubes, and the low-density fluid flows in the shell.

FIGURE 7.17
(*a*) Typical cryogenic cold box for a liquefaction plant. (*b*) Typical liquefaction plant.

FIGURE 7.18
Tubular heat exchanger for liquid-natural-gas plant.

FIGURE 7.19
Tubular heat exchanger for liquid-natural-gas plant.

One type of tubular heat exchanger thought to be limited to cryogenic applications is shown in Fig. 7.20. It consists basically of a fine-bore tube coiled on a large-diameter former. Then a close-fitting insulated sleeve is slid over the whole assembly. High-pressure fluid flows inside the tube, and the return stream of low-pressure, low-density fluid passes in counterflow and crossflow over the tube in the annulus between the former and the outer sleeve. In the large unit shown in Fig. 7.20, many tubes are coiled around the former, with the tube ends welded to toroidal headers at inlet and outlet.

This type of coiled heat exchanger was first used toward the close of the last century by Hampson, the English pioneer of air liquefaction. Later it was improved by Giauque, a professor at the University of California, Los Angeles, who introduced the punched tube-coil spacers shown in Fig. 7.21. The spacers maintain the tubes in the proper positions as specified by the design, to the great advantage of uniform flow distribution and heat-transfer effectiveness.

A somewhat similar type of coil-tube exchanger, used by Parkinson for small Joule-Thomson cryocoolers, is shown in Fig. 7.22. Another original type of heat exchanger for cryogenic systems was devised by Collins at the Massachusetts Institute of Technology. This consists essentially of a coil of thin copper foil closely wound onto a copper tube and soldered thereon to provide an external finned surface. The exchanger element is formed by two or more similar tubes,

FIGURE 7.20
Hampson coiled-tube heat exchanger. (*After Scott [1966].*)

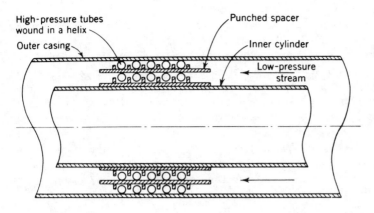

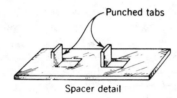

FIGURE 7.21
Giaque spacers for Hampson coiled-tube heat exchanger. (*After Barron [1966].*)

one inside the other, with a close-fitting outer sheath. A section of a two-stage element is shown in Fig. 7.23. Figure 7.24 shows a triple-annulus element presently available commercially.

The same foil-fin tube was used by Collins in the Hampson-type exchanger shown in Fig. 7.25. A novel feature was the introduction of cotton cord baffles

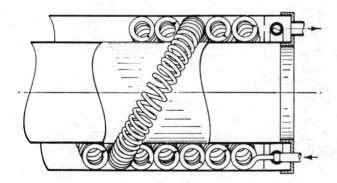

FIGURE 7.22
Parkinson coiled-tube heat exchanger. (*After Scott [1966].*)

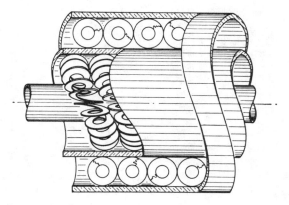

FIGURE 7.23
Collins heat-exchanger element. (*After Scott [1966].*)

between the tube coils to increase the path length and hence the velocity of the low-density flow over the tubes.

Panel-coil heat exchangers are also used extensively in cryogenic engineering systems. This type of exchanger, described in more detail in Chap. 4, is fabricated from embossed light-gauge metal plates welded along the edge seams and at the points of contact. The embossed pattern provides the conduit through which the heat-transfer fluid flows.

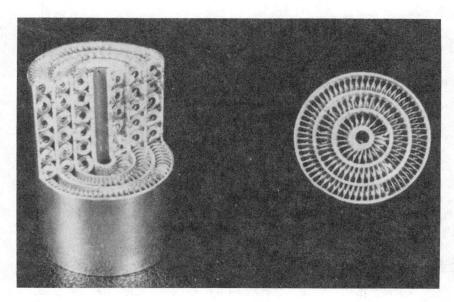

FIGURE 7.24
Triple-annulus Collins heat-exchanger element. (*After Barron [1966].*)

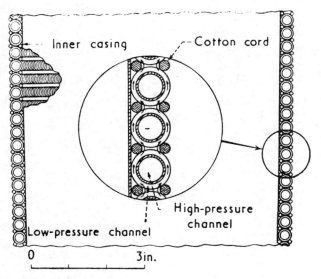

FIGURE 7.25
Collins Hampson-type heat exchanger with cotton and baffle inserts. (*After Collins [1957].*)

Figure 7.26 shows a large panel-coil cryogenic heat exchanger fabricated from stainless-steel sheet and intended for use as a liquid-nitrogen coiled thermal shroud in a space-environment test chamber. Figure 7.27 shows a panel-coil condenser heat exchanger for cryogenic service.

More extensive discussion of the different systems for generating cryogenic refrigeration and aspects of the system design have been given by Walker [1982].

Some Aspects of Cryogenic Service

Many aspects of cryogenic heat-exchanger service are similar to those of higher-temperature systems. In addition, for cryogenic service there are special considerations relative to design, fabrication, and operation.

The production of refrigeration at cryogenic temperatures requires large quantities of work as mechanical input to the gas compressors. This increases dramatically as the temperature decreases. At cryogenic temperatures, it is necessary to practice energy conservation more stringently than at higher temperatures. It is thermodynamically more effective to minimize the temperature difference between fluids in cryogenic fluid-to-fluid heat exchangers. Consequently, fluid conduits should be arranged so the two fluids are progressively matched on the counterflow principle.

In cryogenic heat exchangers, heat leaks from the surrounding low-temperature region will occur naturally by conduction, convection, and radiation in a process obverse to heat leakage from high-temperature systems. Environmental heat leaks to cryogenic systems are more critical than those from

FIGURE 7.26
Large panel-coil cryogenic heat exchanger intended as the liquid-nitrogen thermal shroud in a space-environment test chamber. (*Courtesy Bemco Inc., Pacoima, Calif.*)

FIGURE 7.27
Condensing heat exchanger for cryogenic service. (*Courtesy Bemco Inc., Pacoima, Calif.*)

high-temperature systems because of the large amount of work required to generate cryogenic refrigeration to compensate for the leakage loss. Whereas thermal insulation is often useful for high-temperature systems, it is crucially important to cryogenic heat exchangers.

Substantial efforts are therefore made to thermally isolate the low-temperature system and minimize environmental heat leaks. The measures adopted include the (1) use of long, thin-section fluid conduits and structural members to form tortuous low-conductivity paths, (2) elimination of convective losses by shrouding low-temperature regions in a high-vacuum enclosure, and (3) use of spectrally selective surfaces to minimize or maximize radiation transfers as required.

A familiar example of good thermal-isolation practice is the common vacuum-insulated flask used for picnics to maintain the temperature of hot or cold fluids. This type of flask was invented by Sir James Dewar in conjunction with his pioneer work in the liquefaction of the permanent gases and in cryogenic work is still called the Dewar flask. The interior is a closed double wall of glass or other low-conductivity material. The space between the walls is evacuated to eliminate convection effects, and the internal walls are silvered to provide a polished, low-emissivity, low-absorptivity surface to minimize thermal radiation effects. The flask is contained in a protective plastic or metal enclosure. That such a complicated structure can be obtained for a dollar or two at any hardware store is one of the miracles of modern technology.

For cryogenic service the simple evacuated enclosure is often enhanced by the use of many alternating layers of aluminized Mylar film and a cotton or nylon fiber mat to separate the film. This is called "superinsulation," and it provides the effect of multiple radiation shields to virtually eliminate radiation heat transfer from the cold region. A similar, but less effective, insulant is gained by filling the enclosure with perlite or Min-K granular material, sometimes containing a proportion of metal powder (opacified powder) to interrupt the radiation process.

The key element of all these systems is the very high vacuum in the space between the inner and outer enclosures. Extraordinarily low pressure must be sustained to effectively eliminate convective action. This is achieved in large, welded metal structures only by the unusually painstaking efforts of experienced and competent workers. The production of long, continuous welds, with no pinhole leaks is particularly difficult. Leaks that would be ignored for other applications are the Achilles' heel of cryogenic systems.

Foams are often used as an alternative to vacuum insulation in large systems. Urethane foam has remarkably low thermal conductivity and is much favored as a cryogenic insulation. The conductivity is a virtually linear function of the density, a characteristic also of its strength. If the design of a system utilizes the foam as part of the support structure, it is likely that the density and hence the thermal conductivity will be high.

To further decrease the conductivity, the urethane is sometimes foamed in an environment of low-vapor-pressure gas such as the refrigerant Freons. The Freon is trapped in the foam-bubble interstices, and on cooling to low temperatures condenses to provide a low-pressure multicellular insulation of exceptionally low conductivity.

The choice of materials for low-temperature heat exchangers is somewhat limited. The metals used must be selected from those not subject to the ductile-brittle transformation described in Chap. 6. This eliminates low-carbon steels, the conventional structural material for most applications. Stainless steel is widely used for cryogenic systems, and copper, nickel, and the nickel-steel alloys are also favored.

Plastic materials in general have very low thermal conductivities and may appear attractive on this account. However, many plastics become brittle and have low impact strength at cryogenic temperatures. Furthermore, they are rarely compatible with low-vacuum systems because of the phenomenon of *out-gassing*. Under high vacuum, many plastic materials release low-pressure solvent vapors or gases that make it impossible to maintain a permanent high vacuum without periodic attention. Sometimes the gases are baked out by operation at elevated temperatures and high vacuum for several hours or days. Adsorbing materials called *getters* are then sited in the vacuum space to absorb subsequent out-gassing vapors.

Fluorocarbon polymers are one family of plastic materials commonly found in cryogenic systems. Polytetrafluoroethylene (PTFE, TFE, Fluon, Teflon), fluorinated ethylene, propylene copolymers (FEP), and polychlorotetrafluoro-ethylene (PCTEE, Kel-F) are in this family. They are good electrical and thermal insulators, are inert to most chemicals and solvents, and do not become brittle at low temperatures. They have exceptionally low friction characteristics, stimu-lating their use for dry bearings and seals. They can be used as gaskets for static seals and as the sliding elements in dynamic seals. The mechanical properties of fluorocarbons can be improved by combination with various fillers such as glass fiber and powders of graphite, silver, and bronze. Rulon is a proprietary glass-filled, fluorocarbon plastic much used for piston rings and other dynamic seals in unlubricated low-temperature systems.

Glass is another material often used in cryogenic systems. Some varieties have a low—virtually zero—coefficient of thermal expansion, and so are capable of sustaining severe and repeated thermal shocks or rapid changes in temperature. Glass also has an attractive low thermal conductivity and good mechanical strength, particularly in compression.

Pressure vessels of glass-fiber filament wound with epoxy resin have astounding strength-to-weight characteristics that invite their use as pressure vessels. In cryogenic service the epoxy tends to craze, but there is no apparent loss of strength. Leakage of the contents is prevented by plating the interior of the vessel with a thin, flexible nickel, copper, or gold membrane. Fiber-reinforced composite pressure vessels will likely find increasing use in the future, not only for cryogenic service but more generally.

MINIATURE CRYOGENIC HEAT EXCHANGERS

We have seen that heat exchangers are used over the widest range of temperatures. They are also used over a very wide range of sizes. By a curious coincidence, the largest heat exchangers are also the hottest, while the smallest heat exchangers are found at the lowest (cryogenic) temperatures.

The largest heat exchangers are the giant combustion heated boilers or nuclear heated steam generators found in base-load power stations to raise steam for expansion in the power turbines.

There is no apparent intrinsic upper limit to the size of heat exchangers. The upper limit of size arises in practice from the logistical problem of transporting the unit from the manufacturers to the power station site. Road, rail, and air transportation have specific size limitations. Things are somewhat easier if the heat exchanger can by moved by water. Some extremely large structures have been floated from a manufacturer's coastal locations to a site located somewhere on a sea coast.

Large boilers or steam generators are commonly transported to the site in subassemblies for final construction at the site. This procedure is not possible with the other large heat exchangers used in power stations, namely, the water-cooled steam condensers. These must be fabricated to operate on the steam side at the very low pressure of the steam condensing on the outside of the many water tubes in the condenser drum. It is difficult, at a field location, to accomplish the very difficult welding required for high vacuum service. Therefore, steam condensers are generally completed at the manufacturer's and transported to the site. Although they are, in general, smaller than the steam boiler or generator, it is generally the feasibility of transportating the condenser that fixes the upper size limit.

Turning now to the other end of the scale, we find the smallest heat exchangers are those used in cryocoolers. These are miniature refrigerators capable of achieving cryogenic temperatures used to cool the infrared sensors of night vision and missile guidance systems. Others are used to cool the miniature electronic chips used for a variety of electronic devices, both superconducting and semiconducting, now coming into use. Some believe the era of "cold electronics" presages a technical revolution as profound as that following the invention of the transistor soon after World War II.

One type of heat exchanger widely used for this cryogenic electronic service is the Giaque-Hampson coiled tube heat exchanger shown in Fig. 7.28. An expansion nozzle is located at the end of the tube coil, and the whole assembly is contained in a close-fitting enclosing chamber (the cold finger) to contain the low pressure return gas. High pressure gas (up to 400 atm pressure) is supplied to the tube coil and expands to a low pressure through the expansion nozzle at the other (cold) end of the tube coil.

Nitrogen or argon expanding from high pressures in this way experience a cooling effect (called Joule-Thomson isenthalpic cooling) to temperatures between 77 and 90 K. (See Walker [1982, 1989].) The cold gas or liquefied gas is used to

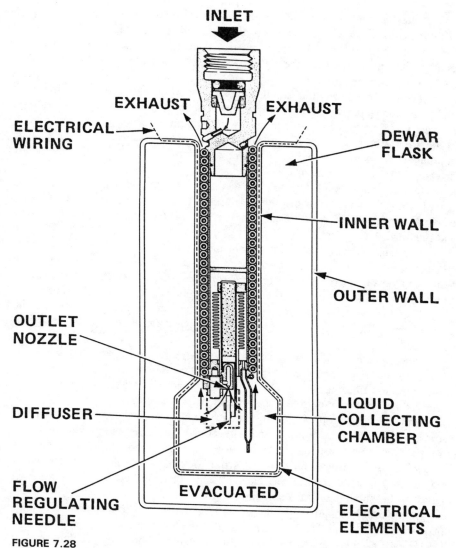

INLET

EXHAUST

EXHAUST

ELECTRICAL WIRING

DEWAR FLASK

INNER WALL

OUTER WALL

OUTLET NOZZLE

DIFFUSER

LIQUID COLLECTING CHAMBER

FLOW REGULATING NEEDLE

EVACUATED

ELECTRICAL ELEMENTS

FIGURE 7.28

Miniature Joule-Thomson expansion nozzle and giaque-Hampson coil tube heat exchanger. (*Courtesy Hymatic Engineering Co., Ltd.*)

cool the infrared sensor or cold electronic device located on the outer end of the chamber enclosing the tube coil.

The sensor and enclosing chamber are themselves contained in a glass or metal "encapsulation" envelope or dewar flask with a high vacuum in the flask to minimize convective heat leakage to the cooled tip of the "cold finger."

After the high pressure gas has expanded the cold low pressure gas exhausts from the cold finger at the inlet end of the tube coil. In passing over the outside of the finned tube coil, the cold gas is warmed to ambient temperature near the inlet end by heat transferred from the high pressure gas flowing through the tube coil en route to the expansion nozzle.

The tube coils are quite small, typically 7 mm in diameter and 50 mm long and attain their operating temperatures very quickly, typically within 1–2 minutes of start up, They are usually supplied with high pressure gas, air, nitrogen, or argon from a high pressure bottle. Some closed cycle systems are available with continuous circulation of the same gas compressed to 200 atm in an electrically driven multistage compressor.

Despite their small size the refrigeration capacity is substantial, ranging from 0.25 to 5 watts at nitrogen temperatures. The rate of gas flow is also appreciable (typically 10–20 liters of gas per min at STP (standard temperature, 0 °C, and pressure, 1 bar).

This high rate of gas consumption makes the open cycle operation of the coolers both expensive and inconvenient. Typically, an operating time of only 3–4 hours can be obtained from a high pressure gas bottle. This can be extended significantly by the use of a more complicated demand flow expansion nozzle.

The alternative is to use a compressor and operate on a closed cycle, but the development of a satisfactory miniature multistage compressor has proved difficult.

Most systems operate with a relatively large air compressor operating periodically to recharge several compressed gas bottles. There are difficult problems in cleaning and filtering the air to prevent the expansion nozzle of the miniature refrigerator from becoming blocked by condensed contaminant.

Although in widespread use for military applications, these miniature Joule-Thomson cryocoolers have not found widespread application for civil use. They are expensive to purchase and relatively expensive to operate. Moreover, the continued reduction in physical size of electronic devices limits the cooling requirements to only a small fraction (perhaps 1/100th) of that available from the even smallest of the tubular systems shown in Fig. 7.28.

Little [1984] recognized in the mid-1970s that it was simply not feasible to contemplate microminiature Joule-Thomson heat exchangers.

Instead, he investigated the possibility of using a photolithographic process as a means for fabricating the very fine channels and nozzles required for the microminiature systems. The system used is similar to that used for the fabrication of very large-scale integrated circuits (VLSI).

Located at the University of Stanford at the heart of Silicon Valley in California, Little was well placed to be familiar with the needs and requirements as well as the technology for manufacturing electronic chips.

In his development of planar microminiature Joule-Thomson refrigerators similar to that shown in Fig. 7.29, Little and his coworkers at MMR Technologies, Inc., Palo Alto, California, were completely successful.

FIGURE 7.29
Microminiature plenar Joule-Thomson refrigerator.

A large number of these microminiature refrigerators have now been made and are in use in laboratories all over the world. They are extremely small in size, of rugged construction, and their freedom from vibration and fast cool down makes them the system of choice for many cryogenic sensors and instruments.

SUMMARY

Heat exchangers of many different types and sizes are used all across the temperature spectrum. Most are used in the middle range of temperatures, not too significantly different from ambient conditions. Great numbers are also used at high temperatures and at low temperatures. In this chapter we have discussed briefly some applications and some aspects of the design and operation of heat exchangers at extreme temperatures.

The principal application of high-temperature heat exchangers is steam raising in fossil-fuel or nuclear-heated systems, primarily for power production and secondarily for process heating. Power boilers operate at temperatures and pressures fronting the limits of technology and are among the largest, most complex, and most expensive engineering creations of man. Smaller, lower-pressure process steam and heating boilers are also widely used as gas- or oil-fired heaters where steam heating is not available or desire. Exhaust heat recovery in recuperative exchangers is another increasingly important area of application of heat exchangers for energy conversation and improved fuel efficiency.

Low-temperature heat exchangers are also widely used. The primary concentration is in the cryogenic range (less than 120 K), for the liquefaction and reevaporation of the industrial gases oxygen, nitrogen, helium, hydrogen, and liquid natural gas, primarily methane. Miniature cryogenic heat exchangers are widely used in infrared night vision and missile guidance systems. Their use in cold electronic devices, both semiconducting and superconducting, is expected to increase rapidly in the future.

Loss of strength and accelerated mechanisms of corrosion and deposition are important consequences of high-temperature operation. At low temperatures, it is important to avoid materials experiencing the ductile-brittle transformation. for both high- and low-temperature systems it is thermodynamically advantageous to utilize the counterflow principle wherever possible, and to design the system to be thermally isolated to minimize extraneous heat leaks.

REFERENCES

Babcock and Wilcox: "Steam—Its Generation and Use," 38th ed., Babcock and Wilcox Co., New Orleans, 1972.

Barron, R.: Cryogenic Systems," McGraw-Hill, New York, 1966.

Collins, S.: "Low Temperature Expansion Machines," Oxford University Press, Oxford, 1957.

Little, W. A.: Microminiature Refrigeration, *Rev. Sci. Instrum.*, vol. 55, no. 5, pp. 661–680, May 1984.

Scott, R. B.: "Cryogenic Engineering," Van Nostrand, Princeton, N. J., 1966.

Walker, G.: "Cryocoolers," International Monographs in Cryogenics, Plenum, New York, 1982.

Walker, G.: "Cryocoolers," vol. 1 and 2, Plenum, New York, 1983.

Walker, G.: "Miniature Refrigerators for Cryogenic Sensors and cold Electronics," Oxford University Press Monographs in Cryogenics Series, Oxford University Press, New York, 1989.

8   Degradation of performance

INTRODUCTION

The purpose of a heat exchanger is to facilitate the transfer of thermal energy from a hot fluid to a cold fluid. It is designed to transfer a specific amount of heat with known fluid inlet temperatures and mass flows and given pump or compressor input, while preserving and containing the fluids whole, separated, and uncontaminated. We shall define degradation of performance as any failure to accomplish this duty. Degradation includes failure to achieve the desired thermal load and leakage of one fluid to the other or to the outside. Ultimate degradation of performance occurs when the exchanger is no longer able to contain the fluids because of corrosion or rupture.

Degradation of performance in heat exchangers is somewhat akin to aging in humans. It is an inevitable process for everyone but affects some to a greater extent, depending on their antecedents and the duties they are called upon to perform. Some heat exchangers never achieve their design objective. Their degradation stems from inadequate design or improper execution and poor workmanship. They are born crippled. Others achieve their design objectives but then deteriorate progressively in performance as time wears on. Deterioration may be due to fouling, the generic name for the accretion of deposits that increase the thermal resistance to heat transfer. This diminishes the heat transfer while, simultaneously, increasing the compressor and pump work inputs because of partial blockage of the flow conduits.

Degradation of performance due to fouling may be overcome by cleaning, with the potential for restoration of the exchanger to its pristine performance. Alternatively, the buildup of corrosion products may be inhibited by power treatment of cooling water, but this is not always economically feasible.

Corrosion is another principal source of heat-exchanger degradation. Corrosion of the heat-exchanger structural materials arises from a variety of mechanisms and progressively weakens the elements to the point where failure by rupture or leakage occurs or is imminent.

Various measures can be taken to eliminate or reduce the rate of corrosion, but it is an irreversible process. Once it has occurred, nothing can be done to restore the part to its original quality. The only recourse is repair by replacement. This is generally limited to the occasional replacement or plugging of a few tubes in a tube-and-shell heat exchanger or plates in a plate-and-frame exchanger.

In addition to ultimate mechanical failure, corrosion also contributes to the degradation of thermofluid performance. The oxide films generated in the course of some forms of corrosive action introduce an additional thermal resistance. The corrosion products will likely occupy a larger volume, partially blocking the flow conduits and increasing the input pump work or inhibiting the mass rate of flow.

In this chapter we shall consider the various aspects of design and execution that commonly lead to low performance, the principal mechanisms of fouling, and the various forms of corrosion. Measures to combat or repair degradation of performance are briefly discussed.

CORROSION IN HEAT EXCHANGERS

Corrosion was defined by Fontana and Greene [1978] as "the degradation of a material because of reaction with the environment." It is a part of the cycle of growth and decay that is the natural order of things. Today's tanks, battleships, and automobiles are the iron oxide deposits of tomorrow.

Except in a very few cases (for example, chemical milling and aluminum anodizing), corrosion is undesirable. It is something we feel we could do without. Then we realize that corrosion in its various forms is nature's way of disposing of garbage. Imagine the world today if there had never been corrosion or decay. There would be no growth, for there would be no room and no nutrients returning to the soil. Our homes would be old decrepit places filled with ancient, broken-down items that are impossible to get rid of. In this context, corrosion is a good thing—in fact essential. However, like many things in nature, its course can be directed to our advantage.

Corrosion is the principal cause of failure for all engineering systems. Estimates of the annual cost of corrosion in the United States were attributed, by Fontana [1977], to run as high as $30 billion, greater than the costs of floods, hurricanes, tornadoes, lightning, and earthquakes. These corrosion costs did not include the loss of production, plant downtime, and the cost of the legal cases that are sometimes consequent to corrosion failure.

Corrosion can be categorized into eight forms [Fontana and Greene, 1978]:

1. Uniform or general corrosion
2. Galvanic or two-metal corrosion
3. Crevice corrosion

4. Pitting corrosion
5. Intergranular corrosion
6. Selective leaching or dealloying
7. Erosion corrosion, including fretting and cavitation
8. Stress-corrosion cracking (scc), including corrosion fatigue

Uniform or General Corrosion

Uniform or general corrosion is the most common form of corrosion. It is characterized by a chemical or electrochemical reaction that proceeds uniformly over the entire exposed surface or a substantial portion of that surface. The metal becomes progressively thinner and eventually fails because of the stress loadings imposed on it.

This type of corrosion is the easiest to handle. The rate of decomposition can be determined by comparatively simple immersion tests of a specimen in the fluid. The life of equipment can therefore be predicted and extended to the degree required by the addition of a *corrosion allowance* to the metal wall thickness as necessary to sustain the pressure or other stress loading applied.

Prevention of Uniform Corrosion

Uniform corrosion can be prevented or reduced by the selection of appropriate materials (including internal coatings), the addition of corrosion inhibitors to the fluid, treatment of the fluid to remove corrosive elements, and the use of sacrificial cathodic protection or impressed electrical potentials. Other forms of corrosion are more difficult to predict. They tend to be localized and concentrated with consequent premature or unexpected failure.

Galvanic or Two-Metal Corrosion

When dissimilar metals are immersed in a corrosive or electrically conductive solution, a voltage (potential difference) will become established between them. If the metals are then connected by an electrically conducting path, a small current or electron flow will pass continuously between them. The principle is illustrated in Fig. 8.1. Corrosion of the less corrosion-resistant metal is accelerated and that of the more resistant metal is decreased, as compared with their behavior when they are not coupled electrically. The less resistant metal is described as *anodic,* and the more resistant metal as *cathodic.* Usually corrosion of the cathode is virtually eliminated.

The combination of dissimilar metals and a corrosive or electrically conductive medium constitutes a galvanic cell similar to the dry battery commonly used for flashlights and radios. The various metals and alloys, along with other materials of interest, can be arranged in order of decreasing corrosion resistance, as shown in Table 8.1. This listing, called the *galvanic series,* refers to materials immersed in seawater, a common and highly corrosive fluid.

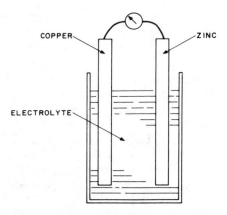

FIGURE 8.1
Galvanic cell comprised of dissimilar metals in an electrolyte and coupled electrically.

The noble metals leading the list are cathodic and least subject to corrosion. Those at the bottom are anodic and most subject to attack. The combination of a metal from the upper half of the table with any other further down the table will establish a galvanic cell with the potential to accelerate the rate of corrosion of the anode, lower in the table, while decelerating the corrosion rate of the cathode. The effect increases for metals that are further apart in Table 8.1. Magnesium would rapidly corrode in seawater in conjunction with a titanium cathode, but much less rapidly in combination with aluminum or zinc.

Brackets in Table 8.1 embrace groups of alloys similar in base composition. There is little danger of galvanic corrosion if metals in a given bracket are coupled electrically in an engineering structure.

An appreciation of the galvanic series is important in the design selection of materials. Fontana and Greene [1978] quote many examples to illustrate the point. One relevant to heat exchangers concerns the use of thin copper tubes and heavy cast-iron or steel tube sheets. Galvanic corrosion occurs, and the ferrous corrosion is accelerated while the thin copper tube is preserved. However, long life is obtained through the use of thick tube sheets, and the more expensive bronze tube sheets are not required. In more severe corrosion conditions, e.g., dilute acidic solutions, bronze tube sheets would be required.

Various factors make the phenomenon of galvanic corrosion more complex and less predictable than would appear from the above discussion. Readers are urged to consult the text referenced above.

Prevention of Galvanic Corrosion

Recommendations for preventing or minimizing galvanic corrosion are:

1. Use a single material or a combination of materials that are close in the galvanic series.

TABLE 8.1

Galvanic Series for Some Commercial Metals and Alloys in Seawater
(*After Fontana and Greene [1978].*)

↑ Noble or cathodic	Platinum Gold Graphite Titanium Silver
	⎡ Chlorimet 3 (62 Ni, 18 Cr, 18 mo) ⎣ Hastelloy C (62 Ni, 17 Cr, 15 mo)
	⎡ 18-8 Mo stainless steel (passive) ⎢ 18-8 stainless steel (passive) ⎣ Chromium stainless steel, 11–30% Cr (passive)
	⎡ Inconel (80 Ni, 13 Cr, 7 Fe) (passive) ⎣ Nickel (passive)
	Silver solder
	⎡ Monel (70 Ni, 30 Cu) ⎢ Cupronickels (60–90 Cu, 40–10 Ni) ⎢ Bronzes (Cu-Sn) ⎢ Copper ⎣ Brasses (Cu-Zn)
	⎡ Chlorimet 2 (66 Ni, 32 Mo, 1 Fe) ⎣ Hastelloy B (60 Ni, 30 Mo, 6 Fe, 1 Mn)
	⎡ Inconel (active) ⎣ Nickel (active)
	Tin Lead Lead-tin solders
	⎡ 18-8 Mo stainless steel (active) ⎣ 18-8 stainless steel (active)
	Ni-Resist (high-Ni cast iron) Chromium stainless steel, 13% Cr (active)
	⎡ Cast iron ⎣ Steel or iron
Active or anodic ↓	2024 aluminum (4.5 Cu, 1.5 Mg, 0.6 Mn) Cadmium Commercially pure aluminum (1100) Zinc Magnesium and magnesium alloys

2. Avoid the use of a small ratio of anode area to cathode area. Use equal areas or a large ratio of anode to cathode area.
3. Electrically insulate dissimilar metals where possible. This recommendation is illustrated in Fig. 8.2. A flanged joint is equipped with bolts contained in insulating sleeves with insulating washers under the head and nut. Paint, tape, or asbestos gasket material are alternative insulants.
4. Local failure of the protective coating, particularly at the anode, can result in the small anode-to-cathode area syndrome marked by accelerated galvanic corrosion. Maintain all coatings in good condition, especially at the anode.
5. Decrease the corrosion characteristics of the fluid where possible by removing the corrosive agents or adding inhibitors.
6. Avoid the use of threaded or riveted joints in favor of welded or brazed joints. Liquids or spilled moisture can accumulate in thread grooves or lap interstices and form a galvanic cell.
7. Design for readily replaceable anodic parts or, for long life, make the anodic parts more substantial than necessary for the given stress conditions.
8. Install a sacrificial anode lower in the galvanic series than both the metals involved in the process equipment.

Crevice Corrosion

This type of corrosion is also called *deposit* or *gasket* corrosion and sometimes *concentration-cell* corrosion. It is characterized by intense local corrosion in crevices and other shielded areas on metal surfaces exposed to *stagnant* corrosive liquids. It can occur where any undisturbed liquid film exists, such as at a small hole, gasket-flange interfaces, lap joints, surface deposits, and the crevices under bolt and rivet heads (Fig. 8.3). Relative to heat exchangers, it is important to note that nonmetallic deposits (fouling) of sand, dirt, or crystalline solids may

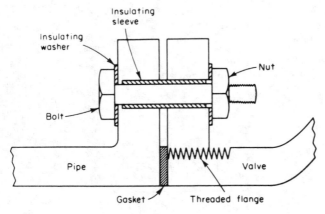

FIGURE 8.2
Isolation of dissimilar metals by electrical insulation. (*After Fontana and Greene [1978].*)

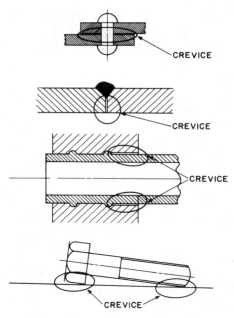

FIGURE 8.3
Crevice corrosion.

act as a shield and create the necessary stagnant condition—the essence of crevice corrosion.

The mechanism of crevice corrosion is a complex electrochemical process whose details are beyond the level of our interest. It is well described by Fontana and Greene [1978], with references to the research papers. Sufficient to say it is associated with the depletion of oxygen in the stagnant liquid pool, which results in corrosion of the metal walls adjacent to the crevice.

This type of corrosion occurs with many fluids but is particularly intense with those containing chloride. Common salt is sodium chloride, and consequently seawater is highly corrosive. The nature of the electrochemical process is such that the corrosion attack is localized within the stagnant or shielded area, while the surrounding surfaces over which the fluid moves suffer little or no damage.

Some time is required between the initial establishment of the conditions for crevice corrosion and the occurrence of visible damage. This "incubation period" is necessary to establish the oxygen-depletion situation referred to above. Sometimes a year elapses before the attack begins, but, once initiated, it proceeds at an accelerating pace.

Metals and alloys that depend on oxide films or passive surface layers for corrosion resistance are most susceptible to crevice corrosion. The films are often destroyed by the chloride or hydrogen ions resulting from the electrochemical process in the stagnant film. Stainless steel and aluminum, usually thought to be corrosion-resistant, are in this category.

Contact between a metallic and any nonmetallic surface can result in crevice corrosion of the metal part. Wood, plastic, rubber, glass, concrete, asbestos, wax, and fabrics, when associated with a metal, can all result in crevice corrosion of the metal. So long as the crevice between the surfaces is wide enough to allow liquid to enter (via surface-tension forces or other means) and is sufficiently narrow to prevent convective action, the conditions for crevice corrosion are established. Fibrous gaskets have a wick action and can form a completely stagnant solution in contact with a flange face, an ideal site for crevice corrosion.

Prevention of Crevice Corrosion

Recommendations for preventing or minimizing crevice corrosion are [Fontana and Greene, 1978]:

1. Use welded butt joints instead of bolted or riveted joints. Good welds with deep penetration are required to avoid porosity and crevices on the inside if the joint is welded on one side only.
2. Eliminate crevices by continuous welding, by solder or braze filling, and by caulking.
3. Design to eliminate sharp corners, crevices, and stagnant areas and for complete drainage.
4. Minimize the accretion of deposits, or clean at regular intervals.
5. Eliminate solids suspended in the fluids, if possible.
6. Use solid soft-metal or PTFE gaskets in favor of absorbent ones.
7. Weld tubes to tube sheets instead of rolling.

Pitting Corrosion

Pitting corrosion is the phenomenon whereby an extremely localized attack results in the formation of holes in the metal surface that eventually perforate the wall. It is illustrated in Fig. 8.4. The holes or pits are of various sizes and may be isolated or grouped very closely together.

Once established, the mechanism of pitting is very similar to crevice corrosion and occurs as the result of a similar complicated electrochemical process. The pits correspond to the stagnant films necessary for crevice corrosion. Pits usually grow in the direction of gravitational action, i.e., downward from horizontal surfaces. They sometimes develop on vertical surfaces, but only in very exceptional circumstances do pits grow upward from the bottoms of horizontal surfaces.

As with crevice corrosion, an incubation period is required before pitting corrosion starts; thereafter, it continues at an accelerating rate. Furthermore, once below the surface, the pits tend to spread out, undermining the surface as

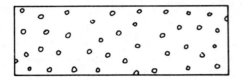

FIGURE 8.4
Pitting corrosion.

shown in Fig. 8.4. This is particularly unfortunate, for the small surface pits can easily become obscured by corrosion products or other sediments and deposits. Failure as a leak resulting from complete perforation of the metal wall therefore occurs suddenly and unexpectedly.

Most pitting corrosion emanates from the action of chloride or chlorine-containing ions. These are present in most solutions made with water and are particularly evident in brackish water and seawater. Pitting also occurs with many other chemical solutions, with varying degrees of aggressiveness. It may often be reduced by the addition of hydroxide, chromate, or silicate salts, but unfortunately in small concentrations these very substances tend to increase the propensity for pitting.

The mechanism for initiation of pitting at any particular point is not clear. It is generally attributed to a momentary local increase (for one reason or another) in the rate of metal dissolution by general corrosion. This attracts chloride ions to the point of rapid metal dissolution and stimulates a further increase. Once established in a crater, the process continues at an accelerating rate while the surrounding surface remains cathodically protected.

The process of establishing a pit site is unstable and is interrupted by any movement of fluid over the surface. Thus, pitting corrosion is rarely found in metal surfaces over which fluids move constantly. Even in these few cases it can be reduced if the fluid velocity is increased. Often a heat-exchanger pump or tube carrying a corrosive fluid shows no sign of pitting corrosion when in service but deteriorates rapidly if the plant is shut down and the fluid not drained from the system.

Stainless-steel alloys are particularly susceptible to pitting-corrosion attack. According to Fontana and Greene [1978], even carbon steel is more resistant to pitting than stainless steel. Pitting of stainless-steel condenser tubes carrying brackish water or seawater can often be alleviated by the substitution of carbon-steel tubes. Of course, the rate of general corrosion is much higher, but rapid perforation of the tubes by pitting does not occur.

Prevention of Pitting Corrosion

In general, the recommendations given above with regard to crevice corrosion apply also to pitting corrosion. The principal measure is to use materials that are known to be resistant to pitting. These include:

Titanium
Hastelloy C or Chlorimet 3
Hastelloy F, Nionel, or Durimet 20
Type-316 stainless steel
Type-304 stainless steel (pits badly in chloride solutions)

Intergranular Corrosion

Metals and their alloys are crystalline solids, atoms of metal arranged in a regular, repeating array. There are various possible crystal structures, but most metals can be categorized as:

1. Body-centered cubic (iron and steel)
2. Face-centered cubic (austenitic stainless steels, copper, aluminum)
3. Hexagonal close-packed (magnesium)

Metals melt when heated to a sufficiently high temperature, and in the liquid state the atoms are randomly distributed. As the liquid cools, the atoms arrange themselves in their characteristic crystalline array. The crystal growth is initiated at innumerable sites in the liquid, and the crystals or grains grow independently until, eventually, the process is limited by the proximity of adjacent grains. The boundaries between individual grains tend to be more chemically active than the grains themselves and are attacked more rapidly when exposed to a corrosive.

The extent to which this happens depends on a number of factors. In many cases the grain boundary effects are of little consequence, and the metal corrodes uniformly. In other cases and under certain conditions, the grain interfaces are very reactive. Preferential corrosion occurs with localized attack at the grain boundaries, and the alloy simply disintegrates or loses strength.

Austenitic Stainless Steels

The phenomenon of intergranular corrosion, discussed in detail by Fontana and Greene [1978], is beyond our level of interest here. Sufficient to say the matter is of particular interest with austenitic stainless steels of the common type-304 18-8 composition (18 percent chromium, 8 percent nickel). Numerous failures of 18-8 stainless steels have occurred because of intergranular corrosion in environments where excellent corrosion resistance was anticipated. The explanation is that when these steels are heated in the temperature range 800 to 1000 K, they become *sensitized* and highly susceptible to intergranular corrosion. In this temperature range chromium precipitates out of solution in areas adjacent to the grain boundaries, leaving a chromium-depleted metal with insufficient resistance to corrosion attack.

The matter is of particular significance with welded structures of 18-8 stainless steels. The heat associated with the welding process is normally dissipated by conduction in the base structure and can result in an area of sensitized alloy, called the *weld-decay zone,* near the weld. This is illustrated in Fig. 8.5, which shows the temperature-time history of the joint area during the electric-arc welding of type-304 stainless-steel alloy. All the metal between points A and C is in the sensitizing temperature range (1200 to 1600°F) for some time and may become susceptible to intergranular corrosion in severely corrosive environments.

Prevention of Intergranular Corrosion in Austenitic Stainless Steels

The problem of intergranular corrosion in austenitic stainless steels is resolved by use of heat treatment or special alloys. The high-temperature heat treatment called *quench annealing* or *solution quenching* consists of very rapid cooling, by water quenching from a temperature of 1340 to 1400 K. At these temperatures chromium carbide is in solution in the metal, and rapid quenching "freezes" the alloy in this homogeneous state. Most austenitic stainless steels are supplied in this condition. Subsequent welding during fabrication can create a heat-affected

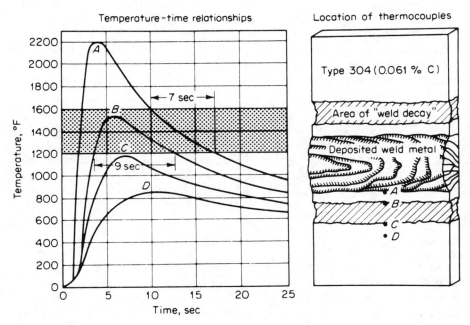

FIGURE 8.5
Temperatures during electric-arc welding of type-304 stainless steel. (*After Honnaker [1958].*)

zone near the weld, subject to intergranular corrosion. This can be eliminated if the weldment can be quench-annealed, but this treatment is clearly impractical for large vessels and for add-on welds or repairs to an existing installation.

The alternative is to use special alloys that are immune from intergranular corrosion. Stabilizing elements such as columbium, tantalum, and titanium are added, as in type-347 and type-321 stainless steel. These elements have a greater affinity for carbon than chromium and are added in sufficient quantity to combine with all the carbon in the steel. Such steels do not exhibit chrome depletion and the consequent corrosion-susceptible zones in the grain boundary regions.

Another approach is simply to lower the carbon content of the alloy to below 0.03 percent (from the 0.2 percent commonly found in the original 18-8 alloy steels). These alloys are called low-carbon or extra-low-carbon steels and are characterized by the designation 304L.

The low-carbon steels have a high solubility for carbon in the molten state. As Fontana and Greene [1978] point out, the whole purpose of using low-carbon steels is lost when the welder carefully cleans the beveled edge of the joint with an oily rag before welding. Similarly, pitting corrosion of stainless steels has been noted where the surface was cleaned by brushing with a carbon-steel wire brush that left fragments embedded in the stainless-steel surface.

Stabilized austenitic steels containing columbium and titanium sometimes exhibit a type of intergranular corrosion called *knife-line attack* because of its distinctive appearance as a straight, narrow band, a few grains wide. It occurs immediately adjacent to welds and arises from the solubility of columbium and titanium in the stainless steel (see Fontana and Greene [1978]). Low-carbon alloys are recommended for use where knife-line attack of stabilized steels has occurred.

Intergranular Corrosion of Other Alloys

Intergranular corrosion is not confined to the austenitic stainless steels, but occurs with other complex alloys that depend on precipitated phases for strengthening. Fontana and Greene [1978] mention the susceptibility of high-strength aluminum alloys (aluminum-copper Duraluminum type) and many other precipitates, including magnesium- and copper-based alloys and the die-cast zinc-aluminum alloys.

Selective Leaching

Selective leaching is the term used to describe a corrosion process wherein one element is removed from a solid alloy. The phenomenon occurs principally in brasses with a high zinc content (dezincification) and in other alloys from which aluminum, iron, cobalt, chromium, and other elements are removed. There are two characteristic forms of selective leaching—a general, uniform process and a local, concentrated or *plug-type* process.

The mechanism for selective leaching is likely a multistage electrochemical process in which the metal first dissolves in the corrosive fluid. The more noble element plates back on the surface, while the anodic ions remain in solution. The process is characterized by soft, porous surface deposits of the noble element.

The presence of oxygen or chloride in solution increases the rate of attack, temperature has some effect. It is usually associated with stagnant conditions and the formation of scale deposits or other accretions also favoring the development of crevice corrosion. There are virtually no dimensional changes as a consequence of selective leaching. If a surface is heavily fouled, there may be a sudden unanticipated failure because of the lowered strength of the selectively leached substrate.

Prevention of Selective Leaching

The only effective method of preventing corrosion by selective leaching is to avoid the use of materials known to be subject to it in association with the fluids concerned. Brasses with a high zinc content (> 35 percent) in acid environments are particularly susceptible. Red brass (15 percent zinc) is virtually immune from selective leaching. Dramatic improvements may be gained by the addition of small amounts of extra alloying elements. Admiralty metal, formed by the addition of 1 percent tin to 70-30 (copper-zinc) brass is noted for its resistance to corrosion. Further improvement may be gained by the addition of arsenic, antimony, and phosphorus. In severe environments the cupronickels, 70 to 90 percent copper and 30 to 10 percent nickel, are favored.

Grey cast iron is subject to a selective leaching known as *graphitization* whereby the iron is dissolved, leaving behind a weak, porous graphite network. The surface shows some rusting that appears to be superficial or has a layer that is apparently a deposit of soft graphite. It is, in fact, a residue from which the metal has gone with consequent loss of strength. This is a relatively slow process that occurs more or less uniformly.

White cast iron and the nodular or malleable cast irons are not subject to this form of corrosive attack.

Erosion Corrosion

Erosion corrosion is the term used to describe corrosion that is accelerated as a result of an increase in the relative motion between the corrosive fluid and a metal wall. The process is usually a combination of chemical or electrochemical decomposition or dissolution and mechanical wear action. Erosion corrosion therefore differs markedly from most other forms of corrosion, where the rate of attack is highest under stagnant or low-velocity conditions.

Erosion corrosion can be recognized by the appearance of grooves, gullies, and waves in a directional pattern, similar to sand formations on a shoreline. Figure 8.6 is a sketch of the erosion-corrosion pattern on a condenser tube wall. Failure by erosion corrosion can occur in a relatively short time (a matter of

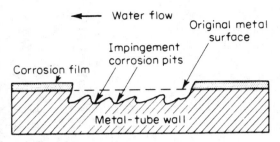

FIGURE 8.6
Erosion corrosion of condenser tube wall. (*After Fontana and Greene [1978].*)

weeks or months). It often comes as a surprise, following satisfactory tests for the corrosion susceptibility of specimens submerged in the corrosive fluid under static conditions.

Metals that depend for their corrosion resistance on the formation of a protective surface film are particularly susceptible to attack by erosion corrosion. Aluminum and stainless steels are in this category. The protective film is eroded by mechanical scrubbing, exposing the soft core to chemical or electrochemical attack in addition to the continued mechanical wear.

Many fluids that are not normally considered aggressive corrosion agents can promote erosion corrosion. High-velocity gases and vapors at high temperature may oxidize a metal and then physically strip off the otherwise protective scale.

Impingement baffles are provided downstream of the inlet nozzles of tube-and-shell heat exchangers to prevent damage to the tube bundle by erosion corrosion due to the high-velocity streams. Similarly, many erosion-corrosion failures in heat exchangers occur in tube-side flow, particularly at the tube inlet; the process is frequently called *inlet-tube* corrosion. It arises essentially from the highly turbulent flow ensuing as a consequence of the sudden change in section as the fluid leaves the inlet bonnet and enters the reduced flow section of the tubes.

The effect of velocity on erosion corrosion is very important but not entirely predictable. An increase in the velocity does not always result in increased erosion-corrosion attack. It may produce no significant change or even a decrease in the rate of attack, depending on the mechanism involved and the material-fluid combination. The general effect, however, is an increase in the rate of erosion corrosion as the velocity increases. For many materials there appears to be a critical value above which the rate of attack increases rapidly. Table 8.2 reproduces some results showing the effect of velocity on the corrosion of different metals by seawater.

Prevention of Erosion Corrosion

Five methods of preventing or reducing damage due to erosion corrosion are recommended by Fontana and Greene [1978]:

1. Use materials with superior resistance to erosion corrosion.
2. Design for minimal erosion corrosion.
3. Change the environment.
4. Use protective coatings.
5. Provide cathodic protection.

Substitution of better materials is the self-evident and invariably most economical solution to an erosion-corrosion problem that becomes manifest in existing equipment.

Solution to problems of erosion corrosion by design is favored for equipment yet to be constructed, either as original or replacement units. Fontana and Greene [1978] quote many examples from their experience of design solutions to problems of erosion corrosion in heat exchangers. Most solutions follow the pattern of changing the shape or the geometry to reduce or divert the impact of high-velocity streams. The use of robust sacrificial impingement baffles to protect tube bundles near inlet nozzles was already mentioned above. Tubes designed to project about 0.1 m beyond the tube sheet provide a sacrificial extension that can be eroded while preserving the integrity of the tube–tube-sheet joint. Similarly, ferrules or short lengths of bell-mouth flared tubing inserted in the tube entrance may avoid the ravages of inlet-tube corrosion.

TABLE 8.2

Corrosion of Metals by Seawater at Different Velocities
(*Courtesy International Nickel Co., New York.*)

	Typical corrosion rates, mdd*		
Material	1 ft/s[†]	4 ft/s[‡]	27 ft/s[§]
Carbon steel	34	72	254
Cast iron	45	–	270
Silicon bronze	1	2	343
Admiralty brass	2	20	170
Hydraulic bronze	4	1	339
Aluminum bronze (10% Al)	5	–	236
Aluminum brass	2	–	105
90-10 Cu-Ni (0.8% Fe)	5	–	99
70-30 Cu-Ni (0.5% Fe)	2	–	199
70-30 Cu-Ni (0.5% Fe)	< 1	< 1	39
Monel	< 1	< 1	4
Stainless steel, type 316	1	0	< 1
Hastelloy C	< 1	–	3
Titanium	0	–	0

*Milligrams per square meter per day.
[†]Immersed in tidal current.
[‡]Immersed in seawater flume.
[§]Attached to immersed rotating disk.

Bakelite and other plastic ferrules are sometimes used in condenser tubes. The use of plastic inserts avoids the possibility of galvanic corrosion that might occur with metallic inserts.

Thoughtful design to facilitate replacement of worn parts is important. Damaged or worn tube bundles that are replaced by spares can be repaired at leisure. One interesting example quoted by Fontana and Greene concerns a vertical evaporator. The life of the tubing was doubled by turning the evaporator upside down when the inlet or bottom ends of the tubes became thin. The outlet ends, in a virtually pristine state, then become the inlet ends and a new cycle began.

Alteration of the environment is another effective, but usually economically impractical, way of reducing the effects of erosion corrosion. Deaeration and the addition of inhibitors are effective. A reduction in temperature can also dramatically reduce erosion corrosion as well as other types of corrosion. Filtration and the removal of solids are also helpful but not usually practical.

Coatings are sometimes useful in combating erosion corrosion. The approach may follow two paths:

1. Application of a soft, resilient liner (rubber) for heavily contaminated slurries
2. Application, to areas subject to attack, of very hard armor, either tungsten carbide shields brazed in position or Stellite inserts welded in place

Cathodic protection to help reduce erosion-corrosion attack in heat exchangers takes the form of sacrificial steel or zinc anodes and is most generally used in seawater and brackish-water equipment.

Cavitation Erosion

Cavitation erosion is a special class of erosion corrosion that is associated with the periodic growth and collapse of vapor bubbles in liquids. It is of interest in heat exchangers, particularly evaporators, handling liquids at pressures and temperatures near saturation.

Cavitation erosion occurs when vapor bubbles at the wall suddenly collapse as a result of local hydrodynamic pressure differences. The local shock wave created by sudden collapse of the vapor bubbles can result in an enormous impact pressure on the wall, estimated as high as 400 MPa. The bubble collapses are equivalent to hammer blows and sufficient to mechanically dislodge particles of metal from the surface. This is a serious matter for those metals that depend on a protective surface layer for corrosion resistance. Removal of the surface layer exposes the core to chemical corrosion before the protective layer is reestablished. Furthermore, the presence of a small irregularity in the surface provides a nucleation site for the growth and collapse of other bubbles, to continue the destructive process.

Surfaces on which boiling or evaporation occurs rapidly assume a dull, diffusely radiating, roughened surface even if originally highly polished. This is caused by the mixed mechanical and chemical action of cavitation erosion described above.

Cavitation erosion also occurs in nonboiling situations where sudden changes in velocity can result in sufficient hydrodynamic pressure differences in the process flow streams to result in the creation, albeit momentary, and collapse of vapor bubbles. This situation may occur in local high-velocity regions in the reversing shell-side flow of tube-and-shell exchangers and in some plate-exchanger flows.

Under steady-state conditions the vapor bubbles tend to form and collapse in the same region of the exchanger, so the cavitation erosion is confined to particular areas.

The techniques used to prevent cavitation erosion are generally similar to those for erosion corrosion. Smooth finishes or polished surfaces generally inhibit the formation of bubbles. Similarly, the application of resilient coatings of rubber or plastic have proven beneficial. However, neither the polished nor the resilient surface would be helpful if the objective was, in fact, to promote evaporative action.

Cavitation action can be reduced by redesign to moderate velocity changes. This results in a reduction of the hydrodynamic pressure differences giving rise to the process of vapor-bubble formation and collapse. In general, the best approach to reducing cavitation-erosion damage is the use of erosion-corrosion-resistant metals. The stainless steels and hard nickel alloys are particularly favored in this regard.

Fretting Corrosion

Fretting corrosion occurs at the contact points of stressed metallic joints that are subject to vibration and slight movement. It is also called *friction oxidation, wear oxidation, chafing,* and *false brinelling.* Fontana and Greene [1978] consider fretting corrosion to be a special case of erosion corrosion occurring in air rather than aqueous conditions. The classification of fretting as a special case of erosion corrosion is not universally accepted; many regard fretting as a separate category of corrosion.

Fretting is principally of interest with heat exchangers of the tube-and-shell variety, particularly with regard to deterioration of the tube–tube-sheet joint and the tube-baffle contacts, as shown in Fig. 8.7. Problems associated with fretting corrosion arise at these points as a result of tube vibrations (see Chap. 6). One approach to minimizing tube vibration is to increase the tube hole clearance in the baffles, with the result that periodic rubbing contact and consequent fretting corrosion are much less likely. This increases the shell-side baffle-bypass flow and so cannot be carried too far.

The essential elements for incipient fretting corrosion include:

1. A loaded interface. Tube–tube-sheet joints are heavily loaded by the strains induced in rolling the tubes in the tube-sheet.

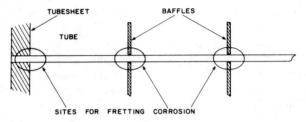

FIGURE 8.7
Tube–tube-sheet and tube-baffle fretting corrosion.

2. Vibration or repeated relative motion between the two surfaces.
3. The load and relative motion of the interface must be sufficient to produce slip or deformation on the surfaces.

The relative motion necessary to produce fretting is very small; displacements of the order of 10^{-9} m can result in damage. Fretting is detrimental in heat exchangers because destruction of the metal results in loosening of the tube–tube-sheet joint, permitting increased movement and an accelerating rate of wear including failure by metal fatigue. The production of oxide debris is detrimental because the debris accumulates in the interstices of the joint and acts as an abrasive, causing further wear. Severe fretting can result in the production of pits or grooves in the metal surfaces of the interface, sufficient to reduce the tube wall thickness to a dangerous level. The pits formed in fretting corrosion act as stress raisers so that local stresses may be two or three times the general level of stress.

Fretting corrosion is generally explained on the basis of two theories: wear/oxidation and oxidation/wear. The two processes are illustrated in Fig. 8.8.

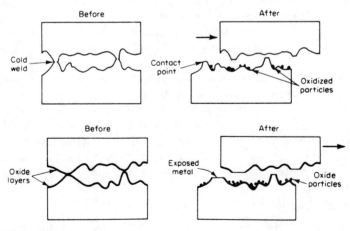

FIGURE 8.8
Theories of fretting corrosion. (*After Fontana and Greene [1978].*)

The mating surfaces are apparently flat but at the microscopic level are a rough series of mountains and valleys as shown. The wear/oxidation theory presumes that the mountain peaks are in contact and so highly stressed that cold welding or fusion occurs. During subsequent vibrations or slip these fused contact points are ruptured, and small metallic particles are removed. The particles are very small and are heated to high temperatures because of the friction effects and so are immediately oxidized. The process is continuously repeated, with subsequent loss of metal and the accumulation of wear debris. The process is therefore one of frictional wear causing the damage, with subsequent oxidation as a secondary effect.

The reverse oxidation/wear theory is based on the concept that most metal surfaces are protected from oxidation by a surface oxide layer. In a fretting-corrosion environment the metal surfaces are put into contact under load and subject to vibration or slight movement. The oxide layer is ruptured at the points of contact, and pieces accumulate in the valleys. The exposed metal peaks reoxidize, and the process continues. This theory is therefore based on accelerated oxidation due to frictional effects.

In practice it is likely that both mechanisms apply. Which one is dominant depends on the surfaces involved and the surrounding environment.

Fretting corrosion can be prevented or reduced by the following measures:

1. Eliminate the source of vibration or movement of the two interfaces. In tube-and-shell exchangers this likely involves reduction of the shell-side fluid velocity. This can be done by reducing the shell-side mass flow or redesigning the fluid flow distribution as discussed in Chap. 6.
2. Eliminate the high-stress interface of the moving surfaces. This might be done by welding the tube–tube-sheet joint instead of rolling. Even light loads can result in fretting corrosion under vibratory conditions.
3. Lubricate the joint with low-viscosity oils or greases. Phosphate coatings are sometimes used because they are porous and act as oil reservoirs.
4. Use hard surfaces on one or both interface elements. Experimental observation has indicated that unlubricated interfaces generally experience a lower rate of fretting corrosion when one or both surfaces are hard. The surface hardness may be gained by cold-working, as the rolling of tubes in tube sheets, by carburizing, or by shot peening.
5. Increase the friction at the interface by roughening the surface.
6. Use soft metallic or nonmetallic interface gaskets.
7. Relax the design to *increase* the relative motion at the interface (not generally practicable for heat exchangers).

Stress Corrosion

Stress corrosion is the name given to the process whereby cracks appear in metals subject simultaneously to a tensile stress and specific corrosive media. The metal is generally not subject to appreciable uniform corrosion attack but is penetrated by fine cracks that progress by expanding over more of the surface and proceeding further into the wall. The cracks may be unbranched or may be many-branched. They may proceed along the grain boundaries only or may be transgranular and advance with no preference to follow the grain boundaries.

Stress-corrosion cracks develop in specific metal-fluid combinations when the stress level is above a minimum level that depends on the temperature, alloy structure, and environment. In some materials minimum stress levels for crack formation are as low as 10 percent of the yield stress. In other cases the critical value may be as high as 70 percent.

For stress-corrosion cracks to initiate, the stresses must be tensile in character and exceed the critical level referred to above. They can be induced from any source, including residual welding stresses. Stress corrosion often occurs in lightly loaded structures that are not stress-relieved after fabrication.

Not all metal-fluid combinations are susceptible to cracking. Stainless steels crack with fluids containing chloride but not with ammoniacal fluids, whereas brasses crack in ammonia but not in chlorides. For a given material the number of fluids that give rise to cracking is generally small. For example, stainless steels crack in chloride and caustic solutions but not in sulphuric, nitric, or ascetic acid or nonchloride aqueous solutions.

Temperature and the presence of oxygen and chloride are significant environmental characteristics affecting stress-corrosion crack initiation and propagation. Stress level, alloy composition, orientation of the grains, and the distribution of precipitates and dislocations appear to be significant metallurgical characteristics. No generalized pattern for stress-corrosion cracking has been identified, but some alloy-fluid combinations known to be susceptible are given in Table 8.3.

It is likely that stress-corrosion cracks are initiated at a corrosion pit or other surface irregularity. The base of the pit acts as a stress raiser so the local stress concentration is very high. Once a crack is started, the stress at the tip of the crack is very high and fosters continuing development of the crack. As the crack penetrates further into the metal, the remaining wall section assumes the whole load. The general stress level is therefore raised and is further magnified at the tip of the crack, so the rate of propagation is accelerated. Eventually the metal fails suddenly and catastrophically when the stress in the remaining metal exceeds the ultimate strength.

Corrosion products are sometimes the source of stresses that result in stress-corrosion cracking. This occurs when corrosion products are generated in a constricted or confined region. The volume of solid corrosion products formed is greater than that of the metal consumed; this gives rise to a wedging action that exerts an outward pressure, with consequent high tensile stress at the tip of a crack and propagation of the stress-corrosion crack.

A similar phenomenon, called *denting,* has been observed in nuclear steam generators. The Inconel tubes are deformed or dented inward where they pass

TABLE 8.3

Metal Alloy and Fluid Combinations Known to be Susceptible to Stress-Corrosion Cracking (*After Fontana and Greene [1978].*)

Material	Environment	Material	Environment
Aluminum alloys	$NaCl-H_2O_2$ solutions $NaCl$ solutions Seawater Air, water vapor	Ordinary steels	$NaOH$ solutions $NaOH-Na_2SiO_2$ solutions Calcium, ammonium, and sodium nitrate solutions Mixed acids ($H_2SO_4-HNO_3$) HCN solutions Acidic H_2S solutions Seawater Molten Na–Pb alloys
Copper alloys	Ammonia vapors and solutions Amines Water, water vapor	Stainless steels	Acid chloride solutions such as $MgCl_2$ and $BaCl_2$ $NaCl-H_2O_2$ solutions Seawater H_2S $NaOH-H_2S$ solutions Condensing steam from chloride waters
Gold alloys	$FeCl_3$ solutions Acetic-acid–salt solutions		
Inconel	Caustic-soda solutions	Titanium alloys	Red fuming nitric acid Seawater N_2O_4 Methanol–HCl
Lead	Lead acetate solutions		
Magnesium alloys	$NaCl-K_2CrO_4$ solutions Rural and coastal atmospheres Distilled water		
Monel	Fused caustic soda Hydrofluoric acid Hydrofluosilicic acid		
Nickel	Fused caustic soda		

243

through the carbon-steel supports, because the annulus between the tube and the inner surface of the hole in the support is filled with corrosion products. In addition to reduction of the tube diameter, this results in dilation of the holes in the tube support plates, with associated distortions and some tube leakage. Malinowski and Fletcher [1978] report that denting has been observed extensively in units "that operated for some time in phosphate water chemistry then subsequently switched to all volatile treatment (AVT) and employed seawater cooling in the condensers."

Prevention of Stress Corrosion

Fontana and Greene [1978] recommend the following measures to reduce or alleviate stress corrosion:

1. Lower the stress level below the critical threshhold level by reducing the fluid pressure or increasing the wall thickness.
2. Relieve residual stresses by annealing at 860 to 920 K for carbon steels and 1080 to 1200 K for stainless steel.
3. Change the metal alloy to one that is less subject to stress-corrosion cracking in the given environment. Carbon steel is more resistant than stainless steel to corrosion cracking in a chloride-containing environment, but less resistant to uniform corrosion. Replacing type-304 stainless with an alloy of higher nickel content is often effective.
4. Modify the corrosive fluid by process treatment or the addition of corrosion inhibitors such as phosphates.
5. Apply cathodic protection with sacrificial anodes or an external power supply. This should be used with caution only in cases where stress corrosion has
5. been positively identified; hydrogen damage is accelerated by cathodic protection.
6. Use shot peening to induce compressive surface stresses.
7. Use venting air pockets to avoid the concentration of chloride in the cooling water.

Corrosion Fatigue

In Chap. 6 the failure of material by metal fatigue was discussed briefly. Fatigue is the process whereby a metal subjected to a fluctuating load fails at stress levels substantially below the yield point, depending on the number of stress reversals to which it is subjected. The conventional correspondence of stress level with cycles to failure is represented by the upper curve in Fig. 8.9.

The fatigue characteristics of metals are drastically affected when the fluctuating load is imposed in the presence of a corrosive medium. The fatigue life is reduced, and there is no threshold value below which the number of stress reversals becomes insignificant. The lower curve in Fig. 8.9 represents the fatigue characteristic of a metal in a corrosive environment. In seawater, aluminum, bronzes, and austenitic stainless steels retain 70 to 80 percent of their normal fatigue life. High-chrome alloys fare worse, retaining only 30 to 40 percent of their normal fatigue resistance.

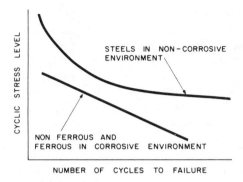

FIGURE 8.9
Corrosion-fatigue effects in the *S-N* diagram.

Corrosion fatigue appears prevalent in metal-fluid combinations that are susceptible to pitting corrosion. The fatigue crack likely starts at the base of a corrosion pit, and the crack propagates via simultaneous mechanical and electro-chemical action. It has been found that the *frequency* of stress fluctuation is important with corrosion fatigue, whereas it is not significant in conventional fatigue failures. Corrosion fatigue is accelerated at low stress frequencies.

In conventional applications, an increase in fatigue life can be gained by increasing the tensile strength of the metal or alloy. This resists the formation of the original crack, but thereafter the crack tends to progress more rapidly than in a material with lower strength. In a corrosive medium the crack is usually initiated by corrosive action. Therefore the fatigue resistance of a high-strength material may be quite low compared with that of a material of lower strength, but less subject to corrosive attack.

Corrosion fatigue can be eliminated or reduced by decreasing the stress levels through design change to increase the metal thickness, stress-relief by heat treatment, or shot-peening to induce compressive surface stresses. Corrosion-fatigue resistance can also be improved by the application of corrosion-resistant coatings—zinc, chromium, nickel, copper, and nitrides—or by the addition of inhibitors to the fluid.

Hydrogen Damage

Hydrogen damage is the generic term applied to the variety of unfortunate consequences following exposure of metal to hydrogen. Fontana and Greene [1978] categorize hydrogen damage as four distinct types:

1. Blistering
2. Embrittlement
3. Decarbonizing
4. Hydrogen attack

Hydrogen may exist in the monatomic form (H) or the diatomic form (H_2). Atomic hydrogen can diffuse through many metals. Molecular hydrogen cannot do this, nor can any other chemical species. There are various sources of atomic hydrogen, including high-temperature atmospheres, corrosion, and electrochemical processes. Corrosion and cathodic protection, electroplating, and electrolysis all produce hydrogen ions, which reduce to atomic hydrogen and hydrogen molecules. Some substances (sulfide ions, phosphorous and arsenic compounds) inhibit the reduction of hydrogen ions, leading to a concentration of atomic hydrogen on the metal surfaces.

Hydrogen Blistering

The production of hydrogen ions will, in some way, result in the aggregation of hydrogen ions, atomic hydrogen, and molecular hydrogen on the metal surface of a pressure vessel or heat exchanger. Some of the atomic hydrogen will diffuse into and through the metal before reducing to molecular hydrogen on the outer surface.

The atomic hydrogen diffusing through the metal will enter any voids in the metal. Some will then reduce to molecular hydrogen, which cannot permeate the metal wall. The equilibrium pressure for atomic and molecular hydrogen is several hundred thousand atmospheres so the one-way accumulative process continues, giving rise to very high pressures—far exceeding the yield stress of the material. The voids grow to appear as "blisters" on the wall of the pressure vessel or tank.

Hydrogen Embrittlement

Hydrogen embrittlement arises from the same source as blistering—the penetration of apparently solid metal by atomic hydrogen. In some metals the hydrogen reacts to form brittle hydride compounds. In others the mechanism of embrittlement is not known. Alloys are most susceptible to cracking from hydrogen embrittlement at their highest strength levels. The tendency to embrittlement increases with the hydrogen concentration in the metal.

Decarbonization and Hydrogen Attack

Decarbonization and hydrogen attack are associated with metals exposed to high-temperature gas streams containing hydrogen and a variety of other gases. Decarbonization is the removal of carbon from a steel alloy on exposure to hydrogen at high temperature. It results in a reduction of the tensile strength and an increase in the ductility and creep rate. Hydrogen attack is the interaction of metals or an alloy constituent with hydrogen at high temperatures. Fontana and Greene [1978] quote as an example the disintegration of copper containing oxygen at high temperatures in the presence of hydrogen.

Prevention of Hydrogen Damage

A reduction in hydrogen damage follows adoption of the following measures:

1. Use of voidfree steels. Killed steel, has few voids as compared with rimmed steel.
2. Use of metallic, inorganic, and organic coatings and liners in steel vessels. The liner must be impervious to hydrogen penetration and resistant to other media in the vessel. Carbon steel clad with nickel or austenitic stainless steel is sometimes used. Rubber, plastic, and brick liners are also used.
3. Addition of inhibitors to reduce corrosion and the rate of hydrogen-ion production. These are economically feasible only in closed recirculating systems.
4. Fluid treatment to remove hydrogen-generating compounds such as sulfides, arsenates, cyanides, and phosphorous-containing ions.
5. Substituting alternative alloys. Nickel steels and nickel-base or molybdenum alloys have low hydrogen diffusion rates and are less susceptible to embrittlement.
6. Use of low-hydrogen welding rods and the maintenance of dry conditions during welding operations. Water and water vapor are major sources of hydrogen.

Recommended Reading

The above is a superficial summary of the many complex corrosion processes that occur in heat exchangers. It is based almost entirely on Chap. 3 of the book "Corrosion Engineering" by Fontana and Greene [1978]. This excellent text of 500 pages is not sufficiently well known and should be on the desk of every engineer and technician. It is strongly recommended to all readers of this book.

Relative Incidence of Corrosion

Fontana [1977] presented data, reproduced in Table 8.4, showing the relative incidence of the various forms of corrosion discussed above. The study, based on a 2-year survey carried out by the duPont Co., involved a total of 313 failures. Of that total, 178 (57 percent) were attributed to corrosion, and 135 (43 percent) to mechanical causes. The most important corrosion failures, in order of frequency, were: general, cracking, pitting, intergranular, and erosion-corrosion.

This pattern of distribution is probably representative of the highest standards of engineering practice and typical of chemical companies employing a well-trained staff. Fontana considered this to be the reason for the virtually zero incidence of two-metal galvanic-corrosion failures.

FOULING

In service most heat exchangers suffer progressive loss of heat-transfer performance due to fouling. The term originated in the oil industry and is now used generally for any deposit on heat-exchanger surfaces that increases the resistance

TABLE 8.4

Relative Incidence of Various Forms of Corrosion
and Metal Failure (*After Fontana [1977].*)

Form	Percent
General	31.5
Stress-corrosion cracking	21.6 ⎫ 23.4
Corrosion-fatigue cracking	1.8 ⎭
Pitting	15.7
Intergranular	10.2
Erosion corrosion	7.4 ⎫
Cavitation	1.1 ⎬ 9
Fretting	0.5 ⎭
Crevice corrosion	1.8
Selective leaching	1.1
Two-metal (galvanic) corrosion	0.1
High temperature	2.3
Weld corrosion	2.3
Cold wall	1.8
End grain	1.1
Hot wall	0.5
Hydrogen embrittlement	0.5

to heat transmission. Taborek et al. [1972] reviewed the fouling literature and
methods for predicting fouling behavior. The following is a summary of the
introductory part of their work.

Fouling Factor

In a heat exchanger the process of heat transfer from a hot fluid to a cold fluid
involves various conductive and convective processes. These can be individually
represented in terms of thermal resistances. The summation of these individual
resistances is the total thermal resistance, and its inverse is the overall heat-
transfer coefficient U. That is,

$$\frac{1}{U} = \frac{1}{h_o} + \frac{A_o}{A_i}\frac{1}{h_i} + R_{fo} + \frac{A_o}{A_i}R_{fi} + R_w$$

where U = overall heat-transfer coefficient, based on outside area of tube wall

A = tube wall area

h = convective heat-transfer coefficient

R_f = thermal resistance due to fouling

R_w = thermal resistance due to wall conduction

and the suffixes i and o refer to the inner and outer tubes, respectively.

It is customary in design work for the heat-transfer coefficients h_o and h_i to be determined from complicated relations involving the Nusselt, Prandtl, Reynolds, and Grashof numbers. Similarly, the conduction thermal resistance is determined from calculations involving properties and dimensions of the material of the tube walls. Such detailed processes are not involved in determining the fouling resistance, the so-called fouling factors R_f and R_{fo}. The uncertainty is such that one simply includes arbitrary values of the fouling factors selected from sources based on experience. The less experience one has, the less confidence one will have in the eventual result. There are various compilations of experience-based fouling factors, showing appropriate values for different metal-fluid combinations. An extensive listing is contained in the TEMA standards [1978].

In heat exchangers involving liquid or phase-change processes, the heat-transfer coefficients h are very high, typically 600 to 1400 Btu/(h·ft^2·°F). The tube walls are relatively thin and made of metal with a high thermal conductivity, so the conductive wall resistance is negligible.

In such cases the typical fouling factors of 0.002 will account for half the total thermal resistance. In other words half the exchanger heat-transfer surface area determined to be necessary from the equation $A = Q/U \, \Delta T$ will arise from the arbitrarily selected fouling resistances. Taborek et al. [1972] suggest that the uncertainty of fouling resistance is "optimistically estimated as ±50 percent." Such uncertainty casts doubt on the worth of time and effort spent on computing the heat-transfer coefficient h, an activity beloved by those of academic bent.

Fouling as a Function of Time

The assumption of constant values for the internal and external fouling factors implies that, when put in service, the new heat exchanger instantaneously deteriorates to the fouled condition. Of course it does not do this, but instead deteriorates progressively. Considerable time—years, perhaps—may elapse before it arrives at the condition where it can no longer perform adequately and must be cleaned.

The buildup of fouling resistance as a function of time may follow various forms as indicated in Fig. 8.10. Curve A describes a process starting with clean surfaces having zero fouling resistance, which then develops at a constant rate with time. Curve B describes a process where the fouling resistance develops at a progressively diminishing rate, asymptotic to a constant value.

The family of curves C, D, and E all share a lengthy incubation or induction period in which there is little or no buildup of fouling resistance, followed by a period of rapidly increasing buildup. For curve C the rate of increase in fouling resistance develops progressively with time. For curve D it moderates to a linear increase, and for curve E it further moderates to a progressively diminishing rate asymptotic to an ultimate constant value.

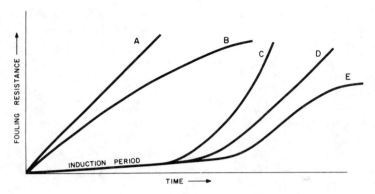

FIGURE 8.10
Development of fouling resistance as a function of time.

There is therefore a substantial time lapse before the heat-exchanger fouling resistance approaches the design value arbitrarily selected from some experience-based source. When first put into service, the heat exchanger will operate with a reduced thermal resistance and therefore with a surplus of heat-transfer area. In many cases involving boiling, the fouling resistance is the principal resistance. Thus, when the heat exchanger is new, the available temperature difference may be so great as to carry the process into the film-boiling region, with the possibility of enhanced surface corrosion and consequent accelerated development of fouling resistance.

In other cases the new heat exchanger with zero fouling resistance may be so effective as to overcool the process stream. To compensate, the cooling-water flow may be reduced, with the result that the water velocity is decreased and the water temperature increased. Both these factors are highly conducive to fouling on the water side. The provision of excess allowance for fouling or an excess of heat-transfer area "just to be on the safe side" does not automatically increase the interval before cleaning is necessary; quite likely it has the reverse effect. The excess area results in reduced flow velocities and elevated temperatures, so the exchanger deteriorates in performance at a drastic rate.

Mechanisms of Fouling

Various mechanisms of fouling have been recognized and can be categorized as follows:

1. Crystallization
2. Sedimentation
3. Polymerization and chemical reaction
4. Coking
5. Organic growth
6. Corrosion effects

The concentration of salts in solution as a function of temperature is generally of the two forms shown in Fig. 8.11. Normally the concentration increases with temperature, but some salts typically found in brackish water or seawater exhibit the inverse solubility characteristic above a certain temperature. Liquids with direct solubility characteristics will precipitate crystals on cooled surfaces, and those with inverse solubility characteristics will precipitate on heated surfaces.

Solutions containing a single salt precipitate strong crystalline formations with substantial adherence forces to the surface. These deposits rarely break off under the action of the fluid velocity passing over the surface. They tend to grow as a linear function of time, but as the deposit thickens, the surface temperature and hence the rate of deposition decrease. There is an induction or incubation period before the deposition starts, which depends largely on the material and surface finish of the wall. The development of fouling resistance due to single-salt crystallization therefore follows the form of curve D in Fig. 8.10.

Solutions containing a mixture of salts precipitate crystals in a more complicated fashion. The deposits tend to form an irregular pattern and are less adherent and strong than the single-salt structures. As a consequence the deposits are relatively easily dislodged by the fluid flow over the surface. They tend to accumulate over time in accordance with the form of curve E in Fig. 8.10, asymptotic to a fixed value. The rate of deposition of new crystals equals the rate of removal of established structure. This type of crystallization tends to be relatively independent of the nature and finish of the surface.

Fouling by sedimentation occurs with fluids containing suspended solids. It may assume several forms, usually in association with other fouling processes. Water streams often contain mud, sand, dust, or rust fragments that aggregate in regions of low fluid velocity. The particles sometimes act as catalysts precipitating chemical reactions and combine with crystallization processes occurring simultaneously. Sediments combine with crystal deposits to sometimes strengthen

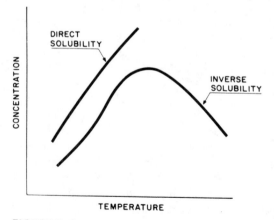

FIGURE 8.11

Typical solution-concentration characteristics of salts as a function of temperature.

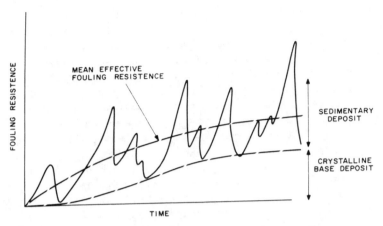

FIGURE 8.12
Typical fouling-resistance development history with combined sedimentary and crystallization processes.

and sometimes weaken them. Fouling-resistance histories for combined sedimentation-crystallization processes may be of the form shown in Fig. 8.12. The sedimentary aggregations accumulate rapidly to increase the fouling resistance. They are easily torn away in large fragments by the movement of fluid over the surface, with consequent sharp decreases in the fouling resistance. This sawtooth variation in fouling resistance is accompanied by a steady accumulation of base fouling resistance, resulting from crystal formation on the surface. This is represented in Fig. 8.12 by the lower dashed curve, shown asymptotic to a final constant value.

High-temperature gas streams containing dust or combustion products will cause sedimentary or particulate fouling with no accompanying crystallization, although there may be some condensation and solidification of the combustion products. The surface roughness and temperature, the gas turbulence and temperature, and the specific nature of the deposits (dry, gummy, etc.) will all affect the fouling process. Quantitative predictions are difficult without previous experience with a similar case.

Fouling of surfaces by polymerization will occur in many petroleum and chemical streams. Critical parameters include the surface temperature and the presence of oxidation promoters. To the extent that surface temperature decreases with the fouling deposit, the process is essentially self-limiting and will likely follow the form of the asymptotic curve B in Fig. 8.10. A special case of this type of fouling is the hard, crusty coke deposits often formed on high-temperature surfaces by combined chemical-reaction and sedimentation fouling processes. Not enough is known to generalize about the rate of deposit formation.

Nuclear reactors with organic-fluid coolants are subject to fouling with products generated by exposure of the moderator to radiation. The surface temperature and radiation levels are important parameters, and an asymptotic fouling history similar to curve B in Fig. 8.10 is to be anticipated.

Organic growth, both animal and vegetable, is sometimes a problem in untreated, once-through river and seawater streams. Growth occurs in the warm downstream discharges rather than in the exchanger itself and is usually combined with other types of fouling. It is primarily a problem in the discharge conduits. Crustacea grow exponentially in such agreeably pleasant streams with plentiful food supplies passing by, although the opportunities to capitalize on this source of oyster, shrimp, and lobster production have not yet been commercially developed. Underwater acoustic ultrasonic generators are sometimes used to combat the luxuries of the environment and inhibit the growth of crustacea.

Surface corrosion contributes to fouling in different ways. On one hand, corrosion products form a crusty, heat-resisting deposit and act as a catalyst, stimulating other fouling processes. On the other hand, as a result of corrosion the surface will be roughened. This increases the number of nucleation sites on previously smooth surfaces for both crystallization and bubble growth in nucleate boiling.

Effect of Heat-Transfer Process

Fouling occurs to some extent in all systems where liquids, gases, and vapors are being heated or cooled. The process may involve boiling, condensing, or heat transfer without phase change.

The greatest source of fouling, principally inverse solubility crystallization and chemical reactions, occurs on hot surfaces in heating processes without phase change. Cooling processes without phase change also result in appreciable fouling as the result of particulate deposition, sedimentation, and chemical reaction.

Heat-transfer processes involving phase change may be either (1) cooling where the fluid is condensing or (2) heating where the fluid is evaporating. Fouling is rarely significant in condensing processes, for vapors are relatively clean, but corrosion of the condensing surface can be appreciable. Moreover, condensate flow velocities tend to be rather low so that fouling may become important in the condensate region.

Fouling in boiling or evaporative processes is a serious problem, particularly in large-scale steam generation, to the extent that specially demineralized, deaerated, and corrosion-inhibited feedwater is often used in the closed circulatory evaporative or condensing loop. With other fluids the nucleate-boiling process is often enhanced by minor fouling on otherwise smooth, polished surfaces. It increases the density of nucleation sites. When the process proceeds too far, the fouling deposit becomes so heavy as to diminish the rate of heat transfer. Much depends on the surface temperature, the principal parameter for raising or suppressing the rate of chemical-reaction fouling. Fouling is not readily predictable with evaporative processes, and experience with a similar situation is the best guide.

Effects of Surface Material and Structure

By the time the fouling deposit has covered most of the surface, the material or finish of the wall has become irrelevant; the primary effect is during the incubation or induction stage. Different materials have different catalytic actions with various fluids and may promote or inhibit the reactive processes responsible for the initial fouling. Figure 8.13 shows typical fouling-resistance development histories during the induction period for carbon-steel, stainless-steel, and admiralty-brass surfaces exposed to brackish water streams under constant flow conditions.

Polished surfaces resist the growth of fouling deposits but are highly susceptible to corrosive action that roughens the surface and increases the potential crystallization sites. Improperly cleaned heat exchangers with residual fouling deposits on the surfaces will degrade by fouling more readily than those restored to the "as-new" clean condition.

Effect of Fluid Velocity

There is much evidence suggesting fluid velocity as the most important parameter affecting fouling. In most cases, an increase in velocity decreases both the rate of fouling-deposit formation and the ultimate level attained, as shown by the typical development histories given in Fig. 8.14. Improvement tends to be at a progressively diminishing rate. Doubling the fluid velocity from a low value may

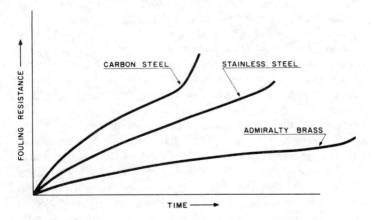

FIGURE 8.13
Typical fouling-resistance development histories for the induction period with carbon-steel, stainless-steel, and admiralty-brass surfaces exposed to brackish-water streams.

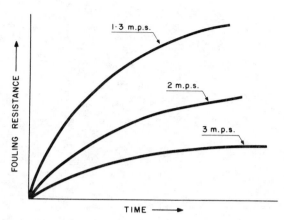

FIGURE 8.14
Typical fouling-resistance development history as a function of fluid velocity.

halve the fouling resistance. Doubling it again may halve the remaining resistance. However, the second doubling requires an increase to four times the original velocity and gains only a reduction of one-quarter the original thermal resistance.

In addition to decreasing the fouling, the higher velocity increases the heat-transfer coefficients so that a double-barreled reduction in the size and cost of the heat exchanger might be anticipated. With reduced fouling there will also be a decrease in the maintenance requirements and cost. However, it must be recalled that the pressure drop is a function of the *square* of the fluid velocity. Doubling the velocity increases the pressure drop by four times, increasing both the capital cost and the operating cost for pumping.

Effect of Temperature

Temperature has a pronounced effect on fouling that can be generalized as shown in Fig. 8.15. The rate of development of fouling resistance and the ultimate stable level both increase as the temperature increases. Temperature refers to either or both of the surface temperature and the fluid bulk temperature. The rates of chemical and inverse crystallization, including catalytic effects, are strongly dependent on temperature, which explains the increase in fouling rate. The rate of removal of fouling deposits is less a function of temperature than of fluid velocity. Therefore, an increase in the rate of deposition with no concomitant increase in removal will result in a higher ultimate stable level.

Effect of Exchanger Type, Operation, and Maintenance

The relative propensity to fouling and the ease with which cleaning can be accomplished are important factors in selecting the type of exchanger for a given application.

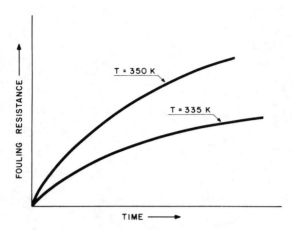

FIGURE 8.15
Typical fouling-resistance development history as a function of temperature.

Manufacturers claim that plate-and-frame and spiral-plate heat exchangers (see Chap. 4) are less susceptible to fouling deposits than tube-and-shell heat exchangers. Users of such equipment generally support these claims, but there are no directly comparable data that have been obtained under controlled conditions. In both types of exchanger, a low fouling resistance might be expected because of the characteristic high velocity and turbulence levels of the fluid flows.

The superiority of the plate-and-frame heat exchanger with regard to cleaning is unquestioned. When simple chemical flushing is not sufficient, the entire unit can be readily dismantled and cleaned in situ without "breaking" any pipework.

Another form of tubular exchanger claimed to be remarkably resistant to fouling is the flexible-tube immersion heater shown in Fig. 8.16. The tube bundle is made up of many small-diameter Teflon tubes, with tube spacers located at intervals along the tube bundle. The assembled bundle is quite flexible and can be arranged in a variety of forms and shapes. Teflon has a low tendency to initiate and retain fouling deposits, and the flexibility of the tube bundle tends to shatter and dislodge any deposits that do accumulate.

The low thermal conductivity, relatively low strength, and poor high-temperature properties of Teflon plastic do limit extensive use of this type of exchanger. However small-diameter tubes reduce the required wall thickness to a point where the conduction thermal resistance is tolerable. It is much favored for tank heating and cooling of acid and plating baths and other highly corrosive fluids.

Tube-and-shell heat exchangers exist in the widest range of sizes, shapes, and types. Fouling and the corrosion characteristics of the fluids are important concerns affecting the choice of material and the type of exchanger. Another important factor is the facility for cleaning.

The tube-side and the shell-side flows should be considered separately. Traditionally, the petroleum industry has allocated process fluid to the shell side, and cooling water to the tube side. In chemical plants the reverse situation is favored; process fluid is put into tube-side flow, with cooling water on the shell side. It is not uncommon to find a refinery and a chemical plant side by side on the same river, with water used as the tube-side flow in one case and as the shell-side flow in the other.

Situations where brackish water is used as coolant to condense a petroleum or chemical vapor favors the use of the heat exchanger shown in Fig. 8.17, a split-ring, floating-head heat exchanger with removable channel and cover, two tube passes, and a single shell pass. The alternative fixed-tube-sheet design has the advantage of relatively low cost, but the shell side is inaccessible for cleaning. A condensing vapor is unlikely to result in appreciable fouling and could be put on the shell side to allow the use of a fixed-tube-sheet design. The brackish water, containing miscellaneous salts and other contaminants, would have a high propensity to foul surfaces. It would be best located on the tube side, with high velocity in the tube to minimize fouling.

Removal of the front channel cover and rear bonnet allows direct access for straight-through mechanical cleaning of the tube internal surfaces. Use of a two-tube-pass channel with removable cover provides for tube access without breaking any pipe connections. Precisely the same result could, of course, be

FIGURE 8.16
Flexible-tube-bundle heat exchanger composed of many small-diameter Teflon tubes. (*Courtesy duPont Corp.*)

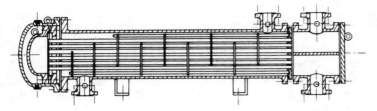

FIGURE 8.17
Split-ring floating-head heat exchanger with removable channel and cover, two tube passes, and single shell pass.

gained with a single-pass tube bundle with front and rear channels having separate covers. In this case the exchanger would have half the number of tubes but would be twice the length.

An alternative possibility is the exchanger shown in Fig. 8.18. This has a U-tube bundle with the front channel integral with the tube sheet and a removable channel cover. The tube bundle can be withdrawn from the shell to provide good access to the tube exterior for cleaning, but it is necessary to break the tube-side flow piping to allow removal of the channel.

Access to the tubes in situ can be gained simply by removal of the channel cover. Cleaning U-tube internal surfaces is less convenient than cleaning straight-through tubes, but it is negotiated relatively easily by the power, mechanical, and very high pressure water-jet cleaners commonly used today.

On the shell side, baffle design and tube arrangement are influenced by fouling and cleaning considerations. Because high velocity is important to minimize fouling, it is clear that the baffle arrangement shown in Fig. 8.19a would lead to many stagnant areas in the shell-side flow, with consequent high fouling. The baffle arrangement shown in Fig. 8.19b has fewer stagnant areas and a longer mean flow path. If the shell-side mass flow were the same in both exchangers, the velocity in Fig. 8.19b would be much greater than that in Fig. 8.19a. Of course the pressure drop and cost of pumping increase as the square of the fluid velocity.

Tubes are generally arranged in the triangular or square pattern shown in Fig. 8.20. Triangular arrangements allow for inclusion of the greatest number of

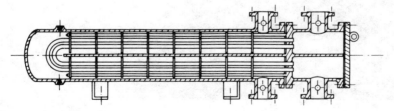

FIGURE 8.18
U-tube-bundle heat exchanger having the channel integral with the tube sheet, a removable channel cover, and two-pass shell with longitudinal baffle.

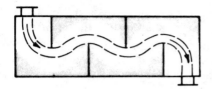

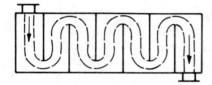

FIGURE 8.19
Baffle designs affecting fluid velocity at the creation of stagnant areas in shell-side flow. (*After Taborek et al. [1972].*)

tubes in a given shell diameter and for the strongest tube-sheet ligaments. However, they are much more difficult to clean with mechanical scrapers and brushes than square tube arrangements. Exchangers likely to require periodic cleaning on the shell side should therefore have square tube arrangements. Of course there may be other compelling reasons to override this general rule, so as to increase the tube count or take advantage of the stronger tube-sheet ligaments of triangular arrangements.

Fins are used to increase the effective area for heat transfer in situations where the inside and outside tube heat-transfer coefficients are substantially different. The increased area is, of course, put on the side with the low coefficient. Usually the high heat-transfer-coefficient fluids are put on the tube side, and the low coefficient fluids on the shell side. Therefore finned tubes are very much more likely to have external fins than internal fins. Exceptions are sometimes made, and it is not uncommon in some applications to find tubes having both internal and external fins.

External fins on the tubes in tube-and-shell heat exchangers are usually of the low-radial-fin type or are longitudinal fins welded onto bare tubes. Manufacturers claim that finned tubes are less susceptible to fouling than bare tubes, and many users find that finned-tube exchangers require cleaning less often than

TRIANGULAR   **SQUARE**

FIGURE 8.20
Triangular and square tube patterns.

those with bare tubes. On the other hand, mechanical cleaning on the shell side is more difficult and tedious for finned-tube exchangers than for their bare-tube counterparts, for it is essential that the cleaning process be performed thoroughly. The degree of reduced fouling resulting from the use of finned tubes varies so much that the only reliable guides are experience and test data.

Practical Fouling Factors

It is customary for the purchaser to specify the fouling resistance used in the thermal design of the exchanger. The above exposition will do little to increase users' confidence in the value of the fouling resistances marked on exchanger specification sheets; however they should now have a clearer understanding of the uncertainties prevailing in the specifications. Many users have their own private collections of fouling factors, based on past experience with similar equipment under equivalent conditions. These are the most reliable data. However, the indiscriminate application of these factors to equipment larger in size and operating under more arduous conditions is of questionable validity. The uncertainty increases the more one departs from past experience.

The proprietary organizations have conducted fouling research over a broad range of different conditions. Their recommendations for design are often applicable to new situations confronting a user. In the public domain the most complete listing of fouling factors recommended for design is contained in the TEMA standards [1978]. This compilation was generated from the accumulated experience of the 28 tubular exchanger manufacturers constituting the association and responsible for publication of the TEMA standards. The value and applicability of this listing to any specific situation is highly questionable. Nevertheless, it will continue to be much quoted in the absence of anything better. Users should simply be aware that the accuracy of single-value fouling factors is probably no better than ±50 percent.

MALDISTRIBUTION OF FLUID FLOW

One of the principal reasons for heat exchangers failing to achieve the expected thermal performance is that the fluid flows do not follow the idealized paths anticipated from elementary considerations. These departures from ideality can be very significant indeed. As much as 50 percent of the fluid can behave differently from what is expected based on a simplistic model. *Maldistribution of flow* is a term often used to describe unequal flow distribution in the several parallel flow paths found in most heat exchangers. We retain that usage here and extend the term further to embrace all other departures from idealized flow.

The Tinker Diagram

Flow on the shell side of a tube-and-shell heat exchanger was classified by Tinker [1951] into a number of separate streams, as represented diagrammatically in Fig.

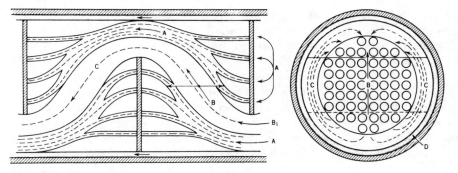

FIGURE 8.21
Bypass and leakage paths in shell-side flow. (*After Tinker [1951].*)

8.21. The A stream represents flows that occur in the clearance space between the baffle tube holes and the tubes. Flow is due to the pressure drop between the upstream and downstream sides of the baffle. The B stream is the true crossflow stream, passing through the tube bundles and performing the real function of the shell-side fluid.

The C stream bypasses the tube bundle and flows in the annulus between the shell and the tube bundle. This is a highly ineffective use of the fluid. If the tube-bundle shell clearance is greater than the tube pitch, it is advisable to include a sealing device to inhibit bypass flow. The sealing device can be strips, rods, or dummy tubes as shown in Fig. 8.22.

The F stream includes other bypass streams that arise when the tube partitions of multiple-pass tube bundles are arranged parallel to the direction of the main crossflow stream. The D stream is leakage flow that occurs in the clearance space between the edge of the baffle and the shell. This represents a direct loss of fluid, for it serves no useful heat-transfer function.

Parallel-Path Flows

Shell-side flows in a tube-and-shell heat exchanger follow a complicated, irregular path, and it is not at all difficult to appreciate that there may be significant

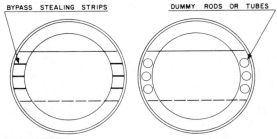

FIGURE 8.22
Methods to reduce bypass flows on the shell side.

differences between theory and practice. It is less easy to understand that very significant differences in flow distribution will arise on the tube side also. Ostensibly the situation is very simple: We have n tubes, all identical, and the total mass flow of fluid is $\dot{M}$. Thus, the mass flow of fluid per tube is $\dot{M}/n$. In practice this is by no means the case, and fluid flow in the apparently identical paths differs by as much as 50 percent. Flow paths cannot be made absolutely identical, and fluid flows are incredibly sensitive to apparently trivial differences between one path and another. When the number of parallel paths is limited to two or three and the paths are highly restricted, the difference in channel mass flow rates may be as high as 90 percent. The flow is then a function of some power of the principal flow-resistance parameter, e.g., the third power of the width of a slit or the square of the cross-sectional area of a flow aperture.

The length of the flow path is also a factor in the resistance to flow, and differences in flow length contribute very significantly to different mass flow rates in various parallel channels. Because of this, plate heat exchangers are not immune to some measure of maldistribution, even though they have a multiplicity of small, precision flow ducts in their die-formed plate elements. Flow in the tubes in U-tube bundles is also affected by variations in length. Tubes at the periphery of the bundle are considerably longer in length than those at the heart. Partial compensation is achieved by the tendency to increased flow in the small-radius bends associated with the shorter tubes and the increased propensity for tube distortion in bending with the small-radius bends.

Tube distortion in bending, or the squashing resulting from improper handling in fabrication, can contribute appreciably to flow maldistribution, as shown in Fig. 8.23.

A difference in the mass rate of flow through the tube carries the implication that the flow velocity is significantly different. The heat-transfer rate depends on the fluid velocity, and the tube-wall and fluid temperatures depend on the heat transfer. Low mass flow and fluid velocity in some tubes may give rise to high fluid and wall temperatures, with accelerated corrosion and fouling deposition rates. The fouling deposits and products of corrosion exacerbate the difference in flow resistance between one tube and another and further

TUBE DEFORMATION INCREASES FLOW RESISTANCE.
TUBE SUBJECT TO EROSION CORROSION AT THE
SITE OF DEFORMATION

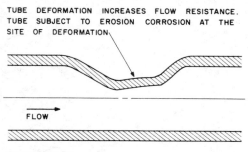

FLOW

FIGURE 8.23
Tube damage effects.

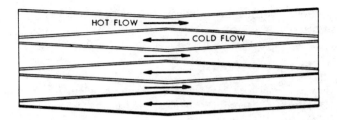

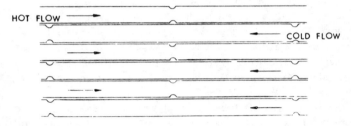

FIGURE 8.24
Counterflow heat exchanger with variable-geometry flow section provides a compensating feedback mechanism to overcome flow maldistribution. (*After Cowans [1972].*)

diminish the mass flow in tubes already starved of fluid. The process is a cancer feeding on itself.

In most industrial heat exchangers a surprising degree of maldistribution of fluid flow can be tolerated—perhaps not happily, but at least without noticeable interference with the process. Most users are even unaware that there are significant differences in the mass flow rates of different flow channels.

This is not true for the high-performance heat exchangers found in miniature cryogenic cooling systems operating on the Joule-Brayton and Claude cycles. These counterflow heat exchangers are required to be of such high performance that even high-precision manufacturing tolerances can result in sufficient maldistribution of flow to prevent the attainment of adequate performance. Alternative solutions to this exacting problem were explored by Cowans [1974]. He described special heat exchangers, shown diagrammatically in Fig. 8.24. The flow channel was of variable geometry designed to incorporate a compensatory feedback mechanism, acting to adjust the duct geometry to ensure uniform distribution of flow in the various channels. The system described by Cowans operated with outstanding success. The miniature high-performance heat exchanger was designed to achieve a very high NTU (see Chap. 2) of 200 (the NTU of most industrial heat exchangers is less than 5). Even with great attention to manufacturing detail, the early high-performance heat exchangers described by Cowans were unable to exceed an NTU of 33. With the compensating feedback geometry, values of 167 were achieved. It is unlikely that such complexity will ever be considered routinely worthwhile in industrial heat exchangers, but it may find application in special circumstances.

Maldistribution of Flow in Air Coolers

Maldistribution of flow is sometimes a particularly severe problem in air-cooled heat exchangers. The nature of the problem is illustrated in Fig. 8.25. The liquid or condensate tube flow typically occurs through multiple parallel paths coupling inlet and outlet distribution headers. There may be a single row of tubes, but more likely the tube bank will be several rows deep. Air passes in crossflow over the tubes, and the tubes are usually finned to increase the surface area so as to compensate for the low heat-transfer coefficient on the air side. To further improve heat transfer, fans are used to increase the velocity of flow over the tubes. The fan may be upstream or downstream of the tube bank, pushing or sucking the air across the tubes (described as forced flow or induced flow, respectively).

It is most convenient to construct tube banks that are square or rectangular in section, and propeller fans are favored on grounds of convenience and economy. It then becomes very difficult indeed to ensure proper flow distribution, particularly in the corners of the tube banks (shaded in Fig. 8.25). With forced-flow systems a transition diffuser or plenum is usually interposed between

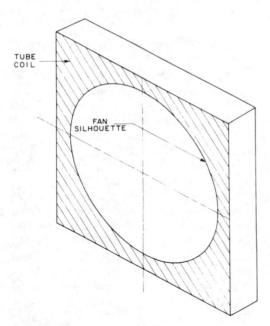

FIGURE 8.25
Maldistribution of flow in air-cooled heat exchangers.

the fan and the tube bank. For the sake of the economy and compact construction, these circular to rectangular transition sections are made as short as possible, with the result that they are remarkably inefficient diffusers. Often their greatest effect is in the appearance they give that something is being done to regulate the flow.

Air coolers are often deficient in performance because they fail to provide the airflow that one would expect after reference to the fan manufacturer's literature. This is because fan characteristics are derived from model tests in wind tunnels under near-perfect conditions. In actual service the fan entry sections, crucial to high performance, are often reduced to negligible proportion or even discarded entirely. Furthermore, the fan drive system and various support stays are often located close to the plane of fan rotation, where they cause significant disturbance to the smooth flow of air.

Distortion of the fan shroud ring during fabrication on site, in service, or during maintenance operations is another factor frequently responsible for appreciable deterioration of fan performance. Finally, the cumulative buildup of dirt and miscellaneous environmental deposits on the fan blades and other aerodynamic surfaces, including the heat-exchanger tubes, contributes to a progressive deterioration of airflow characteristics.

An overall loss of up to 40 percent of published fan-capacity characteristics would perhaps be a realistic assumption for design purposes.

Stagnant Areas

Disappointing heat-exchanger thermal performance often arises from the creation of stagnant areas in the fluid-flow circuits. In stagnant or semistagnant areas the fluid velocities are, by definition, zero or negligibly low. The consequences are often very serious. The obvious effect is that with low fluid velocity the area for heat transfer is not effectively utilized. Less obvious but of greater importance is the fact that corrosion and fouling processes are highly accelerated under stagnant conditions. Sediments in slurries or contaminated streams aggregate preferentially in the low-velocity areas. Surface temperatures in the low-velocity areas may be appreciably higher than the mean design condition, which further accelerates chemical reactions exacerbating the corrosion and fouling processes.

A common location of semistagnant fluid zones in tube-and-shell heat exchangers is the region on the shell side between the tube sheet and the inlet and outlet nozzles (Fig. 8.26). It is necessary to establish the centerlines of the inlet and outlet nozzles some distance from the tube sheets so as to accommodate the nozzle flanges and to provide sufficient shell strength in the high-stress areas near the tube sheets. The existence of some low-velocity regions on the shell side near the ends of the tubes is then virtually inescapable but is frequently overlooked by inexperienced thermal designers. They fail to add an extension to the calculated tube length to compensate for the "dead area."

High fluid velocities at the inlet nozzles frequently require the provision of

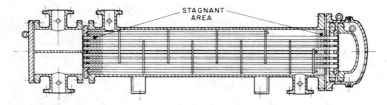

FIGURE 8.26
Stagnant fluid areas in tube-and-shell heat exchangers.

impingement plates over the tubes that are below the inlet nozzle. The
impingement plates eliminate erosion corrosion of the tubes near the inlet
nozzles and the stimulation of tube vibration by the highly turbulent inlet flow.
Careful design of the impingement plates can go far in stimulating fluid motion
in the dead areas near the ends of the tube. It is equally possible to extend and
enhance the stagnant region with improperly designed impingement plates. A
number of different impingement-plate concepts are illustrated in Fig. 8.27.

Baffle design and placement are the principal means by which to ensure
adequate fluid velocities on the shell side and a well-regulated, dispersed flow.
Even good designs can be hopelessly compromised if they are improperly or
inadequately executed. Excessive clearance of the baffle in the shell will certainly
facilitate loading the tube bundle in the shell during construction. However, that
clearance will lead to substantial bypassing of the fluid at the periphery of the

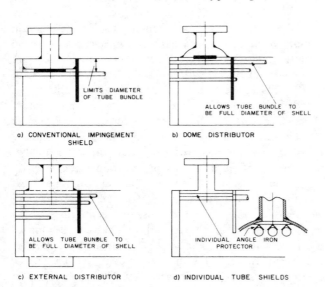

FIGURE 8.27
Impingement plates for tube protection below the inlet nozzles in tube-and-shell heat
exchangers.

baffle, so that little of the fluid actually traverses the tube bundle. Excessive clearance of tubes in the baffle tube holes will greatly facilitate construction, but again will result in a high proportion of the fluid not passing through the tube bundle as intended.

A hazard of interchangeable quality-controlled production is that tube bundles can be inserted upside down with catastrophic consequences. This apparently unlikely situation was discussed by Smith [1979], who asserted that while there are "numerous reasons why that should not happen, the facts are that it can and does happen." He illustrated with the example reproduced in Fig. 8.28. This is a floating-head, two-tube-pass, single-shell-pass exchanger with horizontal-cut segmental baffles, TEMA type AES, commonly found in refineries and chemical plants. The upper diagram shows the tube bundle correctly installed. In the lower diagram the tube bundle has been reversed. It is immediately clear that the compartments between the tube sheets and the first and last baffles are completely stagnant and virtually useless for heat transfer. The effectiveness of the tube bundle is reduced by as much as 40 percent. Moreover, the tubes now located below the inlet nozzle are not protected by the impingement shield and extend over the long span between the tube sheet and the *second* baffle. Therefore erosion corrosion and severe vibration of the tubes might be anticipated.

The final and perhaps most decisive failing of the inverted tube bundle is that at the floating-head end the support plate is no longer supporting the heavy floating-head assembly. Instead, the load is carried by the penultimate baffle, a relatively fragile element situated a long distance from the floating-head tube sheet.

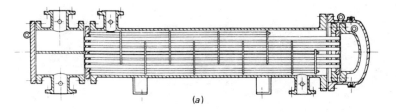

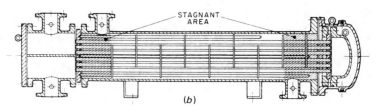

FIGURE 8.28
(*a*) Correct and (*b*) inverted placement of tube bundle in floating-head tube-and-shell heat exchanger. (*After Smith [1979].*)

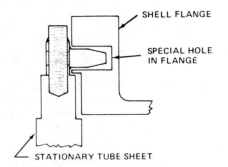

SHELL FLANGE

SPECIAL HOLE
IN FLANGE

STATIONARY TUBE SHEET

FIGURE 8.29
Dowel locating device in tube-sheet–shell-flange joint. (*After Smith [1979].*)

Reversal or inversion of the tube bundle during fabrication or replacement after cleaning can be eliminated by use of a dowel locating pin in the tube-sheet–shell-flange joint as shown in Fig. 8.29. With such a pin, the tube bundle can be installed in only one position—the correct one.

With a floating-head exchanger, the need to provide a support plate increases the size of the stagnant region in the final baffle compartment. Smith attributes 85 percent of the failures from external tube corrosion to the two dead areas of the first and last baffle compartments, with the great majority in the support-plate floating-head tube-sheet stagnant region. This has led to the development of improved designs to eliminate the dead areas in the first and last compartments. One concept found to result in dramatically improved heat transfer and extended tube life is illustrated in Fig. 8.30. This arrangement has

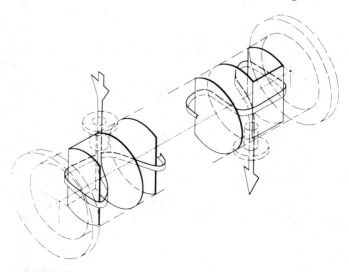

FIGURE 8.30
Improved baffle arrangement and floating-head support plate for tube-and-shell heat exchanger.

vertical-cut baffles, but a similar system with horizontal-cut baffles can be devised. Fluid enters the second compartment through the inlet nozzle and is led from there into the first compartment. Fluid leaves the first compartment by passing *behind* the second compartment and enters the third compartment to begin its routine transit of the exchanger. At the floating-head end, the fluid passes through the last compartment before entering the penultimate compartment and exiting the exchanger. The last baffle is made sufficiently robust to serve also as the support plate for the floating-head assembly. This ingenious arrangement overcomes entirely the stagnant-region deficiencies of conventional designs.

Summary

Maldistribution of flow is responsible for many of the failures of heat exchangers to attain the anticipated thermofluid performance and extended operational life before failure, as well as the need for substantial maintenance. In tube-and-shell heat exchangers the tube and baffle clearances provide paths for fluid to bypass the tube bundle. Stagnant fluid areas adjacent to the tube sheets are virtually inescapable, unless special baffle designs are adopted. Improper fabrication or maintenance replacement can create additional bypass or stagnant zones.

Nearly all heat exchangers of every type contain a number of parallel flow paths. It is virtually impossible to secure absolutely identical flow resistances in every path, and fluids naturally follow the path of least resistance. As a consequence the mass flow rates can be radically different in ostensibly identical flow paths. Corrosion and fouling processes are accelerated in the flow paths of highest resistance, to further accentuate unbalance in the flow. A compensating feedback mechanism to adjust the geometry of the duct may be incorporated in the design, but its use is presently limited to heat exchangers of very high performance in miniature cryogenic systems.

Air-side flow and distribution problems account for much of the disappointing thermofluid performance commonly experienced with air coolers. Judicious reduction of predicted fan ratings will assure more realistic estimates of air-cooler performance in practical situations.

APPENDIX 8A: TUBING DEFECTS

The photographs with annotations contained herein were contributed by Mr. R. E. Smallwood of E. I. duPont de Nemours and Co. They were included in Mr. Smallwood's presentation entitled, "Heat Exchanger Tubing Reliability, Flow Detection and Causes of Failures," at the American Society of Metals Symposium, "Shell and Tube Heat Exchangers," September 17 to 19, 1979, in Houston, Texas.

FIGURE 8A.1
Selective weld attack on zirconium tubing in boiling 70 percent H_2SO_4.

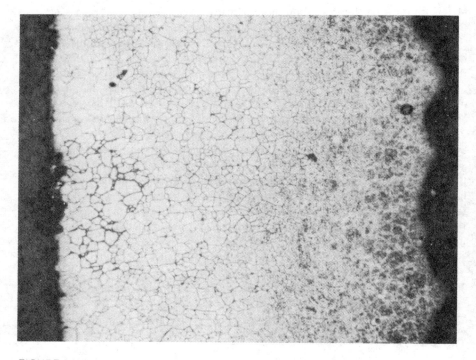

FIGURE 8A.2
Inside diameter of the tube was very heavily sensitized by carbon pickup in the tube mill. The material was sensitized and experienced intergranular attack on the shell side. (Inconel 600; 100×; oxalic acid).

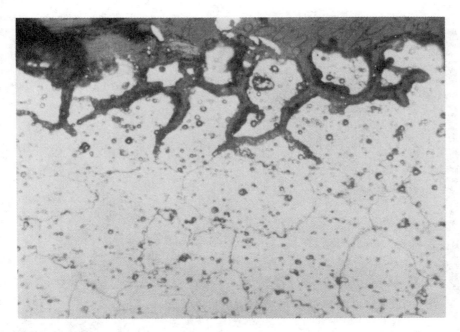

FIGURE 8A.3
Hastelloy C tube that failed the Streicher test for 24 h owing to improper heat treatment. (200×)

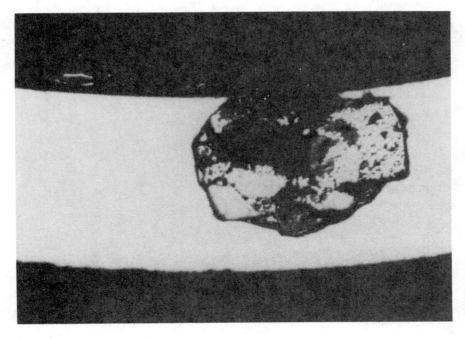

FIGURE 8A.4
Tramp metal and slag found in type-304 stainless-steel weld. (Oxalic acid; 25×)

FIGURE 8A.5
Preferred grain orientation at outside-diameter surface (white area) of Ti–0.15-Pd tubing. In
some acid chloride salts, there may be quite a large difference in corrosion resistance between
the inner and outer surfaces of a tube. (HF–HNO$_3$ etch; 50×)

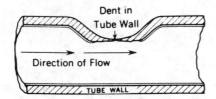

Impinging stream of water reduces wall thickness
at point where tube has been dented.

FIGURE 8A.6
Erosion corrosion of a dented tube.

FIGURE 8A.7
Through-the-wall defect in a skelp near a weld in type-304 stainless-steel tubing. The defect probably was caused when embedded metal in the skelp was pickled out. (50×; oxalic acid)

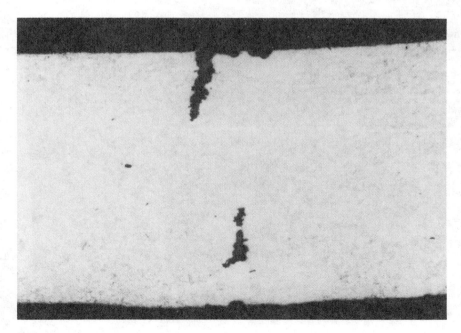

FIGURE 8A.8
Partial through-the-wall defect in a weld of type-316 stainless-steel tubing. The surface crack was about 3 in long. The tube did not leak during pressure testing. (50×; oxalic acid)

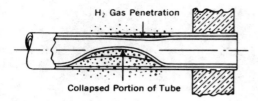

H₂ Gas Penetration

Collapsed Portion of Tube

Hydrogen cannot pass through joint,
hence pressure builds up causing collapse.

FIGURE 8A.9
Collapse of bimetallic tube via H_2 gas penetration. Similar failures have occured in tubes
containing delaminations.

8.A.10

FIGURE 8A.10
Crack in the weld was readily apparent by even casual visual inspection.

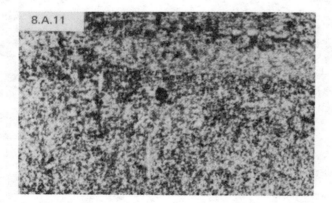

8.A.11

FIGURE 8A.11
Hole in the tubing was picked up only with difficulty.

APPENDIX 8B: MATERIALS SELECTION FOR HEAT EXCHANGERS

The photographs with annotations contained herein were contributed by Mr. W. G. Ashbaugh of the Union Carbide Corp. They were included in Mr. Ashbaugh's presentation entitled, "Materials Selection for Heat Exchangers," at the American Society of Metals Symposium, "Shell and Tube Heat Exchangers," September 17 to 19, 1979, in Houston, Texas. The following paragraphs were taken from the abstract of Mr. Ashbaugh's presentation.

For the corrosion engineer and metallurgist in the petrochemical industry, the shell-and-tube heat exchanger is one of our greatest challenges. These heat exchangers are complex pieces of equipment requiring the best efforts of all who design, build, operate, and maintain them to achieve the high levels of efficiency and reliability that are required today.

Just how successful are we as an industrial technology in our efforts concerning heat exchangers? The scrap pile behind every plant would indicate that we still have a long way to go.

Through the use of several illustrations, we shall describe some of the problems and pitfalls that we face in attempting to select materials of construction for heat exchangers. We intend to show by example that, in many cases, the corrosion and metallurgical problems that develop in heat exchangers need not occur if the available technology is carefully applied.

The selection of materials for heat exchangers is a complex one because of the many factors that are involved. We have grouped these factors into three major categories:

1. Process conditions
2. Equipment design
3. Maintainability

The first category contains the greatest number of variables and, when extended into the operating situation in the plant, results in the greatest number of problems. The second area is one of a greater degree of specialization, yet problems still develop here if they are not adequately considered at the appropriate time. Finally, maintainability is the most abstract of the factors affecting the selection of materials, but it is a real one that should not be overlooked.

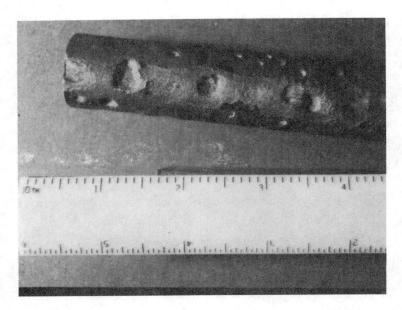

FIGURE 8B.1
Steel tube pitted by inhibited cooling water. Periodically, the water flow was shut off to "thaw" the process side. This operating condition was not described to the corrosion engineer before the tube material was selected.

FIGURE 8B.2
A test heat exchanger was built and operated to determine what materials of construction might be used for retubing an economizer in a process heat-recovery boiler. The process gas being burned contained traces of chlorides that formed HCl and collected on the economizer tubes, which operated below the dew point.

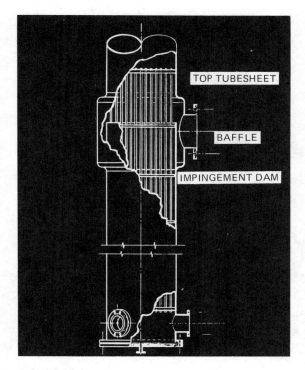

FIGURE 8B.3
A large vertical condenser with seawater inside the copper-nickel tubes failed by corrosion pitting from the water side, owing to hot spots caused by precondensed liquid trapped in the vapor feed zone.

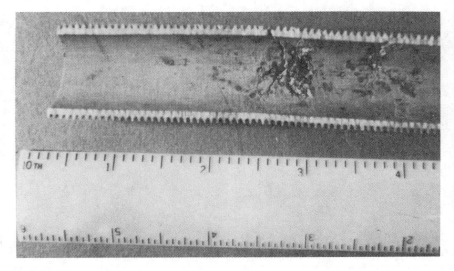

FIGURE 8B.4
A six-pass flash evaporator cooler was starved of water, resulting in boiling and evaporation of the river water inside the stainless tubes and causing rapid pitting and stress-corrosion cracking.

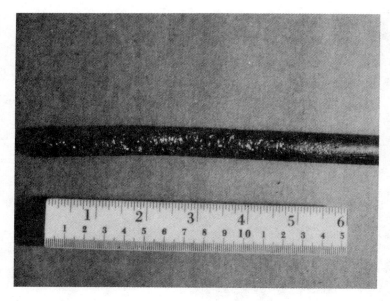

FIGURE 8B.5

This copper-tubed boiler feedwater heat exchanger suffered rapid attack when water-treating chealants were added upstream. The water had not yet been deaerated at this point, thus combining with the chealant to cause the rapid corrosion.

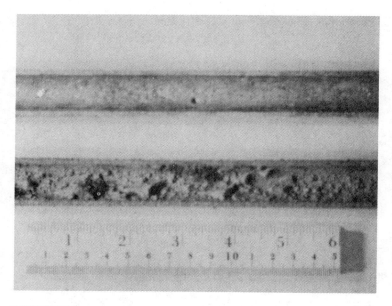

FIGURE 8B.6

A cooling-water system was designed to be inhibited against scale and corrosion so that low-cost steel tubes could be used in the process heat exchangers. Inattention to the water treatment allowed severe scaling and corrosion to occur.

FIGURE 8B.7
Inlet-end velocity corrosion of 90-10 copper-nickel tubes by seawater. This was caused by a very small channel head (water box), which resulted in severe turbulence at the tube inlet end.

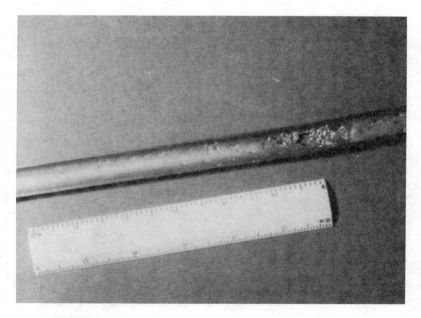

FIGURE 8B.8
Pitting of this $\frac{5}{8}$-in outside diameter copper-nickel tube was caused by a piece of shell or other foreign matter trapped inside the tube. The $\frac{1}{2}$-in inside diameter is too small for clean operation, even in strained seawater.

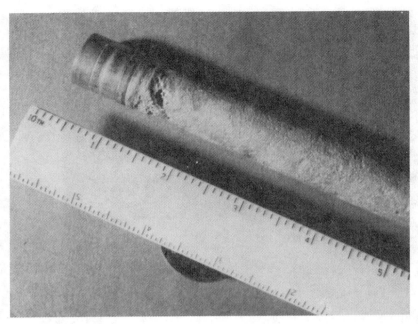

FIGURE 8B.9
Unvented shell-side cooling water on vertical condensers causes corrosion of tubes under the top tube-sheet.

FIGURE 8B.10
A tube-to-tube-sheet seal weld. Not all such welds are as good as they might appear to be from the surface.

REFERENCES

Cowans, K.: A Countercurrent Heat Exchanger that Compensates Automatically for Maldistribution of Flow in Parallel Channels, pp. 437–444 in K. Timmerhaus (ed.), "Advances in Cryogenic Engineering," vol. 19, Plenum, New York, 1974.

Fontana, M.: Alpha Sigma Mu Lecture, American Society of Metals annual meeting, Chicago, October, 1977.

Fontana, M., and N. D. Greene: "Corrosion Engineering," McGraw-Hill, New York, 1978.

Malinowski, D. D., and D. D. Fletcher: Update of Operations with Westinghouse Steam Generators, *Nucl. Technol.,* vol. 37, pp. 103–110, 1978.

Smith, H. E.: "A Users View, The Interrelationships between Codes, Standards and Specifications for Process Heat Transfer Equipment," American Society of Mechanical Engineers, pp. 35–38, 1979.

Taborek, J., T. Akoi, R. B. Ritter, and J. W. Palen: Fouling: The Major Unresolved Problem in Heat Transfer, *Chem. Eng. Prog.,* vol. 68, no. 2, pp. 59–67, 1972.

TEMA: "Standards of the Tubular Exchanger Manufacturers Association," 6th ed., New York, 1978.

Tinker, T.: General Discussion on Heat Transfer, *Proc. Inst. Mech. Eng. London,* pp. 89–116, 1951.

9

Buying a heat exchanger

INTRODUCTION

Heat exchangers are large, expensive items, and their purchase is never a trivial matter. The complexity of the purchasing process depends on the application: The need for a small, simple unit to operate with well-behaved fluids at moderate temperature and pressure may be satisfied by a relatively low-cost, off-the-shelf heat exchanger. At the other end of the scale, very large units operating at extremes of pressure and temperature with toxic or corrosive fluids are individually designed and specially manufactured—custom made.

Many factors have to be considered in the acquisition of a heat exchanger. Types suitable for one case will be unacceptable for another. For high pressures and temperatures, tubular exchangers are the only possibility, and they must conform to safety regulations and manufacturing codes. Even here, there are multiple variations, each with distinct advantages and disadvantages. For other applications the conditions of operation may be less rigorous, even without the need for certified compliance with the ASME code or the standards of API or TEMA. There are many variants and alternatives to tubular systems, and each type has established its niche in the overall mosaic.

In this chapter we summarize some of the more important considerations confronting those responsible for the acquisition of heat exchangers.

SELECTING THE TYPE OF HEAT EXCHANGER

This lengthy section is reproduced verbatim (with minor editorial changes and deletions) and with permission from the "Heat Exchanger Guide" published by the Alfa-Laval Co. This company manufactures a family of compact heat exchangers of the plate, spiral, and lamella types, but not tubular heat exchangers. Given this situation, it would be naive to expect anything other than the most favorable light to be placed on compact heat exchangers. I accept this and ask readers to apply their own judgment to the arguments presented. I offer no apology for including material that could be criticized for overstating the case for the compacts. I am of the opinion that the plate, spiral, and lamella heat exchangers are not sufficiently well known or documented in the classic texts. Who better to state the case than one of the leading manufacturers?

The choice of heat-exchanger type depends on a large number of factors including the duty requirements, fluid properties, temperature and pressure conditions, maintenance possibilities, and comparative costs. The four types of heat exchangers generally considered are:

1. Tubular heat exchanger (shell-and-tube heat exchanger)
2. Gasketed plate heat exchanger
3. Spiral heat exchanger
4. Lamella heat exchanger

The last three are called *compact* heat exchangers.

Operating conditions vary widely, and a broad spectrum of demands are imposed on design and performance. Points to be considered in selection include:

1. Materials of construction
2. Pressure and temperature
3. Performance parameters—temperature program, flow rates, pressure drops
4. Fouling tendencies
5. Inspection, cleaning, addition, and repair
6. Types and phases of fluids
7. Overall economy

Materials of Construction

Tubular heat exchangers can be manufactured in virtually any materials required for corrosion resistance, e.g., graphite tubes in a glass shell. Compact types are limited to materials that can be pressed (plate) or welded (spiral and lamella).

For the same duty, tubular heat exchangers normally require more material than the compact types. Welded heat exchangers (tubular, spiral, lamella) demand more labor-hours in manufacture than plate heat exchangers, but the latter require a large capital investment to manufacture. Initial costs for compact types, particularly plate heat exchangers, compare favorably with those for the tubular type when relatively expensive materials of construction are required. The cost per unit area of compact types is higher than for tubulars, but the increased efficiency, resulting in smaller area requirements, may more than compensate.

Gaskets for spiral and lamella heat exchangers are the same materials used in tubular types and, in special cases, they may be all welded. Plate heat exchangers are normally fitted with elastomeric or compressed asbestos-fiber gaskets and cannot use rigid gaskets, such as metal-asbestos compound gaskets. This limits the upper conditions of application.

Pressure and Temperature

Tubular heat exchangers are designed for almost any combination of temperature and pressure. In extreme cases, limitations are imposed by fabrication problems associated with material thicknesses, and by the weight of the finished unit.

Large tubular units operating at high pressures and temperatures may suffer from fatigue problems associated with sudden changes in cross section (e.g., tube-to-tube-sheet joints) in combination with fluctuating pressure or temperature. Even under steady conditions, differential expansion can lead to severe stress concentrations. Vibration problems are well known, especially at high shell-side velocities.

Compact heat exchangers are constructed of thinner materials with no sudden changes of cross section. The use of thinner materials limits the operating pressure and temperature ranges of these types, but there are no vibration, fatigue, or thermal effects leading to mechanical failure.

Figure 9.1 indicates the general operating range for compact heat exchangers. Working pressure and operating temperature are subject to continuous development, and special designs are often available.

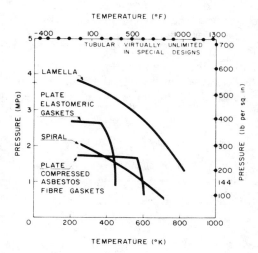

FIGURE 9.1

Pressure and temperature limits for the application of compact plate, spiral, and lamella heat exchangers of conventional design.

Performance Parameters

Thermal Length

The thermal length θ, or the number of heat-transfer units, (HTU) is a useful criterion of the performance of a heat exchanger. It may be defined, for a heat exchanger in single-phase duty, as the ratio of the temperature change of one fluid $t_i - t_o$ to the mean temperature difference between the fluids Δt_m; that is: $\theta = (t_i - t_o)/\Delta t_m$ NTU.

A tubular heat exchanger in water/water duty will achieve a θ value of about 0.5 HTU/pass with moderate pressure drops. Compact heat exchangers, particularly of the plate type, have "thermally long" channels, with θ values up to 4 HTU/pass. For the same duty, a plate heat exchanger is more likely than a tubular to be of the single-pass type. Full countercurrent operation is normally obtained, with neither crossflow nor cocurrent flow reducing the effective value of the temperature difference between fluids.

When multiple-pass plate heat exchangers are used, correction factors applied to the logarithmic mean temperature difference (LMTD) are nearer unity than with tubular heat exchangers because crossflow is absent, and departures from pure countercurrent flow are minimal. Duties involving a high degree of temperature "crossover" (outlet temperature of cold fluid higher than outlet temperature of hot fluid) are therefore feasible in plate heat exchangers.

Values of θ below 0.2 heat transfer units are not as well suited to compact heat exchangers, with the exception of a spiral heat exchanger designed for crossflow on one side (spiral heat exchanger, type 2). It can handle very large flow ratios economically. For single-phase duty, the temperature change on one side will be much less than on the other, resulting in a low θ value for the high-flow side.

One-pass plate heat exchangers have all the pipe connections located in the fixed part of the frame. This is advantageous for maintenance since the unit can be opened without loosening any connections, and the entire heating surface can be made accessible in a few minutes.

Flow Rate

The flow rate is an important parameter affecting selection of the type of exchanger. Limiting volumetric liquid flow rates for commercially available heat exchangers, assuming a 20-kPa (3-lb/in^2) pressure drop, are:

Tubular heat exchangers: unlimited
Plate heat exchangers: 2000 m^3/h
Spiral heat exchangers, type 1 (spiral flow both sides): 400 m^3/h
Spiral heat exchangers, type 2 (crossflow one side): 8000 m^3/h
Lamella heat exchangers: 4000 m^3/h

For fluids other than water or with different pressure drops, the volumetric flow rates are substantially different from these.

The maximum flow rate is determined in the plate heat exchanger by the total flow passing through the channels without excessive pressure drop. For low-viscosity liquids, the pressure drop in the connections is the limiting factor. For viscous liquids, gases, and vapors, the channel pressure drop is critical.

Type-1 spiral heat exchangers are designed for spiral flow on both sides. The maximum fluid flow is a function of the channel flow area (spacing times breadth) and allowable pressure drop. The maximum flow areas for the two sides are similar, and the type-1 spiral is normally best suited to liquid/liquid duties. Spacings for the two sides are independently variable, and relatively large flow ratios can be handled. Full countercurrent flow is maintained.

The type-2 spiral heat exchanger incorporates crossflow on one side. Large flow rates can be handled on the crossflow side, making the type-2 spiral suitable for duties involving large volumes of low-pressure gas and vapor or liquid/liquid applications where the flow rate of one fluid is much greater than that of the other.

The lamella can be designed with comparatively large flow areas. Moreover, flow ratios for liquid/liquid duties up to 5:1 can normally be accepted while maintaining countercurrent single-pass operation on both sides.

Pressure Drop

Pressure drop is important in heat-exchanger design and is the price paid for heat transfer. Limits may be imposed by economic (pumping cost) considerations or by process limitations. For economic design, any imposed pressure-drop limits must be closely approached.

Unproductive pressure drop must be avoided; inlet, outlet, and return-bend losses must be minimized; and the smallest number of tube passes should be used. The plate heat exchanger can develop higher θ values (number of heat transfer units) per pass than other types. Apart from inevitable losses in the connections and the channel inlet and outlet sections, the pressure drop is used effectively for heat transfer.

The specific pressure drop Je is defined as

$$\text{Je} = \frac{\Delta P}{\theta} \quad \text{kPa/HTU}$$

For the normal operating-cost parameters (interest rate, depreciation, electricity cost), optimal overall economy is achieved when Je is in the range 20 to 100 kPa/HTU [3 to 15 lb/(in$^2 \cdot$HTU)] depending on materials of construction; the more sophisticated the material, the higher the economic pressure drop. The above figures refer to water/water duties; for other fluids, the optimal value for Je is considerably higher.

Fouling Tendencies

Fouling is the least predictable phenomenon associated with heat transfer. The fouling characteristics of a given fluid depend largely upon boundary-layer and surface conditions.

The most important factor is the fluid velocity and its influence on shearing force, turbulence, laminar-layer thickness, and residence time close to the surface. The lowest velocity in the exchanger channel system is the most significant. There must be good velocity distribution over the whole channel section, and, with more than one channel in parallel, the flow distribution among various channels must be uniform. Conditions for uniform flow distribution demand the cross section be constant over the entire width, with minimal changes along the flow path. With several heat-transfer channels in parallel, the pressure drop in the channel should be high. The requirements for low fouling are generally satisfied in compact heat exchangers and result in lower fouling factors than in comparable shell-and-tube units.

Shell-and-tube exchangers with high-fouling fluids on the tube side should have small-diameter tubes and a high pressure drop per pass to give better flow distribution over the tubes in parallel. The nozzle arrangement should be chosen so the dynamic pressure directing the flow toward the tubes is low compared with the pressure drop in the tubes. When the velocity head in nozzle inlets, channels, or headers is less than 10 percent of the pressure drop over a whole pass, the difference in flow rate between tubes in parallel is likely to be less than 5 percent. A close pitch prevents stagnant pockets of fluid and deposits on the tube sheets.

With high-fouling fluids on the shell side, it is difficult to achieve a well-distributed flow when the baffles do not cover the main tube-bundle diameter. Regions of low velocity behind baffles are difficult to avoid if the baffles are cut more than 25 percent of the tube-bundle diameter.

Spiral heat exchangers fulfill most requirements for low fouling. The fluid flows in a single channel, so there is no problem of uneven distribution. Spiral heat exchangers are custom made, and the channel dimensions are chosen according to requirements and the permissible pressure drop. When deposits build up in one part of the channel, there is an increase in velocity because of the reduction in the cross section and, hence, a self-cleaning effect.

Plate heat exchangers are little affected by fouling, owing to good fluid distribution and extremely high turbulence. The corrosion-resistant material of the thin plates reduces fouling tendencies. Lamella heat exchangers of single-pass unbaffled design are less liable to shell-side fouling than tubular types.

The channel dimensions of the heat exchanger must allow free passage for the largest particles present in the fluid. For tubular sections the diameter should be three to four times the diameter of the largest particles. Slot-shaped channels in spiral, lamella, and plate heat exchangers should have widths two to three times the maximum particle diameter. Lamella and plate heat exchangers may be used with particles of small size (diameter below 2 mm). Spiral heat exchangers will handle 5- to 8-mm particles so long as the concentration is not too high and the fluid flows easily. Testing is required for fluids with a high concentration of solids and a tendency to stick to the surface.

Spiral heat exchangers are suitable for liquids with suspended fibers. In the pulp and paper industry, spirals with spacing studs in the channel are used for fluids containing 0.5 percent by weight of fibers. Even higher concentrations of fibers can be handled if the unit is made without studs.

When erosive particles are suspended in a fluid, smooth straight channels are used without changes in direction. Spiral heat exchangers are the most suitable compact exchanger for this type of application. Shell-and-tube units require that tube joints and tube inlets be protected.

When severe fouling is expected, it is best to design for high velocity and good flow distribution with regard to pressure drop rather than to add extra surface area to allow for fouling. Additional surface results in lower fluid velocities, and the gain is lost through higher fouling rates.

Fouling deposits may be removed by mechanical means or by chemical means referred to as "cleaning in place." Requirements for mechanical cleaning are that the affected surfaces are readily accessible. This is a characteristic feature of plate heat exchangers. Cleaning-in-place methods require that the cleaning fluid be well distributed and in a high degree of turbulence. Plate and spiral heat exchangers (particularly the latter with severe fouling) offer significant advantages over other types of exchangers in this regard.

Inspection, Cleaning, Repair, and Additions

Table 9.1 shows, for various heat exchangers, the comparative facility of inspection, cleaning, repair, and additions. The comparison is relative and approximate, for some types offer great flexibility in design. For example, spiral heat exchangers can be made with channels 5 to 25 mm wide with or without studs. Tubular heat exchangers can be made with fixed tube sheets, with a retractable tube bundle, and with small or large tubes on close or open pitch. Lamella heat exchangers are similar, with a retractable lamella bundle, easy to clean on the shell side but not accessible for inspection or mechanical cleaning internally. Plate heat exchangers are easy to open and clean, especially when all the nozzles are located on the stationary end plate. The plate arrangement can be changed for other duties within the frame and nozzle capacity.

Guidelines for Selection

Heat Transfer between Two Low-Viscosity Fluids

For the purpose of this discussion, *low viscosity* is defined as less than 10 cP at mean bulk temperature.

TABLE 9.1

Relative Ease of Inspection, Cleaning, Repair, and Additions for
Different Types of Heat Exchangers

	Heat-exchanger type			
Procedure	PHE (plate)	SHE (spinal)	LHE (lamella)	THE (tubular)
Inspection for fouling:				
One side	a*	b	b	b
Both sides	a	b or d	c	b or d
Inspection for leakage:				
One side	b	a or b	a	a
Both sides	b	a or b	a	a
Inspection for corrosion:				
One side	a	a or c	b	b
Both sides	a	b or d	d	b or d
Chemical cleaning:				
One side	a	a	b	b
Both sides	a	a	b	b or c
Manual cleaning:				
One side	a	b or c	b	b
Both sides	a	b or d	d	b or d
Additions	a	d	d	d
Repair	a	c	c	b

*a = very good; b = acceptable; c = poor; d = impossible.

Plate heat exchangers require the smallest surface area and are the first choice, provided the duty lies within the temperature and pressure limitations. The initial cost of a stainless-steel plate heat exchanger with elastomeric gaskets will be within 25 percent of that for a carbon-steel tubular heat exchanger designed for the same duty. For water/water duty, plate heat exchangers offer technical and economic advantages, particularly if the water quality is such that stainless steel or better is necessary. The cost of a titanium plate heat exchanger is double that of a similar unit in stainless steel.

If one or both fluids are such that conventional elastomeric gaskets are likely to fail, compressed asbestos-fiber gaskets may be suitable.

Closed-Circuit Cooling Water

Direct use of cooling water from natural sources (sea, river, lake, etc.) or from a cooling tower often leads to corrosion and fouling. This is aggravated by decreasing availability and increasing pollution of natural waters. Cooling-tower water is increasingly corrosive due to atmospheric pollution.

It is becoming standard practice to install a circuit of clean noncorrosive demineralized cooling water to cool process equipment. This circuit requires recooling, and selection of the heat exchanger to accomplish this depends on a number of factors. Corrosion and fouling problems will be confined to the central cooler, thus protecting the process equipment from fouling and corrosion, and the environment from leakage of the process fluids due to corrosion.

With ample supplies of natural cooling water available, plate heat exchangers are preferable. If the water is not brackish, the unit should be in stainless steel. However, if the secondary coolant is brackish (sea, estuary, tidal river, etc.), the use of plate heat exchangers in titanium is preferred. Tubular units in copper alloy are comparable in basic initial cost, but such materials are likely to corrode. Replacement costs plus the need to install standby units, will generally turn the overall economy to the plate heat exchanger's favor. With inadequate supplies of natural cooling water, central cooling can be achieved by a water/water exchanger in combination with a cooling tower, or an air cooler.

The choice from among the different cooling systems depends on legislation on thermal pollution, meteorological conditions, energy costs, economic parameters, and the availability of makeup water (for cooling towers).

Low-Viscosity Liquid to Steam

If carbon steel is suitable, tube-and-shell exchangers will normally be the first choice. If higher-grade material or hygienic construction is required with regular manual cleaning, plate heat exchangers are preferred. The plate heat exchanger offers advantages where floor space is limited, and it is the only type that can be extended to meet increased demand.

In certain cases the steam volumetric flow rate makes plate heat exchangers less advantageous, owing to excessive pressure drops; spiral or lamella heat exchangers are normally attractive alternatives to a tubular unit, particularly if the liquid contact surfaces are to be in stainless steel or a better material.

Medium-Viscosity Fluids

Medium viscosity is defined as the range 10 to 100 cP. Flow in straight channels is likely to be laminar, but the plate heat exchanger's plate corrugations induce turbulence at very low Reynolds numbers. With normal pressure drops per unit mass flow, plate heat exchangers are likely to be fully turbulent at viscosities up to 70 to 100 cP. For duties involving such fluids on one or both sides, plate heat exchangers are the first choice, even where carbon steel is adequate as a material of construction.

When plate heat exchangers cannot be used (e.g., because of gasket materials, excessive temperature, or high solids content), the spiral heat exchanger is the first choice. The channel curvature introduces secondary flow effects to cause turbulence at Reynolds numbers of 300 to 500, compared to about 2300 for straight channels. Spiral heat exchangers require a smaller surface area than a straight-channel exchanger in this viscosity range.

Lamella heat exchangers with straight, smooth channels on both sides are less attractive, technically and economically. Tubular heat exchangers with the viscous liquid in crossflow on the shell side are preferred.

High-Viscosity Fluids

High viscosity is defined as being over 100 cP. The flow will be laminar in all types of heat exchangers. Good flow distribution is of prime importance, since maldistribution (particularly in cooling) leads to regions of high viscosity, thus aggravating maldistribution. Plate heat exchangers fulfill this requirement and have been used with fluids having viscosities up to 50,000 cP.

Spiral heat exchangers are preferred for very high viscosities. The single channel involves a perfect flow distribution not achieved in multiple parallel channels. Spiral exchangers have operated with fluid viscosities up to 500,000 cP. Concentric-tube exchangers provide the same advantage but are cumbersome and require more surface area. For this application, serpentine coils are often used to advantage, with a continuous coil of tube immersed in a bath or shell.

Plate heat exchangers offer significant advantages for the transfer of heat between high-viscosity liquids. The flow geometries are normally identical, and this, coupled with good flow distribution, makes plate units the first choice, regardless of material requirements. Spiral heat exchangers are preferred if one or both fluid viscosities are extremely high or if elastomeric gaskets are precluded. Neither lamella nor tubular units are suitable for this duty.

Fouling Due to Suspended Solids

Few heat-transfer duties involve absolutely clean fluids. Some fouling of heat-transfer surfaces is to be expected, and normal design allows for this phenomenon.

Fluids with high fouling tendencies present serious problems. The requirement for good flow distribution leads to the preference for plate and spiral heat exchangers, particularly the latter if suspended solids are fibrous.

Scaling. Scaling is the deposition of insoluble matter on heat-transfer surfaces as a result of thermal decomposition (e.g., calcium bicarbonate, calcium carbonate) or of inverse solubility (e.g., calcium sulfate). All types of heat exchangers will be scaled to some extent. The effects of scaling are limited by good flow distribution (resulting in small variations in local surface temperature). Exchangers with heat-transfer surfaces that are readily accessible for scale removal have obvious advantages. Exchangers should be well suited for cleaning-in-place procedures, and plate heat exchangers fulfill this requirement best. Spiral heat exchangers have also proven to be resistant to scaling of the single channel, and cleaning-in-place methods are effective with this type of exchanger.

Slurries, suspensions, and pulps. These fluids result in heavy sedimentation fouling of the heat-transfer surfaces and are almost invariably non-Newtonian in flow characteristics. The effective viscosity is a function of temperature and local velocity. Good flow distribution is essential to avoid areas of low velocity, with large local variations in viscosity and consequent aggravated maldistribution of flow.

Both spiral and plate heat exchangers fulfill the requirements for good distribution, but the ultimate choice between them depends on a number of factors so that a general recommendation cannot be made. Spiral heat exchangers are the only heat exchangers capable of handling extreme fluids such as rock ore slurries with up to 50 percent solids, owing to the complete absence of maldistribution effects.

Heat-Sensitive Liquids

Successful treatment of heat-sensitive liquids requires that the hold-up volume, time, and surface temperature be limited and well known. These, in turn, require that the exchanger have good flow distribution, high heat-transfer coefficient, and small channel volume—conditions best met by plate heat exchangers. The high degree of turbulence in plate heat exchangers leads to high heat-transfer coefficients, allowing operation at lower terminal temperature differences, and to the breaking up of the boundary layer on the surfaces. The re-formation of the boundary layer contributes to the reduction of burn-on rates.

When heat sensitivity is combined with high viscosity and extreme non-Newtonian behavior, spiral heat exchangers may be preferred for the single channel that gives perfect flow distribution, despite the larger hold-up volumes and lower heat-transfer coefficients.

Burn-on fouling affects the situation, since maldistribution leads to local variations in surface temperature. This applies even to the cooling side of a unit cooling a heat-sensitive liquid; good flow distribution assures troublefree operation.

Cooling with Atmospheric Air

An extended surface exchanger is the first choice. For low-pressure applications, a "car radiator" type is often specified, while for higher pressures, an assembly of finned tubes is used. Such coolers are normally delivered with fan and structural steelwork included.

Gas or Air under Pressure

Heat transfer between gases at approximately the same pressure is similar to heat transfer between two liquids of low viscosity. Plate heat exchangers offer technical and economical advantages over other types. Gas pressures should not be too low (high specific volumes); otherwise, pressure drops will be excessive due to the close plate spacing.

For gases having widely different specific volumes (considerably different pressures) or for duties involving heat transfer between a gas and a liquid, a heat exchanger having extended surfaces on the side with higher specific volume is preferred. Lamella heat exchangers may prove economically attractive if the relative flow areas are compatible with volumetric flow rates; this is often the case for two streams of gas at different pressures. The type-2 spiral heat exchanger often has economic and technical advantages for gas/liquid duties.

Cryogenic Applications

The usual types of exchangers for such duties are the stacked plate-fin of brazed-aluminum "honeycomb" construction and tubular Giaque-Hampson units. Plate, spiral, lamella, and tube-and-shell heat exchangers cannot be considered suitable for cryogenic duties.

Vapor Condensation

In noncorrosive conditions, carbon-steel tube-and-shell heat exchangers (vapor on the shell side) will normally be the first choice. Where conditions require a more exotic material, type-2 spiral or lamella heat exchangers may be used. The type-2 spiral provides condensate at saturation temperature, while yet another variant can be designed for a predetermined degree of condensate subcooling. For vapors at moderate pressure (25 kPa up to the temperature and pressure limit), plate heat exchangers may prove advantageous when hygienic construction or manual cleaning is required; when there is a prospect of future extension; and when the cooling water is brackish, thus requiring titanium.

For vapors at very low pressure, a special type of plate heat exchanger, the box condenser, may be used.

Gas-Vapor Mixture, Partial Condensation

Spiral heat exchangers can be specially constructed for condensation of the vapor component from a vapor-gas mixture. The construction is such that high and calculable degrees of vapor and condensate subcooling are achieved, thereby recovering the maximum possible condensables. This is done by decreasing the flow area, in transition from crossflow to spiral flow, as condensation proceeds.

Heat Exchangers for Distillation Plants

Distillation plants include several heat exchangers: an overhead condenser, reboiler, feed preheater, bottom cooler, condensate subcooler and process feed/effluent exchanger.

Overhead condensing. In general, condensation of a pure vapor is involved. The condensate is returned to the distillation column with little subcooling. Tube-and-shell heat exchangers are most commonly used, but require large-bore vapor pipework and a pumped condensate return line if the condenser is located at the base of the column.

The use of a special spiral heat exchanger, the top condenser, eliminates these requirements. The condenser is flanged directly to the top of the column. The condensate is removed without subcooling and passed to a reflux divider, part being removed as product, and the remainder returned to the column. Alternatively, an overhead condenser can be built with two or more spiral elements. The first provides reflux without subcooling liquid flowing directly back into the column. The other element condenses the remaining vapor, with or without subcooling as required. When the condensate consists of two immiscible liquids, arrangements can be made for phase separation.

Reboiling. The distillation reboiler may be internal or in an external, natural-circulation (thermosyphon) loop. Type-2 spiral heat exchangers are well suited to both duties. As internal reboilers they are bolted directly to the column bottom, with the central part of the spiral body serving as the downcomer. The heating medium is in spiral flow in a closed channel, seam welded at both edges; the boiling liquid is in crossflow in open channels.

Type-2 spiral heat exchangers are also well suited for use as external thermosyphon reboilers, particularly when the operating pressure is atmospheric or less. Such conditions require large flow sections and a small heat-transfer surface height.

At higher pressures, the lamella heat exchangers have proved suitable, with the boiling liquid on the shell side.

Feed preheating and bottoms cooling. The operation of a distillation plant is more economical when the feed is introduced near its boiling point. Heat exchangers are often installed to recover heat from the outgoing bottom product and preheat the feed. This normally involves heat transfer between two low-viscosity liquids, and plate heat exchangers are preferred because high heat recovery is feasible.

Product subcooling. The overhead product emerging from a condenser may require subcooling. This is achieved by built-in subcooling or by a separate subcooler and is a standard application for plate heat exchangers.

Absorption and Stripping Plants

Gases absorbed in liquids are often recovered by stripping. The liquid is recirculated between absorber and stripper, with several heat exchangers involved.

Heat recovery. The absorber operates at low temperatures, while the stripper is essentially at boiling temperature. Heat exchangers are installed to transfer heat between the hot liquid leaving the stripper and the cold liquid leaving the absorber. This is a standard low-viscosity liquid/liquid application for plate heat exchangers. Precooled liquid passing to the absorber may require further cooling, while further heating of the stripper feed liquid (involving the use of a second plate heat exchanger) may be called for.

Condensing. The vapor-phase mixture from the stripper is a mixture of gas, preferentially absorbed in the absorber, plus vaporized absorption media. Partial condensation of the absorption-media vapor from the "product" gas is required, and a special spiral heat exchanger, top-mounted on the column, may be used.

Reboiling. Heat input for the stripper may come from direct injection of steam or a type-2 spiral heat exchanger used as a reboiler.

Scrubbing plant. Certain components, e.g., sulfur dioxide, must be removed from flue gas before discharge to the atmosphere. The waste gas is scrubbed with a suitable liquid-phase reagent, recirculated after regeneration. If the component is to be recovered, the process is one of absorption stripping as discussed above. Absorption of the component removed in the scrubbing liquid involves a heat of reaction-absorption, plus condensation if the waste gas is moist, so the circulating scrubbing liquid must be cooled. In sulfur dioxide scrubbing plants, plate heat exchangers of Incoloy 825, titanium, titanium-palladium alloy, or high-quality stainless steel are often used.

Evaporation Plants

Evaporator. Evaporators operate on the same principle as reboilers in distillation plants. Forced circulation or flash evaporation is used where the liquid to be evaporated has high fouling tendencies. In forced circulation the liquid usually boils on the heat-transfer surface. For noncorrosive environments, tube-and-shell heat exchangers are standard equipment. Where high-grade material is necessary, type-2 spiral or lamella heat exchangers often show significant advantages.

Flash evaporation involves the heating of liquid under pressure to a temperature high enough to cause flash evaporation when the pressure is reduced to the operational level. Heat transfer to low- or medium-viscosity liquid from steam or some other heating medium favors the use of plate heat exchangers.

Condenser. Two basic condensation procedures are used: (1) barometric or contact condensing, where vapor is condensed by spraying with water, and (2) indirect or surface condensing, where vapor is condensed on the surface of a cooled heat exchanger. For surface condensing, spiral heat exchangers, with or without condensate subcooling, and lamella heat exchangers may be used.

High-Temperature, High-Pressure Duties

Heat exchangers for high-temperature, high-pressure applications (over 500°C and 3.5 MPa) are individually designed and custom built. Tubular construction is preferred, provided thermal expansion, vibration, and other effects are taken into account. Hairpin, U-tube, and tube-and-shell units are favored to avoid the problems associated with thermal expansion.

Tube-to-tube-sheet joints are the weak point of construction, and precautions must be taken to protect the end of the tube bundle from direct impingement of hot gas. The tube-bundle support must also be carefully designed to avoid fatigue, creep, and vibration damage.

Extremely Corrosive Fluids

Despite the wide choice of materials offered by the heat-exchanger types discussed above, some fluids are so corrosive that no metal or alloy offers sufficient corrosion resistance within the bounds of economic feasibility. Both tube and plate heat exchangers are available in silver, tantalum, or zirconium, but these can hardly be regarded as economically attractive. Instead, for such applications nonmetallic exchangers are used. They suffer the disadvantage of extreme fragility in conventional tubular construction.

Impervious graphite is sometimes used for nonmetallic heat exchangers. Graphite is weak in tension but relatively strong in compression and does not lend itself to tubular construction. Most successful forms of graphite exchangers are the block type, with the material held under compression at all times.

PURCHASE SPECIFICATIONS

Once the required type of heat exchanger is determined, the next task is to prepare the purchase specifications—a statement of requirements in exact detail with precise definitions. The purchase specifications are used by competing vendors to determine the design and prepare a price quotation for the manufacture and delivery of a heat exchanger meeting the customer's need.

Purchase specifications invariably include a summary sheet similar to the model reproduced in Fig. 9.2. Similar summary sheets may be prepared for air-cooled, tube-and-fin, plate-fin, and regenerative heat exchangers.

Purchase specifications should be written in concise, unambiguous language specifying the minimum detail necessary to yield the designed product at the lowest cost. The temptation to include legalistic circumlocutions that add nothing to the end result should be avoided. All subjective or excess words should be avoided; they lead to misinterpretation and confusion. Restraints imposed on vendors simply limit their flexibility in supplying the most economical product.

Purchase specifications vary greatly in length. Some occupy less than one page; others run to 20 or 30 closely printed pages. Appendix C is a verbatim reproduction of the tube-and-shell heat-exchanger purchase specification of the leading U.S. engineering contractor. It is very comprehensive and has been progressively developed over many years of experience in the design and construction supervision of refineries and chemical process plants worldwide.

Appendix C is a useful model purchase specification. However, most specifications will not include all the material contained in App. C. It is far too voluminous and detailed, except for the heat exchangers used in demanding refinery or process-plant conditions. Readers of this book will probably not have responsibility for purchasing such equipment. It is suggested that App. C be carefully reviewed, and only the necessary elements be included in a purchase specification, along with local requirements specific to the equipment at hand.

Our emphasis is on the word *necessary*. It is easy to incorporate, by reference, model codes or standards such as the ASME code, TEMA standards, ASTM material standards, etc. Incorporation by reference should be qualified by specific definition of the sections or paragraphs intended. Wholesale adoption of the model codes and standards simply leaves vendors doubtful as to which

DATE

#								
1						Job No.		
2	Customer					Reference No.		
3	Address					Proposal No.		
4	Plant Location					Date	Rev.	
5	Service of Unit					Item No.		
6	Size	Type		Horizontal Vertical	Connected	Parallel		Series
7	No. Units	Surf./Unit Gross Net		Sq Ft	Shells/Unit	Surf./Shell Gross Net		Sq Ft

PERFORMANCE OF ONE UNIT

#			Inlet	SHELL SIDE	Outlet	Inlet	TUBE SIDE	Outlet
9	Fluid Allocation							
10	Fluid Name							
11	Fluid Quantity, Total	Lb/Hr						
12	Vapor	Lb/Hr						
13	Liquid	Lb/Hr						
14	Steam	Lb/Hr						
15	Water	Lb/Hr						
16	Noncondensable	Lb/Hr						
17	Temperature	°F						
18	Specific Gravity/°API/Density — Lb/Cu Ft							
19	Viscosity	Cp						
20	Molecular Weight, Vapor							
21	Molecular Weight, Noncondensable							
22	Specific Heat	Btu/Lb°F						
23	Thermal Conductivity	Btu Ft/Hr Sq Ft°F						
24	Latent Heat	Btu/Lb @ °F						
25	Inlet Pressure	Psig						
26	Velocity	Ft/S						
27	Pressure Drop, Allow./Calc.	Psi		/			/	
28	Fouling Resistance (Min.) / Film Coeff.		ro	/ho		ri	/hi	
29	Heat Exchanged			Btu/Hr, MTD (Corrected)				°F
30	Transfer Rate, Btu/Hr Sq Ft°F, Service			Clean			% Over Design	

#	CONSTRUCTION OF ONE SHELL:	Sizes & Materials		Sketch (Bundle/Nozzle Orientation)
31		Shell Side	Tube Side	
32				
33	Design/Test Pressure Psig	/	/	
34	Design Temperature °F			
35	No. Passes per Shell			
36	Corrosion Allowance In.			
37	Connections In			
38	Size & Out			
39	Rating Intermediate			
40	Tube No. OD In.; Thk. — Min/Avg BWG In.	Length Ft; Pitch In.	30 60 90 45	
41	Tube Type	Material		
42	Shell ID OD In.	Shell Cover	(Integ.) (Remov.)	
43	Channel or Bonnet	Channel Cover		
44	Tubesheet-Stationary	Tubesheet-Floating		
45	Floating Head Cover	Impingement Protection	Yes No	
46	Baffles-Cross No. Type % Cut Diam Area	Spacing: Central Inlet In.		
47	Baffles-Long	Seal Type		
48	Supports-Tube Type	U-Bend Type		
49	Bypass Seal Arrangement	Tube-Tubesheet Joint		
50	Expansion Joint	Type		
51	ρv2-Inlet Nozzle	Bundle Entrance Bundle Exit		
52	Gaskets-Shell Side	Tube Side		
53	-Floating Head			
54	Code Requirements ASME	TEMA Class		
55	Weights/Shell & Bundle — Empty Filled with Water	Bundle Only Lb		
56	Remarks:			
57				
58				
59				
60	Sales Engineer Thermal Design /	Mechanical Design /		
61	Revisions ⚠1 ⚠2 ⚠3 ⚠4			

FIGURE 9.2
Model heat-exchanger specification summary sheet for tube-and-shell heat exchangers. (*Courtesy Nooter Corp., St. Louis.*)

sections are applicable. To cover themselves against contingencies, they simply raise the price for which they will undertake manufacture.

Purchase specifications used previously may be revised and carefully edited to remove irrelevant material. Additions are often made to existing specifications to cover the special requirements of unusual and unconventional pieces of equipment. If these additions are not deleted on subsequent jobs, the standard specification grows to unmanageable size and places on the vendor the task of trying to interpret exactly what the client wishes to obtain.

Many applications outside the process and power industries can be adequately satisfied by purchase of standard heat exchangers listed in manufacturers' catalogs. In many cases popular sizes are held in stock. These units are almost always lower in cost than custom units with special features. They can be obtained with minimum delay through routine procedures. The purchase specification can be a brief, concise statement, largely based on the manufacturer's catalog data.

Units with special features need to be custom made and will almost certainly cost more than those purchased from a catalog. Efforts should be made to include standard components or a standard ensemble to gain the economics of volume production.

Special units will likely be designed in detail by the customer or hired consulting engineers. The purchasing specifications will then simply invite vendors to provide a "hardware-only" quotation and performance guarantee. The requirement of a performance guarantee will force vendors to independently verify the design. Often vendors have such broad experience that they can devise alternative designs that are more economical to manufacture, install, operate, or clean. The possibilities for such improved design should not be denied by a rigid purchase specification. Vendors should be encouraged to submit design revisions or alternative designs at the tender or pretender stages. Their suggestions should be given serious consideration and adopted when clearly superior to the original. All flexibility should be concentrated at the early stages. Once the contract has been placed and the purchase order issued, no further design changes should be tolerated unless there are exceptional circumstances.

Many facets of the preparation of purchase specifications are beyond our level of interest here. These include the incorporation of special customer experience, the relationship between specifications and codes and standards, and additions to cover specific deficiencies or generalities in the model codes and standards. This material is well covered by a special publication of the ASME [Rubin, 1979], which is recommended to every engineer concerned with heat-transfer equipment.

TENDER PROCEDURES

The procedures followed in soliciting and obtaining quotations vary greatly from one jurisdiction to another and depend on the magnitude of the order. The simplest situation arises when an existing unit is to be replaced or duplicated. It may be adequate to request the previous supplier to submit a price quotation for supplying the duplicate unit.

More likely, new units will not be exact replicas of existing systems. For undemanding applications three or four manufacturers may be invited to submit price quotations based on products advertised in their catalogs.

For noncatalog items the process is more complicated. The customer or the customer's consulting engineers will have prepared the bid specification documents, including an outline or detailed design of the units required. Public jurisdictions are usually required to solicit bid quotations by advertisement in the approved channels—newspapers, trade magazines, or official government publications such as the *Commerce Business Daily*. The process is quite formal. There may be a preliminary notice inviting companies that are anxious to bid on the contract to produce evidence of their capability and capacity to handle the work. From the respondents a select list is prepared, and these few companies are invited to prepare a detailed bid. This selective policy is open to political manipulation and misunderstanding, so that despite the excess effort invested in preparing many bid proposals for public tender, it is the most visibly fair and unprejudiced method.

Nongovernment operations can be handled more efficiently and with great dispatch. Large customers have established relationships, formal or informal, with a stable of suppliers who have performed adequately in the past. Adequate performance is interpreted to mean delivery, at the price and time stipulated in the original contract, of a unit that meets its thermal and mechanical design objectives without problems. The accolade of adequate performance is not easily achieved. Once a company has found two or three suppliers who can fulfill their promises, the company is happy to remain a good customer. Evaluating competitive bids is a costly, time-consuming process, and there is little to be gained from soliciting and evaluating more than three or four serious bids. Of course, this policy must be practiced circumspectly. There is a corollary of Parkinson's law and Peter's principle that the commercial efficiency of a company is invariably related to its size, history, and level of success. Successful vendors eventually become fat and lazy in their established relationships and are then superseded by young and vigorous companies eager to go the extra mile to get the business.

EVALUATING BID PROPOSALS

Bid proposals received from several vendors should be evaluated in two stages:

1. Technical-proposal evaluation
2. Cost-proposal evaluation

The technical-proposal evaluation should be carried out before, and independently of, the cost evaluation. Only those proposals capable of fulfilling the engineering requirements should be allowed to proced into the cost competition.

The best way to conduct a technical evaluation is to prepare a heat-exchanger summary sheet for all the competing bids, in a common format and with the same system of units. Information similar to that shown in the specification summary sheet in Fig. 9.2 will be required. Some evaluators summarize the information for all bids on the same sheet, to facilitate direct comparison. The arrangement of the comparative summary may be different, but it is good practice to retain the same line designations for the same items of information, as in the model sheet. Most vendors use heat-exchanger data summary sheets closely modeled after the specimen shown or the corresponding version for air-cooled heat exchangers contained in API 641 [1973].

With all the data summarized, the first stage of evaluation is to compare the sizes and types of exchangers offered. The process is facilitated when the types of exchangers, in terms of details of construction and surface area for heat transfer, are all similar. Then the thermal-design engineers have all developed the same solution. Of course, if the customer or the consultant has specified the exchanger in detail, then no significant differences are to be expected.

When significant differences exist among various offerings, a more detailed review is necessary. Discussions with the vendors may be held to establish the basis for their optimistic or pessimistic design approach. When the basis of the design is demonstrated to be sound and a guarantee of performance is offered, the price becomes a significant and decisive factor.

All associated expenses should be calculated in the cost. A prime component will obviously be the actual cost of the unit at the manufacturer's plant. To this must be added a variety of other costs. Transportation, insurance, and other charges may be appreciably different between a local and a remote manufacturer. The cost of sending inspectors or surveyors to the manufacturer during construction may be significant when the unit is made at a remote plant. Variations in credit availability and interest charges may be appreciable in the present era of high inflation and interest charges.

The low total-cost bids are obviously those which command the main interest. Deviations from the bid specifications should be carefully examined, although some variations are usually acceptable. Dimensional changes to the internal system can usually be tolerated. A decrease in the tube length is generally advantageous. Sometimes an increase in the tube length, wall tube diameter, or wall thickness is acceptable; the use of existing stocks of tube or other materials may allow delivery at last year's prices, Similarly, the use of finned-tube instead of bare tube can result in substantial savings in cost, weight, and volume.

Other variations might include a reversal of the shell-side and tube-side fluids. Often a viscous fluid that is laminar in tube-side flow will be turbulent on the shell side, with consequent increase in the overall heat-transfer coefficient. The effect on pressure drop, the fouling tendency, and the ease of access for cleaning must all be considered in regard to such a change.

Higher bids should not be summarily dismissed, but rather examined carefully to determine the reasons for the extra cost. There may be features that make the extra cost worthwhile. Sometimes heavier-gauge or higher-grade materials, available at relatively low cost, could extend the operating life appreciably. An increase in the surface area for heat transfer is sometimes worth a little extra cost; but if the fluids have high fouling tendencies, the increase in area, resulting perhaps in decreased fluid velocities, may actually be a disadvantage.

The temptation to evaluate bids on a cost per unit mass or cost per unit surface area basis should be resisted. It is easy to do but there is no useful purpose served if the mass or surface area is not distributed in the most effective way. Rubin [1980] quoted one example of the folly of this method. Three offers were received, for three similar heat exchangers. The offers were essentially identical except for the information shown below.

Vendor	Size	Surface, ft²	Cost, $	Surface cost, $/ft²	Weight, lb	Weight cost, $/lb
A	21–96	500	9,000	18.0	3700	2.43
B	19–120	600	10,200	17.0	4700	2.17
C	21–120	625	10,000	16.0	5000	2.00

Rubin noted that:

> If the evaluation is based upon Surface Cost in dollars per square foot or upon Weight Cost in dollars per pound, Vendor C would be selected. Prior engineering evaluations showed all three offers were adequate to meet the thermal and mechanical performance requirements. Vendor A should be selected to obtain an acceptable item at the minimum cost.

Preparation of the bid quotation is a difficult task. The effort expended in preparing the quote is rewarded only for the vendor who secures the order. There is rarely enough information or time available to complete the quotation with no uncertainties remaining. Indeed, the whole basis of thermal design for fluids with appreciable fouling characteristics is uncertain and a game played only by those with a gambler's nerve or instinct.

The uncertainties are allowed for by padding the proposal. The degree of padding depends on what the market will stand, the nature of the competition, and the state of the order book. A lean order book may result in lower cost estimates, simply to secure the work necessary to keep the shop going even though there may be little profit in it. A busy shop with years of work can afford to be more selective; however, they need to be careful not to price themselves out of future consideration by the client companies.

ETHICS IN BID ANALYSIS

Vendor confidence in the impartiality and propriety of the bid analysis procedure is important. It is established in the long run by consistent ethical behavior toward the vendor companies.

Sometimes the highly competent heat-transfer engineers employed by vendor companies can originate design improvements resulting in dramatic savings of cost, size, or weight compared with the bid specifications. When this occurs, there is a temptation to request bids from other vendors, using the same innovative design features. This is unethical and unfair to the originator of the design improvement. It is also poor commercial practice, for clients who do this soon become known to vendors. Vendors' innovative responses to a bid proposal are then inhibited in favor of other clients practicing higher standards of business ethics.

The purchase specification should define all the pertinent information intended for use in the evaluation process, including the basis for the comparative evaluation. Evaluation in terms of lowest cost per unit mass or per unit area is frequently used, even though it does not result in the best or most economic design. If evaluation is to be made on this basis, the purchase specification should say so. It will result in modified designs by more alert vendors.

Economic trade-off allowances for savings in pressure drop or adjustments in fluid exit temperatures should be defined in the purchase specification. Rubin [1980] suggests that no credit be allocated for a design with a lower pressure drop unless the type of credit allowance is outlined in the purchase specification—perhaps $1000 for each pound per square inch reduction in pressure drop.

INSPECTION

Heat exchangers are subject to various inspections during fabrication and on completion. The inspectors work for different authorities and have different objectives.

Inspectors from a regulatory authority are government employees concerned with matters of public safety in the use of boilers and pressure vessels. They are not interested in the performance of heat exchangers or internal malfunctions, except where internal fluid leaks create a hazardous situation. Regulatory-authority inspectors have sweeping powers and great responsibility for a broad spectrum of pressurized engineering products. They are usually experienced journeymen or professional engineers but are rarely specially trained in all aspects of their task. The sheer scope of their responsibility makes this impractical. They are invariably too few in number to adequately perform every task within their jurisdiction, and, consequently, they focus their effort on "squeaky wheels." Their first priority is to investigate accidents and matters related thereto, including attention to the manufacturer of the vessel involved in the accident.

The imprint of the ASME-code stamp on a heat exchanger and a certificate of compliance are evidence that the unit was constructed in accordance with the code by a responsible manufacturer using licensed journeymen and welders under the supervision of professional engineers. It is not a guarantee that the unit has been inspected and passed by the regulatory authority or that the unit has been inspected or tested by anybody other than the manufacturer.

The manufacturer applies the stamp indicating compliance with the requirements of the ASME code. The authority to do this is renewed at 3-year intervals by the regulatory agency responsible for the area in which the plant is located. The initial inspection and triennial review are serious business. Manufacturers must demonstrate sufficient competence and capability to construct pressure vessels in accordance with the code. They must be completely familiar with the code, have written design calculation procedures, and show evidence of satisfactory methods of manufacturing, including a quality-control system and the employment of licensed staff.

There is, therefore, every possibility that a unit carrying the ASME-code compliance stamp will not fail or become hazardous to life or property when operated at the design pressure and temperature or less, provided the proper start-up and shutdown procedures are followed. The compliance stamp implies absolutely nothing about the internal arrangement or thermal performance of the unit. These matters are largely covered by industry standards such as the TEMA or API standards and the additional requirements included in the purchase specification.

As part of the quality-control system, the exchanger manufacturer employs inspectors to check various components as they are made and authorize succeeding phases of construction. They also inspect the unit on completion. The quality of their work, their independence, and the managerial support they enjoy are the ultimate basis for the commercial reputation of the manufacturer. Quality-control inspectors are an expensive overhead item and appear often as impediments to completing contracts on time and making a profit. Temptations to dilute the inspectorate must be resisted. Inspectors must be senior, experienced, diplomatic conciliators and, in the end, "made of steel," well trained, highly paid, and given great authority and independence in their areas of jurisdiction. This will ensure the maintenance of high-quality production. Whether that will go hand in hand with a commercially viable and profitable operation is a matter for management, beyond the competence of this author. If such policy is not practiced, the enterprise will surely fail, but unfortunately there is no guarantee of success when it is followed. The dangers of an overly rigid, uncompromising inspectorate are as great as those of a lax and flabby team.

Exchangers of moderate to large size are also examined in the course of construction by the customer's inspector, representing the interests of the purchaser. This person may be a permanent member of the buyer's staff, engaged solely or partly in inspection duties, or work for the consulting engineers representing the purchaser. Again, there are specialists who offer inspection services as their only activity and who are retained specifically for that purpose. Industrial insurance companies maintain staffs of experienced boiler and pressure-vessel inspectors who represent clients as required.

The function of the customer's inspector is to witness the construction and testing of the heat exchanger at critical phases during its manufacture. For example, the best time to inspect heat-exchanger tubing is in the "as received" condition from the mill. At that stage all is visible. Simple visual inspection can often reveal cracks, damage, scale formation, and dimensional and geometric differences.

The customer's inspector is usually present to witness the hydrostatic testing of the tube bundle and the subsequent fully assembled hydrotest. Customer's inspectors keep an eye on general construction procedures and the conduct of business and management at the plant, reporting regularly to the purchaser on matters of interest. On a very large contract the consumer's inspector may be resident at the manufacturer's plant, but this is unusual. In most cases the inspector visits the plant at prearranged times to witness critical operations.

Customer's inspectors are professionals of high integrity and therefore too expensive to be used other than in the most effective way. Their daily fees, subsistence costs, and travel expenses should be included in the total overall costs when evaluating competitive bids. The cost of inspection overseas will be far greater than at a plant a few miles away.

In many applications of heat exchangers, little or no inspection is performed, aside from the manufacturer's quality-control program. Heat exchangers in heating and ventilating systems, water heaters of many types and sizes, lubricating-oil coolers, and automotive radiators are exempt from compliance with the ASME code or other regulatory requirements. They are mass produced or prefabricated in great numbers. The principal responsibility for their quality and durability rests with the manufacturer. Progressively stringent consumer legislation is elevating manufacturer's liability to increasingly higher levels and forcing greater interest in safety and durability.

TESTING

Testing of heat exchangers is almost always taken to mean hydrostatic pressure testing, radiographic inspection of weldments, and inspection for leaks. Tests of thermal performance and fluid pressure drop to validate thermal design are never carried out on custom units by the builder, except on the smallest units or in very special circumstances. The builder simply does not have the capability to generate the large energy and fluid flows characteristic of most custom exchangers.

Pressure Testing

Pressure testing is carried out at various construction stages, depending on the type of exchanger and the rigor of its intended service. It is customary to pressure-test the two fluid circuits independently. In tube-and-shell heat exchangers the tube-side flow circuit is assembled with the tube bundle in or out of the shell, depending on type. Various covers, flange test rings, etc., are provided to seal the system. Then it is filled with water, oil, or some other specified liquid and compressed to 1.5 times the specified design pressure (with a correction made to account for the difference in temperature between the test condition and the operating condition). The pressure is maintained constant for the test period specified in the purchase specification, usually a minimum of 30-min but sometimes as long as 4 h. Periodic visual examinations are made to confirm that there are no leaks from joints or cracked tubes. Joints with low rates of leakage, called *weepers,* are particularly difficult to locate in brief test periods. Sometimes misleading results are obtained because the test-fluid temperature is different from the shop air temperature. Cold fluid and humid atmospheres result in the condensation of atmospheric water on the cold metal surfaces, suggesting a weeper or porous wall. Similarly, hot fluid causes differential thermal expansion that sometimes causes damage in incomplete test assemblies.

Pressure testing on the shell side is accomplished with the tube bundle in place but with the channel and shell covers removed. The tubes are therefore unpressurized and accessible for visual inspection for leaks that become manifest as a trickle of liquid from the tube or tube-sheet joint.

Sometimes heat exchangers are tested pneumatically rather than hydraulically. There are various reasons for this. In some large but lightweight systems the sheer weight of water or oil would impose dangerously high stresses on the incomplete structure. In other cases the presence of liquid contaminants on the surface is unacceptable. For pneumatic testing the pressure is limited to 1.25 times the design pressure.

Hydraulic testing is preferred to pneumatic testing. Liquids are virtually incompressible, so on rupture a small quantity of fluid is released, with very rapid reduction in the pressure. This fail-safe mechanism does not apply to pneumatic systems. Gases are compressible, and the escape of a small quantity does not appreciably reduce the pressure; the propagation of the crack or other rupture mechanism is sustained, with catastrophic consequences. Moreover, minor leaks of gas are less visible than liquids.

Leak Testing

Pressure testing is carried out to confirm the ability of complete or semicomplete systems to withstand pressures greater than those they will be called upon to sustain during normal operations. In the course of pressure testing, leaks from joints, cracks, and other defects in materials will be noted, but this is really secondary to the main purpose. Testing for leaks is normally carried out separately and independently of pressure testing. It may consist of tests applied to single components, subassemblies, or complete heat exchangers.

Heat-exchanger tubing is called upon to perform arduous duties. It is selected not only for its resistance to corrosive attack and its mechanical strength, but also for its inherent reliability based on past experience and the ease with which defects or damage can be detected. The methods used to make the metal and the tubes have a significant effect on the number of defects and the ability to detect certain defects.

Two electrical methods, eddy current and ultrasonic, are frequently used to test for both surface and subsurface defects. Both depend on passing signals through the material, longitudinally or transversely, and receiving signals at a transducer for subsequent display in a strip-chart recorder or cathode-ray oscillograph. Perturbations in the signal indicate inhomogeneities or defects in the material. With these methods, gross defects, large occlusions, blowholes, slag, etc., are easily determined by untrained personnel. Smaller defects require interpretation by trained operators. Typically, at maximum sensitivity, through-the-wall pinholes above 0.065 in in diameter can be accurately located. Holes with smaller diameters tend to be masked in the general background noise of the signal. Unfortunately, hole diameters of 0.005 in result in significant leakage, particularly at high pressure, and a hole diameter of 0.020 in ranks as a massive leak.

An increase in instrument sensitivity, provided to locate the smaller defects, simply increases the number of suspect defects in the component under surveillance. Electrical test methods are therefore best suited to locating major defects and generally sorting the perfect and potentially imperfect components for subsequent visual examination. The sensitivity and usefulness of electrical test methods are greatly diminished when they are applied to a complicated assembly or subassembly such as a tube-bundle and tube-sheet assembly.

The simplest but most effective method of leak detection is to immerse the test unit in a water bath and observe the bubbles. The water should be treated with a small amount of detergent to reduce the surface tension. Air can be used to pressurize the unit to a low pressure, but helium or hydrogen is more effective. Hydrogen is highly flammable, so great care should be exercised in its use. Helium is inert, safe to use, and only slightly less sensitive than hydrogen.

In many instances it is neither convenient nor feasible to immerse large components or subassemblies in a water bath, so leak testing must be practiced in situ. All the various methods depend on charging and pressurizing one side of the exchanger with a test fluid and detecting its presence on the other side. The simplest method is to use a liquid whose presence can be readily detected. Water and oil leaks can be visually observed, and other methods use liquids that fluoresce in ultraviolet light or have readily detected magnetic properties.

Large leaks are easy to find, but the detection, location, and correction of small leaks is one of the most exasperating problems in heat-exchanger fabrication and maintenance. A simple but effective method is to lightly pressurize one side with air or helium and to paint the joints with soap solution or one of the proprietary foaming agents available as liquids or aerosol sprays.

Cryogenic systems are among the most demanding with regard to sealing. Frequently the insulation spaces are evacuated to incredibly low pressures as an integral and vital part of the thermal-insulation process. The space is evacuated and heated to "bake out" the absorbed gases according to a prescribed schedule. Then, with adsorping *getters* in place in the insulating space, the pressure in the space is monitored with highly sensitive pressure-measuring devices to observe the increase per day. The same technique can be utilized (at a less demanding level) with heat exchangers for which a very high degree of leak tightness is required.

Helium leak tests are very sensitive. For this test one side of the exchanger is coupled to a vacuum pump and evacuated to a low pressure. Then, on the other side, a probe from which helium is escaping is moved over the surfaces thought to be responsible for the leak. A small quantity of this helium is drawn, with atmospheric air, through the leak, passes to the pump, and is detected downstream by a mass spectrometer. The mass-spectrometer signal increases sharply as the probe is moved closer to the leak. Such instruments for detecting leaks are incredibly sensitive.

A somewhat similar but less sensitive technique is often used to detect leaks with gaseous halides or halide compounds. One side of the exchanger is lightly pressurized with halide. Then a "sniffer" is drawn over the surface thought to be leaking. The sniffer is coupled to a small pump and gathers a portion of the leaking halide. The pump feeds a gas torch that burns normally with a colorless flame, but with a green flame when halides are present.

Thermofluid Testing

Except for small units or those in regular commercial production, no thermofluid testing is carried out in heat exchangers before delivery to the customer. Builders simply do not have the capability to generate the large energy and fluid flows characteristic of custom exchangers.

Most units are therefore incorporated into process plants with the design unvalidated. Moreover, sufficient instrumentation is rarely included to perform anything beyond a simple heat-balance analysis in operation. During start-up, various deficiencies and irregularities are detected and "fixed" in one way or another. Frequently, changes are made at a late design stage, after the critical components have been ordered, which results in operation off the design point. It is very hard to come by reliable operating data in sufficient depth and detail to provide a sound basis for validating thermal design procedures.

With new types of heat exchangers, for which design data based on previous experience are not available, some form of testing must be carried out. The most effective way is to conduct experimental work in properly scaled models. Many studies have shown that often a better set of tests can be carried out with a well-designed model than with the full-scale heat exchanger. The cost of model testing is very small compared with full-scale testing. Furthermore, the models can be quickly built, are easily accommodated in a laboratory setting, and can be readily modified to evaluate the effects of parametric changes. Fraas and Ozisik [1965] have given an interesting discussion of model testing and the principles of similitude for geometric scaling.

SUMMARY

The process of buying a heat exchanger involves a number of activities. The first and most important task is defining the job that the exchanger is to do and the inherent limitations. Then the type of heat exchanger is chosen and preliminary design work is carried out to define the broad features of the unit. Satisfactory experience with previous units performing similar work is helpful at this stage.

The preparation of the purchase specifications is an important process. The purchase specifications are the clearest possible statement of the purchaser's requirements for a heat exchanger. Great care must be exercised to include all the relevant information and to exclude all the unnecessary requirements and stipulations.

The exchanger requirements may be met by standard products available in the builder's catalog. If they cannot be met in this way, competitive bids must be sought from three or four custom builders experienced in the field and with established performance records.

Evaluation of the bids should be carried out with scrupulous fairness in accordance with the highest standards of business ethics. It is well to maintain a flexible attitude and to modify or relax the design requirements if appreciable advantage can be gained thereby.

Periodic inspection by a representative of the customer is necessary at critical stages in the construction to witness crucial tests of the unit. These consist principally of hydraulic pressure tests and leak tests. Tests of thermofluid performance are rarely performed prior to installation and commissioning of the unit in service, but one can pray hard.

REFERENCES

API: "Heat Exchangers for General Refinery Services," API Standard 660, American Petroleum Institute, Washington, D.C., 1968.

API: "Air Cooled Heat Exchangers for General Refinery Services," API Standard 661, American Petroleum Institute, Washington, D.C., 1973.

Fraas, A. P., and M. N. Ozisik: "Heat Exchanger Design," Wiley, New York, 1965.

Rubin, F. L. (ed.): "The Interrelationships between Codes, Standards and Customer Specifications for Process Heat Transfer Equipment," Heat Transfer Div., ASME Winter Annual Meeting, New York, December, 1979.

Rubin, F. L.: "Shell and Tube Heat Exchangers," course notes, American Society of Chemical Engineers professional education course, Houston, Texas, 1980.

TEMA: "Standards of the Tubular Exchanger Manufacturers Association," 6th ed., TEMA, New York, 1978.

10

Codes and standards

INTRODUCTION

Large numbers of heat exchangers are manufactured with little or no reference to regulations, codes, or standards. Such heat exchangers are simple devices in which no appreciable pressure is involved and thermofluid performance is less significant than cost, size, or weight. They are found everywhere—in cars, trucks, refrigerators, air-conditioning systems, building exhaust and inlet air preheaters, and the like. They are usually used for cooling liquids, oil, water, and the refrigerant fluids with air or, less frequently, water.

Our interest in this chapter is not in these simple exchangers but in the more complicated devices operating at elevated pressures and of sufficient size and significance to warrant conformance to industrial standards. Vessels with internal or external pressures in excess of 1000 kPa (15 lb/in^2) gauge are usually considered to be pressure vessels.

REGULATIONS, CODES, STANDARDS, AND SPECIFICATIONS

Regulations are prescribed by government agencies having jurisdictional responsibility in certain areas. Regulations have the force of law and must be adhered to unless specific exemption is sought from and granted by the regulatory authority involved. All countries have their own regulations, but there is no common format or language.

Codes are bodies of rules that are established to protect the public interest and that evolve in the course of time. They are usually prepared by committees of experts representing manufacturers, users, regulators, and other interested parties, often under the aegis of a disinterested technical or professional organization. They are safety documents prescribing detailed rules for the design and operation of equipment to ensure that the required service will be provided in a safe manner. Government regulations frequently embody the applicable codes as the basis for their regulations, with additional material to cover local conditions. This is called "adoption by reference."

Standards are similar to codes. They give detailed requirements for particular classes of materials or equipment. They define minimum quality levels for materials or operating performance, standardize nomenclature, and prescribe or recommend rules and procedures for design, installation, operation, repair, and replacement. Standards are prepared by committees of experts operating under the aegis of technical, professional, manufacturer, or government-sponsored organizations. Standards are frequently incorporated in codes by reference.

Specifications are written, detailed descriptions of the work to be done on a particular job. They provide data on the quantities and quality of material to be used, the mode of fabrication, and much other information that is not shown on the drawings. In fact, the engineering drawings of components, parts, and systems to be manufactured constitute part of the specifications, although they are not normally regarded as such. Specifications are prepared by the customer or the customer's agent (consulting engineer) for the particular unit required. They almost invariably incorporate references to the applicable standards, codes, and regulations. A very comprehensive specification for tube-and-shell heat exchangers is included in App. C.

It is important to appreciate the clear distinction among the four terms defined above. Regulations and codes are concerned only with public safety, not with thermal performance. Thus, a heat exchanger constructed in conformance with the applicable codes and regulations will safely contain the fluids concerned but may or may not provide the desired thermal performance.

Standards are not concerned with public safety but are largely economic measures fostered by manufacturers and, to a lesser extent, users to provide minimum quality reference levels for materials, nomenclature, designs, and procedures. Standards are exceptionally useful, for they characterize good contemporary engineering practice and permit the prediction of operating performance with reasonable levels of confidence.

PRESSURE-VESSEL CODES

The ASME Code

The American Society of Mechanical Engineers (ASME) Boiler and Pressure Vessel Code has been widely adopted, in whole or in part, by regulatory authorities in North America and many other parts of the world. This code is an extremely comprehensive document in 11 sections (each a substantial volume—or volumes—covering different aspects of its subject matter). Section VIII is applicable to confined pressure vessels and is commonly applied to heat exchangers. Section I is applicable to power boilers of both the finned-tube and water-tube types. Other sections of interest to the heat-exchanger industry include Section II, "Material Specifications"; Section III, "Nuclear Power Plant Components"; Section V, "Nondestructive Examination"; and Section IX, "Welding and Brazing Qualifications."

The ASME code is continuously updated and refined; addenda and revisions are published semiannually, and complete new additions triannually. When urgent needs not covered by the existing code arise, temporary rules are formulated in the form of Code Cases. These are subsequently incorporated into the next edition of the code.

Manufacturers capable of constructing pressure vessels according to the ASME code may apply to be inspected and qualified for a code stamp. Vessels made by the manufacturer then carry the ASME code stamp if the specifications for the equipment require it. Triannual inspections are made to maintain the manufacturer's qualification. Code stamps have been issued to manufacturers overseas as well as in North America.

The Uniform Boiler and Pressure Vessel Laws Society Inc. operates to secure legal status for the ASME code and publishes synopses of the various jurisdictional laws relating to boilers and pressure vessels.

British Standards

In Great Britain safety considerations are the responsibility of the user. Insurance-company requirements therefore dictate the applicable rules. Usually these prescribe construction according to standards of the British Standards Institution.

The British Master Pressure Vessel Standard BS 5500-1976 ("Unfired Fusion Welded Pressure Vessels") was introduced to integrate in one standard the requirements for design, manufacture, testing, and inspection of fusion-welded pressure vessels. It embraces a wide spectrum of pressure vessels from air receivers to nuclear vessels and replaces many previous standards specific to limited areas. The master standard is in five sections, with many supplementary tables and appendixes.

European Pressure-Vessel Codes

In France the mandatory regulations, the A.P.A.V.E., under the authority of the *Service des Mines,* are supplemented by the much more extensive S.N.C.T. code, the *"Code de construction des recipients a pression non soumis a l'action de la flamme."*

The S.N.C.T. (*Syndicat National de la Chaudronnerie, de la Tolerie et de la Tuyanterie Industrielle*) code is a nonmandatory design code developed by a syndicate of French pressure-vessel manufacturers. It is essentially an experience-based guide giving data for the recommended manufacture of pressure vessels. The user is entitled to adopt different approaches, provided they conform to the mandatory regulations; in all cases, manufacturers remain liable for the safety of their vessels.

Design and construction requirements for pressure vessels used in Germany are specified in the *AD-Merkblatter,* Series B. Materials are covered in Series W; manufacture and testing in Series HP; manufacture in Series H; and nonmetallic materials in Series N. Compliance with the rules is mandatory.

In Italy, the A.N.C.C. code covers mandatory design, construction, and operating requirements for pressure vessels and steam generators.

In the Netherlands the Rules for Pressure Vessels are replacing the previous mandatory Grondslagen code.

The Norwegian Pressure Vessel Code specifies requirements for materials, design, construction, inspection, and testing of welded pressure vessels.

The Swedish Pressure Vessel Code covers stress calculations for pressure vessels and is mandatory for all boilers and pressure vessels destined for use in Sweden.

Japanese Pressure-Vessel Code

In Japan pressure vessels are subject to government control, and a permit is required before a pressure vessel may be put into service. The Japanese Industrial Standards embracing the rules for pressure vessels and heat exchangers include:

1. JIS B 8243, "Construction of Pressure Vessels." This is the basic standard for unfired pressure vessels except those exceeding 30 MPa design pressure or if riveted and brazed construction. The standard is in 17 parts with 6 appendixes.
2. JIS B 8249, "Shell and Tube Heat Exchangers for Application to General Use."

There are many other related Japanese Industrial Standards embracing materials, materials testing, and nondestructive testing.

Comparison of Pressure-Vessel Codes

Khella [1980] summarized and compared the regulations for pressure vessels in various parts of the world. He included Fig. 10.1 showing the wall thickness required, as a function of pressure, by the ASME (Section VIII, Division 1) code used in North America and the Japanese, French, and German codes. The particular case selected for comparison refers to a common welded cylindrical carbon-steel shell, 1000 mm in diameter, with 100 percent radiography.

The general conclusion gained from Khella's work is that the ASME and Japanese codes include the most rigorous design requirements. The ASME code is well-known, highly respected, and widely accepted. It is likely that a heat exchanger constructed by a qualified manufacturer and carrying the ASME code stamp would conform to the safety requirements of nearly every jurisdiction. Of

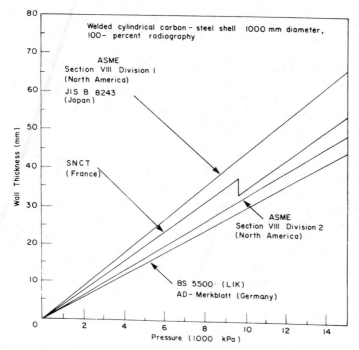

FIGURE 10.1

Wall-thickness requirement as a function of pressure for various national pressure-vessel codes. (*After Khella [1980].*)

course, severe difficulties resulting from meticulous interpretation of the domestic code could arise, depending on the local bureaucracy.

An excellent review of United States and European design standards was given by Dall'Ora [1971]. Interesting commentaries on the ASME Boiler and Pressure Vessel Code were given by Irving [1977] and (with particular reference to heat exchangers) by Yokell [1979]. The recent conference proceedings [Rubin, 1979] of the interrelationships of codes, standards, and specifications for process heat-transfer equipment contains a wealth of relevant detail.

The British Standards Institution, through their Technical Help to Exporters Service, maintains and sells English translations of European codes and standards in many areas, including boilers and pressure vessels.

International Standardization

The plethora of local and national pressure-vessel codes suggests that the introduction of uniform international standards is a matter to be addressed urgently. The need for common standards was recognized early in the century, and in 1946 the International Organization for Standardization (ISO) was established in Geneva. Since then, much work has been done in many areas, and a great variety of international standards have been proposed. Regrettably, pressure vessels and heat exchangers have not been included so far.

In 1968 ISO Recommendation R831, "Rules for Construction of Stationary Boilers," was issued to unify various national codes and establish a basis for countries that did not have their own codes. The draft recommendation was unacceptable to many countries and failed to obtain the necessary majority of voting members. It was returned to the draft committee for revision.

Work continues on the development of international standards for pressure vessels and heat exchangers. It is fraught with difficulties and disagreements that appear to require a long time to reconcile. Agreement on the rules for such matters as safety factors to be applied to the ultimate and yield strengths is hard to achieve, for there are basic differences between the American and European approaches. Furthermore, different countries manufacture materials based on different melting practices and materials chemistry. Therefore, prior materials standards need to be established. Finally, there is, inescapably, a reluctance to yield national control. With the introduction of uniform standards, international competition would be stimulated. Many countries fear their manufacturers would be unable to compete with the large international corporations.

STANDARDS

TEMA Standards

Among the various nonmandatory standards of the heat-exchanger industry, the TEMA standards are probably the most widely known and accepted. TEMA is the acronym of the Tubular Exchanger Manufacturers Association, an organization of 27 U.S. manufacturers of tube-and-shell heat exchangers. The association was formed in 1939, and the TEMA standards were first published shortly thereafter. The standards were established to provide minimum levels of quality and uniformity in the design and construction of shell-and-tube heat exchangers. The standards represent the collective experience of the leading American manufacturers and are widely accepted throughout the free world as the basis for sound engineering practice.

The standards are continuously reviewed and periodically revised in a new edition. The sixth edition was published in 1978. The standards embrace many aspects of shell-and-tube heat-exchanger technology not directly related to safety and therefore not covered by the ASME Boiler and Pressure Vessel Code. These diverse aspects include size designation and nomenclature, manufacturing tolerances, specification sheets, nameplates, guarantees, installation, operation and maintenance, mechanical design of internal components not included in the pressure-vessel code, baffles and support plates, floating heads, gaskets, tube sheets, channel covers and bonnets, nozzles, end flanges and bolting, material specifications, thermal standards including fouling, physical properties of fluids, general information on pipefitting, flange and bolting dimensions, wrench clearances, flange pressure ratings, tubing characteristics, hardness conversions, metal properties, conversion factors, and definitions.

One section of the compendium contains "Recommended Good Practices" on a range of topics including very large units beyond the bailiwick of the standards, seismic design nozzle loading, tube plugging, tube vibration, and tube-sheet design performance.

An important development was the publication by TEMA in 1980 of the "Sample Problem Booklet," containing detailed calculation solutions to exercises in tube-sheet design, channel-cover thickness requirements, longitudinal stress calculations, and tube-vibration calculations. The "Sample Problem Booklet" is an essential adjunct to the TEMA standards. In subsequent editions, the two publications may be combined.

TEMA generally classifies tube-and-shell heat exchangers into three generic groups—classes R, C, and B. Class-R exchangers are those used for the generally *severe* requirements of petroleum and related processing applications. Class-C exchangers are those used for the generally *moderate* requirements of commercial and general process applications. Class-B exchangers are those used for *chemical* process service. Class-R specifications are more severe than those of class B and class C. Rubin [1980] has well summarized the differences and similarities among the three TEMA classes.

EJMA Standards

Another organization, similar and closely related to TEMA, is the Expansion Joint Manufacturers Association (EJMA). This association of manufacturers publishes standards for the design and use of expansion joints in the shells of tube-and-shell heat exchangers and in other applications.

API Standards

The American Petroleum Institute (API) publishes two standards concerned with heat exchangers. API Standard 660, "Heat Exchangers for General Refinery Service," is concerned with shell-and-tube heat exchangers and is supplementary to the TEMA and ASME code requirements. It defines specific requirements above or beyond TEMA requirements in various critical areas and, in general, requires a higher level of manufacturing. The standard is commonly used by the major oil companies for refinery applications.

API Standard 661, "Air-Cooled Heat Exchangers for General Refinery Services," is considerably more extensive because for air coolers there is nothing comparable to the TEMA standards for tube-and-shell exchangers. Standard 661 has therefore become the basic reference standard for many types of air-cooled heat exchangers in rigorous service. The topics covered include nomenclature, types of air cooler, induced versus forced draft, single and multiple fans, coils and bags, heater types, plenums, and drives. Standards are included to cover the design of the process side for internal pressure and temperature and types of fin attachment, and the air side for design of fans, drives, motors, louvers and fabrication, tube-end attachment, welding tolerances, etc.

ANSI Standards

The American National Standards Institute (ANSI), based in New York, is an umbrella organization, similar to the British Standards Institute, that publishes a broad range of engineering standards. The ASME Boiler and Pressure Vessel Code is an ANSI standard. Two others are directly concerned with heat exchangers: ANSI B78.1 (1972), "Tubular Heat Exchangers in Chemical Process Service," and ANSI B78.2 (1974), "External Dimensions for Shell and Tube Heat Exchangers for Chemical Process Service." Neither of these standards has come into general use. In 1979 at the U.S. National Heat Transfer Conference, a panel discussion on the history and evolution of heat-exchanger equipment and standards proceeded without mention of these ANSI standards. In the 1979 symposium volume on codes, standards, and specifications for process heat-transfer equipment, the only mention of ANSI standards was in the chairman's opening remarks drawing attention to their obscurity. Chairman Rubin, possessing encyclopaedic knowledge of the field, went on to say that he knew "of no application for either standard by any user or specifier of heat transfer equipment."

Earlier in 1979, TEMA recommended that ANSI B78.2 (1974) be withdrawn and that ANSI B78.1 (1972) be revised to conform with the latest edition of the TEMA standards. This would grace the widely used industry standard with the propriety of an ANSI standard. No doubt politicking and entrenched personal positions will delay this eminently desirable move, but there appears to be an increasing tide in its direction. It is to be hoped that the new edition of the API standards for air-cooled heat exchangers, now imminent, will eventually be adopted as an ANSI standard.

HEI Standards

The Heat Exchange Institute (HEI), based in Cleveland, is an association of manufacturers of heat-exchange and steam-jet vacuum apparatus including condensers, closed feedwater heaters, steam-jet ejectors, and power-plant heat exchangers. This association publishes the 10 standards listed below, all concerned with the engineering design, nomenclature, operation, and use of heat-exchange-related equipment:

1. "Standards for Steam Jet Ejectors," 3d ed., 83 pages, 1967.
2. "Standards for Field Testing. Addendum Standards for Steam Jet Ejectors," 3d ed., 13 pages, 1975.

3. "General Construction Standards for Ejector Components other than Ejector Condensers," 14 pages, 1979.
4. "Construction Standards for Surface Type Condensers for Ejector Service," 18 pages, 1972.
5. "Code for the Measurement of Sound from Steam Jet Ejectors," 16 pages, 1978.
6. "Standards for Direct Contact Barometric and Low Level Condensers," 5th ed., 28 pages, 1970.
7. "Method and Procedure for the Determination of Dissolved Oxygen," 2d ed., 29 pages, 1963.
8. "Standards for Steam Surface Condensers," 7th ed., 65 pages, 1978.
9. "Standard for Power Plant Heat Exchangers," 90 pages, 1980.
10. "Standards for Closed Feedwater Heaters," 3d ed., 33 pages, 1979.

ASHRAE Standards

The American Society of Heating, Refrigerating and Air Conditioning Engineers (ASHRAE), an exceptionally well-organized and prolific engineering society, has published many standards relating to nomenclature, practices, and methods for heating, ventilating, air-conditioning, and refrigeration systems, many of which include heat exchangers. Many of the ASHRAE standards have been adopted as American National Standards. A complete listing of the ASHRAE standards is given in the society's list of publications [ASHRAE, 1980].

An invaluable compilation of the relevant codes and standards of some 50 organizations and agencies is given in the ASHRAE "Guide and Data Book," a four-volume compendium revised in a four-year cycle—one volume per year. Many of the codes and standards listed therein refer to heat exchangers in one way or another.

SUMMARY

We have briefly discussed some of the codes and standards used to regulate and specify the design, manufacture, and operation of heat exchangers. The list is by no means comprehensive; that would require a volume of this size containing nothing else. Furthermore any comments or interpretations that are included represent only a personal view. Those professionally engaged in this field will, of course, need to have access to and familiarity with the source documents, far beyond the introductory account presented here.

REFERENCES

API: "Heat Exchangers for General Refinery Services," API Standard 660, 3d ed., American Petroleum Institute, Washington, D.C., 1976.
API: "Air Cooled Heat Exchangers for General Refinery Services," API Standard 551, 2d ed., American Petroleum Institute, Washington, D.C., 1978.

ASHRAE: Codes and Standards, chap. 45 of "Guide and Data Book," American Society of Heating, Refrigerating and Air Conditioning Engineers, New York, 1978.

ASHRAE: "List of Publications," American Society of Heating, Refrigerating and Air Conditioning Engineers, New York, 1980.

ASME: "ASME Boiler and Pressure Vessel Code," American Society of Mechanical Engineers, New York, 1980.

BSI: "Unfired Fusion Welded Pressure Vessels," BS 5500-1976 (Master Pressure Vessel Standard), British Standards Institution, Hemel Hempstead, Herts, U.K., 1976 (replaces BS 1500, BS 1515, BS 3915, etc.).

BSI: "T.H.E. Publications Catalogue," British Standards Institution, Hemel Hempstead, Herts, U.K., 1979.

Dall'Ora, F.: European Pressure Vessel Codes, *Hydrocarbon Process.*, June, 1971, pp. 93–96. In the same issue, see also: U.S. Standards for Tanks and Pressure Vessels, pp. 96–99; U.S. Pressure Vessel Codes, pp. 99–101; U.S. Heat Exchanger Standards, pp. 105–106; U.K. and German Material Standards, pp. 113–114.

Irving, R. R.: Why the ASME Code is in a Class by Itself, *Iron Age,* January 17, 1977, pp. 2–11.

Khella, A. I.: "Design Criteria of Pressure Vessels in Canada, the United States of America, the United Kingdom, Germany, Japan and France," M. Eng. thesis, University of Calgary, Calgary, Alberta, Canada, 1980.

Rubin, R. (ed.): "The Inter-relationships between Codes, Standards and Customer Specifications for Process Heat Transfer Equipment," conference publication, American Society of Mechanical Engineers, New York, 1979.

TEMA: "Standards of Tubular Exchanger Manufacturers Association," 6th ed., Tubular Exchanger Manufacturers Associations, New York, 1978.

TEMA: "Sample Problem Booklet for Standards of Tubular Exchanger Manufacturers Association," Tubular Exchanger Manufacturers Association, New York, 1980.

Yokell, S.: Design and Fabrication of Heat Exchangers, pp. 139–167, in "Practical Aspects of Heat Transfer," American Institute of Chemical Engineers, New York, 1979.

11

Regenerative heat exchangers

INTRODUCTION

A regenerative heat exchanger consists of a porous mass of finely divided material (wires, balls, pebbles, powders, etc.) through which hot and cold fluids pass periodically and alternately. During the *hot blow,* when hot fluid is passing through the matrix, heat is transferred from the fluid to the matrix. The fluid cools, and the matrix is heated.

Subsequently the flow is switched to the *cold blow,* wherein the flow of hot fluid is replaced by a flow of cold fluid. Heat is now transferred from the matrix to the fluid. The matrix cools, and the fluid is heated.

The matrix therefore acts as a thermal store, alternately accepting heat from the hot fluid, storing that heat, and then releasing it to the cold fluid. A helpful analogy is to think of the matrix as a *thermodynamic sponge* with water as the energy transfers. The sponge is relaxed during the hot blow and absorbs energy. It is squeezed during the cold blow, thereby discharging heat.

This is in contrast to all other types of heat exchangers, which involve energy transfer between fluids and are collectively called recuperative heat exchangers. In these devices separate flow conduits are provided for the hot and cold streams, normally but not necessarily continuous in nature.

CLASSIFICATION OF REGENERATIVE EXCHANGERS

Regenerative heat exchangers may be generally classified in two types: dynamic and static. The two types are illustrated in Fig. 11.1.

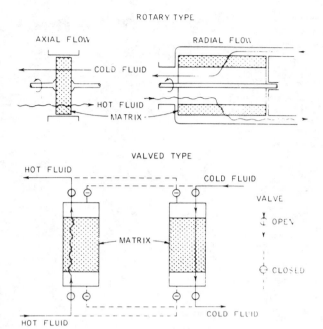

FIGURE 11.1
Static and dynamic types of regenerative heat exchangers. (*After Kays and London [1966].*)

Dynamic Regenerative Exchangers

Dynamic regenerative exchangers have moving parts. In most cases there is a single matrix arranged in the form of a flat disk or hollow drum, as shown in Fig. 11.1 (upper). Elements of the matrix are exposed sequentially to the hot blow and then the cold blow by relative motion of the matrix and fluid ducts. This is accomplished by rotating either the matrix or the fluid headers controlling the fluid distribution to and from the matrix.

The Ljungstom regenerator, shown in Fig. 11.2 and often called the "thermal wheel," is widely used in base-load electric power stations and process industries for inlet-air preheating by hot exhaust gases. In this type of unit, a porous disk matrix rotates slowly, driven by an external power supply, so that any given element of the matrix passes periodically through the hot stream and then through the cold stream.

An alternative arrangement, called the Rothemuhle regenerator, is shown in Fig. 11.3. In this case the disk or drum is stationary, and the headers controlling the fluid flow to and from the matrix are rotated.

Dynamic regenerators have the advantage of a single matrix but offer a severe challenge in the mechanical design of the rotating drives and the fluid seals separating the two streams. This becomes particularly important when there are

FIGURE 11.2
Ljungstrom regenerative heat exchanger. (*Courtesy Combustion Engineering/Air Preheater Corp.*)

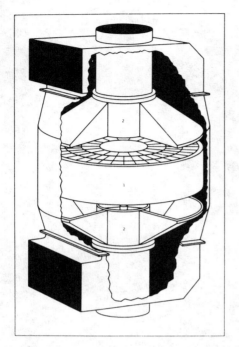

FIGURE 11.3
B&W Rothemuhle regenerative air heater. (*Courtesy Babcock and Wilcox, New Orleans.*)

significant pressure differences between the hot and cold fluids, as in the gas-turbine regenerative inlet-air preheater application.

The fluid-sealing problem is compounded by the effects of repeated heating and cooling of the regenerator matrix. This results in cyclic contraction and expansion of the matrix, depending on the coefficient of thermal expansion of the material involved. In some cases severe distortion may occur or high thermal stresses may be involved. Eventual failure may result from "thermal fatigue," wherein the repeated heating and cooling action causes the matrix elements to become loose and unsecured or, in the case of porous monolithic structures, to fragment.

Static Regenerative Exchangers

Static regenerative exchangers have no moving parts except for the small components used in flow-switching devices. Most applications involve the continuous flow of fluids and so require the provision of dual matrices with periodic flow switching. At any instant one matrix is experiencing the hot blow while the other is in the cold blow. Systems of dual-matrix static regenerators have been used since the early 1920s in steelmaking inlet-air preheaters and large-scale industrial gas liquefaction.

Some applications exist where the flow is discontinuous and a single matrix can be used. Solar heating systems are of this type, as illustrated in Fig. 11.4. Heating is not required when the sun is shining. Therefore a large regenerative store in the basement can be heated by solar energy during the day (the hot blow) and then released heat during the dark hours (the cold blow).

Another application of single-bed regenerators is found in the Stirling engine and similar regenerative machinery used for power production, refrigeration, or cryogenic cooling [Walker, 1980]. The engine consists of a hot space

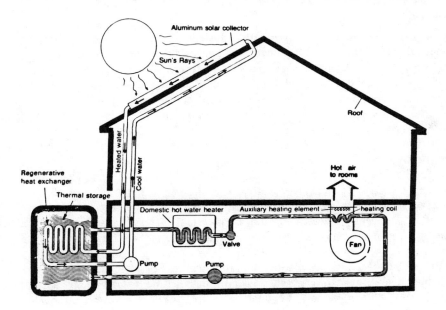

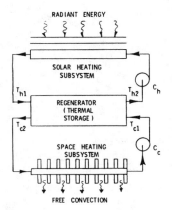

FIGURE 11.4
Solar-heating system incorporating a regenerative heat exchanger. (*After Mondt [1980].*)

and a cold space coupled through three heat exchangers—the heater, cooler, and regenerator. As the piston and displacer move, the fluid in the working space oscillates in a cyclic fashion described as "tidal flow" [Walker, 1981]. The fluid processes involved are very complicated.

ADVANTAGES AND DISADVANTAGES OF REGENERATIVE HEAT EXCHANGERS

Advantages

The principal advantage of regenerative systems is their relative simplicity, particularly those with static-bed matrices. The matrix material can be virtually anything with a high specific heat that can be finely divided to provide a large surface area for heat transfer. A low-cost material can often be found if economy is a principal consideration. In other cases greater expense is justified to allow the use of materials with a higher heat capacity or in finely divided form. Incredibly compact heat exchangers having a very large surface area for heat transfer and exceptional effectiveness values can be provided for situations where space and weight are important considerations.

Another important advantage of regenerative heat exchangers is their self-cleaning ability. In most recuperative exchangers the fluids flow in separate channels and always in the same direction. The inevitable deposits are therefore cumulative and lead to decreased heat transfer and increased pressure drop. In regenerative heat exchangers having the fluids arranged to flow in opposite directions (counterflow operation is preferred for *all* heat exchangers), the matrix is virtually self-cleaning. Any sooty deposits tend to be "blown out" by the succeeding cold blow. In cryogenic systems this feature is particularly advantageous. Any solid or liquid contaminant precipitating in the matrix during the fluid cooling process (the hot blow) will be reevaporated and carried from the matrix after flow switching to the return stream that cools the matrix (the cold blow).

Finally, the fact that only one fluid is entering or leaving the matrix at any one time facilitates the design of the fluid inlet and outlet headers. Systems to separate the fluids into distinct flow paths are simply not required.

Disadvantages

The principal disadvantage of the regenerative heat exchanger is that it is impossible to completely separate the hot and cold fluids. There must always be some mixing of the fluids because the voids in the matrix are filled with one fluid at the instant of flow switching. This void fluid is expelled from the matrix into the downstream flow of the alternate fluid after flow switching has occurred. This is called the *carryover,* and it can become very significant in a large, highly porous matrix with short-duration blow times. The simple fact of inescapable, albeit minor, leakage entirely eliminates consideration of regenerative heat exchangers for the great majority of process heat-transfer applications, where interfluid leakage or contamination cannot be tolerated.

The need for sophisticated fluid seals is another disadvantage limiting the choice of dynamic regenerators, particularly in applications where there is a significant difference in the fluid pressures, e.g., in gas-turbine inlet-air preheaters.

The need for a positive power drive, bearings, and a support system are disadvantages of the dynamic regenerator as compared with the static unit. Sometimes, however, the advantage of compact design with a single matrix having no valves is sufficient to justify the provision of the moving parts.

MATRIX MATERIALS

The material used for the matrix of a regenerative heat exchanger can be virtually anything, provided it has a relatively high heat capacity and can be had in a form that provides plenty of area for heat transfer to and from the fluids passing through. Another important characteristic of the matrix material is thermal conductivity. Ideally there should be a temperature gradient in the matrix along the direction of fluid flow, ranging from a temperature at one end not much *below* the hot-blow inlet temperature to a value at the other end not much *above* the cold-blow inlet temperature. To preserve this gradient, the ideal matrix material should have zero conductivity in the direction of gas flow and infinite heat conduction in a direction perpendicular to the gas flow. Woven wire-mesh screens arranged in stacks with the wire axes perpendicular to the gas flow approximates the ideal condition very closely. Thermal conduction between adjacent screens is minimal because they are in intermittent point or line contact, whereas conduction is maximum along the length of the wires.

Metal spheres or powders are often used for regenerators in Stirling engines, particularly for cryocoolers. Nickel spheres of 0.004 in diameter are common in cryocoolers capable of attaining 80 K. Lead spheres of the same dimension are used for lower temperatures. Cryocooler regenerative heat exchangers are a particularly demanding application, for the specific heat of helium gas *increases* with a decrease in temperature, whereas the available matrix materials decrease in specific heat. This situation becomes acute at temperatures of 10 K or lower and has led to all kinds of esoteric speculation and proposals [Walker, 1981].

For temperatures near ambient, the selection of materials is virtually unlimited. For solar-heating systems, rocks, gravel, and any granular material may be used. Water has a high heat capacity and can serve well as the storage matrix for solar-heating systems, with a set of pipes or tubes through which the heat-transfer fluids pass.

For energy-conservation systems in buildings, Ljungstrom thermal wheels may be made up of low-cost plastic or treated-paper honeycomb materials. Plastic drinking straws with their axes parallel to the flow of fluid have been used in both thermal wheels and fixed-bed regenerators.

For higher-temperature applications in inlet-air preheating systems, regenerator matrices are usually formed from metal strips, rods, wires, or balls and from ceramics. One form of construction for thermal wheels is illustrated in Fig. 11.5. Two strips of metal are helically wound on a former to a disk of the required diameter. At least one of the strips is corrugated or crimped in some way to separate the two strips and provide apertures for the fluid flow. This type of construction is termed "flame trap" after the original application, in large diesel engines and other potentially hazardous or explosive atmospheres, to inhibit the spread of combustion processes.

Ceramics in porous, monolithic form have come to dominate the high-temperature regenerator field for gas turbines. Rotary regenerators with a ceramic honeycomb matrix of lithium aluminosilicate were first used about 1960, but sulfuric acid in the combustion products stimulated a lithium-hydrogen ion exchange resulting in limited life for the core. An aluminous keatite ceramic material, developed to overcome this problem, appears suitable for extremely long lived systems [Grossman and Larning, 1977].

The large dynamic regenerators used in base-load electric power plants as inlet-air preheaters are generally constructed of crimped metal plates contained in a segmental section, as shown in Fig. 11.2. It is likely that the future will see the widespread application of ceramic-matrix regenerators to large systems.

Pulverized brick and other suitable granular masonry or ceramic materials are used for static regenerators operating at high temperatures. Repeated thermal cycling causes eventual fragmentation of many materials, but this only serves to enhance the heat transfer (at the expense of increased pressure drop) until the bed becomes so fragmented that the smallest particles (powder) are carried from the bed, with deleterious consequences for the downstream operation.

Miniature Stirling machines, used for artificial hearts and cryocoolers for infrared night vision and missile thermal tracking systems, use a variety of regenerator systems. The simplest form is the *annular-gap* regenerator illustrated

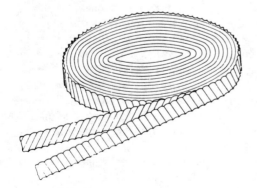

FIGURE 11.5
Flame-trap construction of disk-type axial-flow dynamic heat exchanger. (*After Scott [1966].*)

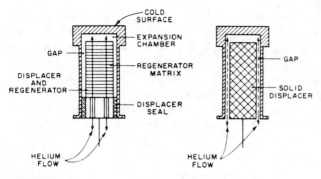

FIGURE 11.6
Comparison of conventional and gap regeneration for miniature cryocooler. (*After Daniels and du Pre [1971].*)

in Fig. 11.6. The right-hand figure illustrates the annular-gap system with a solid displacer reciprocating in a cylinder. The left-hand diagram illustrates the more conventional arrangement with the regenerator located within the moving displacer. Provided the annular gap is small (about 0.02 in), an appreciable measure of regenerative action is achieved from the cylinder and displacer walls adjacent to the surface.

Annular-gap regenerators can be made more effective by the addition of increased surface area in the gap. This may be accomplished with the dimpled-coil foil and straw regenerators shown in Fig. 11.7.

REGENERATOR THEORY

Regenerators are very simple structures compared with recuperative plate-fin or tube-and-shell systems. The simplicity is deceptive, for they are difficult to analyze theoretically. This is because the temperatures of the fluid and the matrix are different from each other and at various points on the matrix. In addition, both the fluid temperature and the gas temperature at all points vary

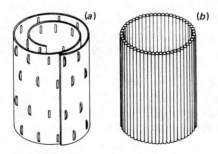

FIGURE 11.7
(*a*) Dimpled-coil foil and (*b*) straw tubular laminar-flow regenerators.

continuously with time in a complex cyclical fashion. Detailed consideration of the theory is beyond our level of interest here. Kays and London [1966] have presented NTU/effectiveness charts and procedures for regenerator design suitable for many purposes. The more substantive theory for regenerators developed in the 1930s by Hausen and other German engineers has been well summarized by Jakob [1957]. An extensive bibliography with a brief commentary on design methods was given by Finegold and Sterrett [1978]. More recently, Mondt [1980] has presented a comprehensive account of many aspects of regenerative heat exchangers.

REGENERATION WITH PHASE CHANGE

In many regenerator applications, the operation proceeds without a change in the phase or state of the fluids or the matrix material. This is not necessarily the case. There are some applications where the fluid wholly or partly condenses in the matrix during the hot blow and reevaporates during the cold blow. This is called a *wet regenerator* process, and it poses highly entertaining mathematical challenges regarding the resolution of its operation.

When water or another more suitable liquid is used as the matrix component of a thermal storage system, it can change phase—freeze—during the cold blow, thereby releasing the latent heat of fusion. The solid melts during the hot blow, and large quantities of energy are absorbed. The system is used to achieve a compact design by utilizing the solid-liquid eutectics having the highest latent heats of fusion near the normal temperature of operation. The water-ice transition is near 0°C, which is not a convenient or comfortable temperature for building heating systems—a great pity, for water has an exceptionally high latent heat of fusion and so is an excellent thermal-storage medium.

FUTURE APPLICATIONS

There appear to be many future applications for regenerative heat exchangers. The greatest opportunities seem to lie in the near-ambient temperature range associated with building energy flows—hotels, restaurants, public buildings, animal houses, greenhouses, etc. There are presently many exhaust systems withdrawing warm, moist air from buildings. This air is replaced by makeup units drawing in ambient-temperature air, which must be heated before distribution. Substantial savings of fuel and utility costs are possible through the use of inlet-air/exhaust-air heat exchangers, either regenerative or recuperative. Similar savings could be gained in refrigerated buildings or cold stores with inlet/exhaust heat exchangers. At these temperatures, low-cost plastic or paper regenerative matrices are entirely feasible, and innovative design and quantity production will surely combine to offer relief from spiraling utility costs. For the same application, low-cost thin metal foil, plastic, and paper recuperative exchangers are also possible, so there will be competition.

SUMMARY

In this chapter we have briefly reviewed the different types and applications of regenerative heat exchangers. There is considerable potential for future applications of these systems as an aid to energy conservation, particularly for the near-ambient temperature conditions associated with building energy flows.

REFERENCES

Daniels, A., and F. K. du Pre: Triple Expansion Stirling Cycle Refrigerator, pp. 178–184, K. Timmerhaus (ed.), "Advances in Cryogenic Engineering," Plenum, New York, 1971.

Finegold, J. G., and R. H. Sterrett: Stirling Engine Regenerator Literature Review, *Jet Prop. Lab Rept. 5030-230,* California Institute of Technology, Pasadena, 1978.

Grossman, D. G., and J. G. Larning: Aluminous Keatite Ceramic Regenerators, *Bull. Am. Ceram. Soc.,* vol. 56, no. 5, pp. 474–477, 1977.

Jakob, M.: "Heat Transfer," vol. II, chap. 35, Wiley, New York, 1957.

Mondt, J. R.: Regenerative Heat Exchangers: Elements of Their Design, Selection and Use, course notes, University of California, Los Angeles, 1980.

Scott, R. B.: "Cryogenic Engineering," Van Nostrand, Princeton, N.J., 1966.

Walker, G.: "Stirling Engines," Oxford University Press, Oxford, 1980.

Walker, G.: "Cryocoolers," Plenum, New York, 1981.

12

Computer analysis of heat exchangers

This chapter, contributed by Andrew Pollard of Queen's University, Kingston, Ontario (formerly of The University of Calgary), outlines the present and future use of digital computers for the design of heat exchangers. Dr. Pollard exemplifies the "new wave" of engineering analysts who are not content simply to mechanize the largely "by gosh and by golly" methods of design as in the presently available commercial proprietary heat-exchanger design programs. Instead he will be satisfied with nothing less than an uncompromising simulation of the system, complete in all thermofluid respects. At present such complete simulation is not possible. However, given the present rate of development of large-capacity computing machines and the numerical processing techniques, its accomplishment lies in the near rather than the distant future. I am most grateful to Dr. Pollard for his enhancement of my own effort.

INTRODUCTION

In Chap. 2 the basic heat-transfer mechanisms associated with heat exchangers were outlined; in Chap. 5 it was shown, in a simple manner, how the principles of the earlier chapter are applied to the design of a two-tube, single-shell heat exchanger. Of course, information of a detailed nature is required to design a heat exchanger that meets a user's requirements. For example, the usual design considerations require the evaluation of the effects of fluid phase change, fouling, variable properties of both tube-side and shell-side fluids, flow maldistribution,

The author expresses his gratitude to Karen Undseth for her excellent typing. Also, thanks go to the University of Calgary for financial support under grant 69-2531. Finally, to G.W., a thank-you for allowing me to contribute this modest effort.

and baffle arrangement on heat-exchanger performance; and there are many other factors to be evaluated. To overcome the tedious trial-and-error solutions necessary for successful design of a heat exchanger, designers have resorted to computers to solve the myriad interlinkages among the various equations.

There are two ways of approaching the final design of a heat exchanger: the integrated approach and the differential approach. The differential approach examines the total exchanger by way of solving the basic equations of fluid mechanics and heat transfer. The integrated approach examines the exchanger in view of overall performance with respect to pressure drop and heat duty rating. The differential method provides details of local variations of fluid velocity, fluid phase distributions, and the like; the integrated approach, while taking local variations into account, provides global information only.

In this chapter, these two approaches will be compared by examining the methodologies used, with particular reference to shell-and-tube heat exchangers. This comparison is not exhaustive; it provides an overview of two ways of designing a heat exchanger by computer analysis utilizing numerical techniques still at an early developmental stage.

Before we proceed, a cautionary note must be made: A computer program is only as good as the physical models it employs, and therefore *the results of computer analysis are not always correct or practical;* good engineering judgment plays a vital role in interpreting the numbers generated by computer analysis.

THE INTEGRATED APPROACH

The integrated approach will be examined first because its methodology is easier to comprehend; indeed, the methodology is basically that given in Chap. 5 of this book.

The integrated approach is supported by various organizations, examples of which are the Heat Transfer and Fluid Flow Services (HTFS) of the United Kingdom and the Heat Transfer Research Institute (HTRI) of the United States. These organizations provide computer programs to dues-paying members (usually companies). The result is that these programs are proprietary, and detailed information about their construction is scarce. The information presented here has been gleaned from the open literature, brochures, etc., and from talking with personnel of companies using codes that originate with HTFS, HTRI, and others.

Some Background Information

The design of heat exchangers using an integrated-approach computer program normally contains two major options: "design" and "performance." The design option provides the physical specifications for a heat exchanger that matches a user-specified heat duty rating. The performance option calculates the heat duty rating for a geometrically specified heat exchanger.

In the preliminary stages of choosing a heat exchanger, the designer progresses through various decision-making processes; these take into account cost, space limitations, and reliability requirements. Here we presume that a shell-and-tube heat exchanger has been chosen. The next step, even before one thinks about approaching a computer, is to ascertain what limits the physical design; fouling or corrosion problems with one stream, pressure-drop limitations, and the sort of temperature differences that are required (if known) are examples of factors that limit the final choice. Additionally, the development cost of the exchanger and the conditions under which it will operate usually influence the final arrangement. A good way of getting first-round (hand-calculation) estimates that aid in heat-exchanger selection is given by Bell [1978]; other methods are also available.

The performance option is the most commonly used portion of a computer program, the input being the results of the designer's conclusion about the initial physical design. Once the heat duty rating, etc., have been determined (an overview of this process will be given shortly), a design-option program may be used to see if the physical layout can be improved; this alternation between the two options can continue until a decision is made to go with a particular exchanger.

Performance Calculations

The performance option in most heat-exchanger computer programs is based on (or is some variant of) the Delaware procedure; the most recent progress report on this procedure can be found in Bell [1979]. The Delaware procedure resulted from an exhaustive experimental program that began by determining pressure-drop and heat-transfer data during forced convection over banks of tubes; over a period of 12 years these measurements were extended to include variations in tube and baffle arrangements, baffle leakage rates, etc. The result is a comprehensive set of correlations that relate heat transfer, pressure drop, leakage rates, etc., to specific geometric configurations through the use of correction factors that are applied against ideal flow situations. These correlations, together with Tinker's diagram for shellside flow-stream distribution [Tinker, 1951] play a pivotal role in any integrated-approach type computer program.

Utilizing the above, the performance calculation starts from given process, fluid-property, and geometry information to obtain the length of the exchanger if the heat duty is not known, the heat duty if the length is known, and the pressure drops across various sections within the exchanger. The process information required usually falls into the following categories: inlet and outlet fluid flow rates, at least one set of stream temperatures, and the fouling characteristics of the fluids (i.e., an estimate of the heat-transfer resistance due to fouling). The physical-property information includes fluid densities, viscosities, and specific heats. The geometry information includes shell diameter, outer-tube-diameter limit, tube layout, baffle spacing and type, and, if necessary, the length of the tubes. It follows from this that usually a shell-and-tube exchanger type can be identified before the performance calculation is performed. Thus, many of the available codes permit various types of exchangers to be rated by specifying the three-letter type.

The rating program can, of course, be rerun to account for the effects of time aging of the surfaces. The result allows the designer to evaluate the exchanger's performance from the as-new condition to, say, the point where the heat transfer becomes unacceptably low due to fouling, corrosion, etc.; this feature highlights one of the most important attributes of computer design, for it permits considerable parametric study to be performed rapidly and cheaply. Of course, the results must be interpreted critically with a view toward ensuring that the correlations employed in the program have not been extended beyond their range of validity.

As was noted above, the performance evaluation of an exchanger usually goes hand in hand with the design option of the program. It is to this latter option that we now turn our attention.

Design Calculations

The design option, within which the performance option can be included, has as its task the selection of a set of geometric parameters that are optimal for the required duty. This option typically starts with the results of the performance option. Then, that which limited the performance must be first determined; furthermore, a method of eliminating this restriction must be found during this process. Of course, little alteration should occur to those characteristics that have been isolated by the designer as being satisfactory.

The design-option programming is very complex, so as to permit a wide variety of adjustments. Indeed, the scope permitted by some programs increases the diligence required of the user; variety introduces possible program logic errors. Consider the following example, which indicates a possible decision-making process for a fixed-duty exchanger.

First, the output of the rating portion of a program provides the required length of the exchanger, the number of tube passes, the allowable baffle spacing, etc. The first decision is to determine if the length calculated is less than the maximum allowable. If the answer is yes, a decision is then made as to whether the tube pressure drop and shell-side pressure drop are greater than the maximum allowable. If so, then a possible solution would be to add another exchanger in parallel to the original; this is perfectly acceptable but may be totally impractical.

THE DIFFERENTIAL APPROACH

The differential approach to heat-exchanger design is more recent than the integrated approach. The methods used in the integrated approach have been available to the heat-exchanger community since the introduction of Kern's classic textbook; a computer wasn't really needed. The differential approach, however, could not even be contemplated until the advent of large digital computers. The reason, as will be made evident shortly, is that there are far too many mathematical operations to perform. Moreover, the advent of the computer required that the heat-transfer specialist devise efficient methods of solving and evaluating complex mathematical relationships; it is only in the last few years that these methods have become an acceptable tool of the fluid-flow and heat-transfer specialist. The reason for the time lag is the need for evaluating and severely testing the computational methods and mathematical models that are used to represent phenomena observed in nature; an example of the latter is turbulence.

In this section, the basic ideas that are embodied in the differential approach are outlined. Detail, as in the previous section, will be kept to a bare minimum. Those wanting details should refer to the growing number of publications in the field of numerical fluid flow and heat transfer, examples of which are to be found in Patankar and Spalding [1971, 1974], Marchand et al. [1980], and Patankar [1980], including the references contained therein.

Some Background Information

In the introduction to this chapter the differential approach was characterized by the need to solve the basic equations governing fluid flow and heat transfer. These equations govern the conservation of mass, momentum, and energy in three-dimensional space and time. Each of these equations has components that transport these quantities via convection and diffusion and includes also provisions for sources or sinks of these three quantities. Moreover, the number of equations for, say, flow that is two-phase in character (i.e., air/water) is normally twice that for describing single-phase flow (i.e., water). The interlinkages that must exist between, say, air bubbles surrounded by water, are ascribed through the source and sink terms of the equations; thus, if the air is higher in temperature than the water, energy loss by the air (a sink term) would be matched by an equal energy gain by the water (a source term). Similar source-sink interactions can be thought of by the reader.

The basic equations that govern any transport process can be cast into the following vector form:

$$\frac{\partial}{\partial t}(\rho\phi) + \text{div }(\rho\mathbf{V}\phi) = \text{div }(\Gamma \text{ grad } \phi) + S_\phi$$

where ρ is the fluid density, $\mathbf{V}$ is the local velocity vector, Γ is a coefficient that governs the transport of a quantity ϕ by diffusion, and S_ϕ is the source-sink term. ϕ can stand for any of three velocity components, temperature, concentration, and the like; furthermore, if $\phi = 1$ and with the right-hand side of the equation being normally zero, the resulting equation governs the conservation of mass. Note that for momentum, the source term contains the pressure gradient. The above equation is valid for single-phase flows; by introducing the notion of a volume fraction, the transport processes of two-phase or even multiphase flows can be described by the resulting set of equations. Furthermore, porosity effects such as the effect of fluid flowing through a tube support can be included by replacing $\rho\mathbf{V}$ by $\beta\rho\mathbf{V}$ in the above equation; $\beta\rho\mathbf{V}$ can then be regarded as the mass flux vector through an unblocked portion of the tube support.

Attention will now be directed toward how the equations are to be embodied into a computer program, and the basic features this program must contain such that it will permit exchangers, among other things, to be designed.

Program Format

In the section dealing with the integrated approach to heat-exchanger design, it was noted that a large variety of standard heat-exchanger geometric arrangements can be accommodated; these are usually based on the standard types. Moreover, the effects of different flow regimes and the like are accounted for by correlation equations derived from experiments. In many instances, the designer may wish to depart from accepted designs and limited correlations. The integrated approach therefore has little to help the designer in this regard. The differential approach, if it is to succeed as a viable method, must permit wide flexibility at costs that are at least comparable to those associated with the integrated approach, and the differential method must be of a form that requires a minimum amount of training of those that are to use it. Unfortunately, both the cost and training factors are at present larger than those associated with the integrated approach; this aspect is discouraging, but cognizance must be taken of the fact that the differential approach is in its infancy.

The features of a computer program that provide the machinery for solving equations such as those presented in the previous section are:

1. A discretization technique that permits the equations to be linked to various subdomains within a particular geometry; these subdomains can be thought of as cells of which there are a finite number.
2. The conservation principles for mass, momentum, energy, etc., must be applied to these cells. Each cell must be connected to its neighbor by rules for interpolation between the values associated with the two cells; the result is a set of algebraic equations or relations between one cell and nearby neighbors. These equations can be referred to as "finite-difference" or "finite-element" equations.
3. The program must feature a provision for the introduction of auxiliary equations, such as those that link density changes with temperature and pressure, and link the equations with other effects such as turbulence. Furthermore, a scheme must be devised to link conservation of mass with momentum in a manner that will ensure satisfaction of these equations in a consistent manner.

4. The program must feature a provision for accommodating complex geometries. This is an obvious need for heat exchangers; however, it is usually only the external shell geometry that must be accounted for. The internal structure can be accommodated in all its complexity by using the distributed-resistance concept that is fundamental to the differential approach. The details of this concept will be discussed shortly.

The above features, when incorporated correctly into a computer program, result in a powerful tool. Open-literature examples of the non-heat-exchanger use of such programs for predicting complex fluid-flow and heat-transfer phenomena can be found in Markatos [1977], Pollard and Spalding [1979] and Serag-Eldin and Spalding [1979].

The Distributed-Resistance Concept

The concept of a distributed resistance is not new; it has been used widely in the fluid mechanics of porous media—that is, a solid material that permits movement of a fluid through it by virtue of the existence of small, interconnected evacuated spaces. A good example of this type of material is a sintered plate. Now, if a sintered plate (or disk) is placed inside a tube, normal to the fluid flow, such that there are no gaps between the edges of the disk and the tube, the disk reduces but still permits fluid flow for the same pressure gradient. The internal piping and baffle network inside the shell of a heat exchanger can be thought of as achieving the same effect as the sintered disk; it resists the movement of fluid between the shell inlet and outlet ports, thereby influencing the local velocity distribution within the exchanger.

This is a brief outline of the basic idea behind distributed resistances. The question remains as to how this is coupled with the equations and the computer program. The answer is given below.

Introduction of Distributed Resistances into a Computer Program

The best way of understanding what follows is to consider a particular example—a shell that is square in cross section, and whose length is longer than its sides. Now, by use of the subdomain idea introduced earlier, this shell is discretized into a large number of small volumes, such that there are 10 (say) in each of the three coordinate directions. One thousand control volumes is within the realm of computational efficiency; any more results in costs that can be considered uneconomical. We now introduce a large number of tubes that run parallel with the long axis of the shell. These tubes cross the control volumes' faces. At each control-volume face there exists but one velocity component, and associated with each control volume there is only one value of temperature, pressure, etc.; this is by virtue of the finite-difference or finite-element formulation noted earlier. With the introduction of so many tubes, much averaging of the effects introduced by the tubes must take place.

To account for the physical presence of the tubes in each control volume, it is best to consider them as occupying a certain fraction of the control volume, thereby reducing the effective area through which the fluid will flow. This is akin to introducing the $\beta\rho V$ term noted previously. It has the effect (for constant-density fluid) of maintaining a constant mass flux through the volume; thus an increase in blockage will increase the velocity, since the area available to the fluid is reduced.

To account for momentum influences, whereby an increase in blockage will tend to decrease the velocity due to higher frictional forces, the notion of sources and sinks of momentum is introduced. These are chosen so as to represent the best available knowledge. Normally these sinks of momentum take the form of friction coefficients.

With respect to heat transfer, the shell-side fluid accounts for the tube heat transfer in a manner similar to that described for the momentum sources and sinks. The tubes, however, are treated differently. Since the tubes interact only with the shell-side fluid, and details of fluid flow within the tubes are not really necessary, the tube flow and heat transfer can be considered separately. This is done by solving (in the example considered here) an energy equation in one-dimensional form. This equation contains the usual provision for sources and sinks of energy that take cognizance of the shell-side temperature and the local tube fluid temperature. The result is that the shell-side equations are influenced by the tube temperature, and vice versa.

SUMMARY

In this chapter two ways in which computers can be used to aid in the design of heat exchangers, and more specifically shell-and-tube heat exchangers, have been outlined.

A brief outline of the integrated approach to performance and design calculations for heat exchangers was first presented. There can be no doubt that these programs can aid in the process of arriving at a suitable exchanger; the most useful of the two, I believe, is the performance option. The programs contain correlation equations that have been found to apply over a wide range of flow conditions and in many types of geometric arrangements. The design option can be invoked either through experience, hand calculation or computer programs (see for example, Bell [1980]).

In the differential approach to heat-exchanger design, the design aspect is totally user-controlled. Performance calculations can be handled, but the user requires intimate knowledge of the program's construction to introduce the desired changes. This is a serious disadvantage; nevertheless commercial groups are becoming established that provide services similar to that of HTFS and HTRI.

It is not expected that the brief overview given here will satisfy the reader's curiosity regarding the differential approach to heat-exchanger design. Little information is at present available that has utilized the distributed-resistance concept. A good review of the status of this concept can be found in a recent publication by Spalding [1980].

What has become apparent is that computer programs are not a panacea for heat-exchanger evaluation. They are to be used only as tools, and, as with all tools, they can be poorly utilized. But, on a positive note, the tools are continually being refined so that the future of computers in heat-exchanger analysis looks very promising indeed.

REFERENCES

Bell, K. J.: Estimate S & T Exchanger Design Fast, *Oil Gas J.,* December 4, 1978, pp. 59–68.

Bell, K. J.: "The Delaware Method for Shellside Thermal-Hydraulic Analysis," D. Q. Kern Award Lecture, 18th National Heat Transfer Conference, San Diego, August, 1979.

Bell, K. J.: Process Heat Transfer Seminar notes, Albuquerque, New Mexico, September, 1980.

Marchand, E. O., A. K. Singhal, and D. B. Spalding: Predictions of Operation Transients for a Steam Generator of a PWR Nuclear Power System, ASME paper 80-C2/NE-5, 1980.

Markatos, N. C. G.: A Theoretical Investigation of Buoyancy-Induced Flow Stratification in the Cylindrical Outlet Plenum of a Liquid-Metal-Cooled Fast Breeder Reactor in Heat Transfer and Turbulent Buoyant Convection, in Spalding and Afgan (eds.), vol. II, Hemisphere, Washington, D.C., 1977.

Patankar, S. V., and D. B. Spalding: Computer Analysis of the Three Dimensional Flow and Heat Transfer in a Steam Generator, *Forsch. Ingenieurwes.,* vol. 44, no. 2, 1971.

Patankar, S. V., and D. B. Spalding: A Calculation Procedure for the Transient and Steady State Behaviour of Shell and Tube Heat Exchangers, in Afgan and Schlunder (eds.), "Heat Exchangers: Design and Theory Sourcebook," Hemisphere, Washington, D.C., 1974.

Patankar, S. V.: "Numerical Heat Transfer and Fluid Flow," Hemisphere, Washington, D.C., 1980.

Pollard, A., and D. B. Spalding: Turbulent Flow and Heat Transfer in a Tee-Junction, ASME paper 70-WA/HT-47, 1979.

Serag-Eldin, M. A., and D. B. Spalding: Computations of Three-Dimensional Gas Turbine Combustion Chamber Flows, *J. Eng. Power,* vol. 101, no. 3, pp. 326–336, 1979.

Spalding, D. B.: Multiphase Flow Prediction in Power-System Equipment and Components, *Int. J. Multiphase Flow,* vol. 6, pp. 157–168, 1980.

Tinker, T.: Shellside Characteristics of Shell and Tube Heat Exchangers, parts I, II, III, Proceedings General Discussion Heat Transfer, I Mech., London, 1951.

13

Heat sinks, sources, and storage

HEAT SINKS

Introduction

A heat sink is, literally, any system, device, process, or procedure for disposing of waste heat derived from any source. In many instances, the heat is not useful for energy conversion, being available only intermittently, at regular or random intervals, or at a temperature that is too low or the amount is too small to be worth attempting at utilize.

The term *heat sink* has come to be particularly applied to those devices and measures adopted to dissipate the heat generated in electrically powered components or systems.

Electrical equipment generally operates with the flow of electrons along an electrically conducting path. In some ways the process corresponds to the flow of water along a pipe. The electrical voltage V corresponds to the pressure of water in the pipe, the current I corresponds to quantity of water flow, and the electrical resistance R corresponds to the frictional resistance of the water in the pipe.

According to Ohm's Law,

$$R = E/I$$

The electrical power Q, lost by the resistance heating resulting from the flow of electrons along the wire or conducting path, is

$$Q = I^2R$$

This power loss is manifest as heat produced internally in the wire or in the conducting path.

In electric water heaters or domestic electric heaters, this resistance heating is useful, the very reason for using such equipment. We shall look at that in more detail in the next section.

In most other electrical applications, the resistance heating is not useful and is minimized where possible by the use of electrical conductors that have a low resistance.

Some resistance heating of the equipment is inevitable, however. If the heat is not dissipated in some way, then the temperature of the equipment will increase above the safe level with consequent reduction in the performance, life, and reliability.

Heat sinks are therefore devices added to electrically powered equipment to facilitate dissipation of the resistance heating so as to maintain the temperature of the device at a safe operating level.

Heat sinks take a variety of forms but generally may be divided into air cooled or water cooled systems.

Cooling Electronic Equipment

The invention of the transistor after the Second World War revolutionized the "light current" electrical engineering. This has continued to the present situation where the world seems full of electronic equipment doing all kinds of measuring, controlling, regulating, processing, computing, and playing records, tapes, television, etc.

Transistors and other semiconductor devices are composed of silicon or germanium chips through which current passes. The junction where current control occurs is the place where the greatest heating takes place so that this junction temperature is the critical factor regulating the design of the heat sink.

A rule of thumb in the trade is that every reduction of 10 °C in the device junction temperature doubles the operating life of the device.

Many heat sinks for electronic units take the form of some type of fin to provide an extended surface for convective air cooling [see Lorenzetti, 1988].

Convective air cooling follows the equation

$$Q_{conv} = hA(\Delta T)$$

where Q_{conv} = rate of heat transfer from the heat sink finned surfaces
h = heat transfer coefficient from the fin to the air

A = area of the fin for heat transfer

ΔT = temperature difference between the fin and the cooling air

It is clear from this simple equation that to minimize the junction temperature of the electronic chip it is necessary to

1. Arrange the thermal conduction path between the transistor silicon chip and the heat sink to be as short as possible with minimal thermal resistance
2. Use the lowest temperature of cooling air possible
3. Seek to maximize the convective heat transfer coefficient
4. Provide the maximum effective area of the fin.

In many cases the heat sink is (1) an extruded aluminum section of complicated shape, (2) a flat metal sheet formed by a slumping or forging process, or (3) a fabricated fin section formed by securing fins to a base plate by using welding, brazing, soldering, or conductive epoxy cement.

Sometimes the heat sink unit itself can serve as the base to which the silicon chips are firmly secured by using metal screws, rivets, or solder to secure a good conducting path. In most instances, however, the heat sink is an add-on unit.

Frequently, the silicon chips are mounted on a plastic composition board of low thermal conductivity. This interposes a high thermal resistance between the cooling unit (the heat sink) and the units to be cooled, a manifestly self-defeating process.

There is very little room for maneuver in the matter of optimizing the heat transfer coefficient h. There are three basic types of convective heat transfer: natural convection, forced convection, and mixed natural/forced convection in which the forced convection aids or opposes the natural convection.

These processes are discussed in some detail in Chap. 2. It is sufficient here to recall the heat transfer coefficient h for natural convection in air is customarily in the range 5–25 W/(m² °C) whereas with forced convection it is customarily in the range 20–500 W/(m² °C).

In normal operation the electrical device and heat sink will be at a temperature higher than the air available for cooling so that the direction of the air flow away from the heat sink fin will be vertically upward.

To facilitate the air motion and hence increase the heat transfer coefficient, the heat sink should therefore be positioned with the fins standing vertically as in Fig. 13.1a with open slots at the bottom.

This is considerably more effective than the alternative shown in Fig. 13.1b. Although in this case the fins are vertical, access to them by the air is inhibited by the solid base.

The least effective orientation for natural convective cooling is to locate the fins longitudinally, one upon the other. There is then virtually no air flow in the interstices between the fins.

Sometimes it is necessary or at least judged worthwhile to gain extra cooling by accepting the cost and complexity of adding a fan. This blows air over the heat

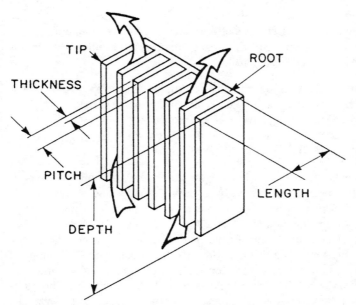

a) VERTICAL FIN HEAT SINK (SIDE MOUNTING)

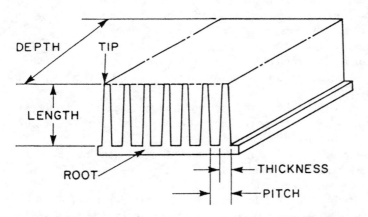

b) VERTICAL FIN HEAT SINK (BASE MOUNTING)

FIGURE 13.1
Fin nomenclature.

sink so that the cooling process is forced convection rather than natural convection. More heat is thereby transferred from the heat sink to the air.

Increase in the rate of heat transfer will be substantial and will increase as the velocity of the air flow increases, more or less in proportion to the square root of the velocity increase (see Chap. 2. Increasing the velocity of the fluid flow increases the Reynolds number of the flow).

If the fan is small and the forced air velocity is low (say, less than 0.5 m per s) the intrinsic natural convection heat transfer process will still likely be significant. In such cases it is important to arrange that the fan blows the air over the fins in a vertical and upward direction so the forced and natural convection heat transfer processes are mutually complementary.

In the unhappy situation that the fan blows the air vertically downward so as to oppose the vertical upward flow of the air in natural convection it could well be that the addition of the fan actually decreases the cooling effect of the heat sink.

Fin design is more of an art than a science; there is no one best shape or size of fin. Some customary fin nomenclature is included in Fig. 13.1. The base of the fin is called the *root*. The distance the fin extends outward from the root is called the *length*. The distance between two adjacent fins is called the *pitch*. Other parameters of interest are the *thickness* of the fin and the *depth*.

The fins may be tapered, being narrower at the *tip* and thicker at the root, as shown in Fig. 13.1*b*. This is easy to achieve with an extruded section. Fabricated fin stock usually has parallel fin (same thickness along entire length) as shown in Fig. 13.1*a*.

The heat sink is usually firmly secured by screws and washers of low thermal resistance to the base or platform that is to be cooled. The root of the fin will then be at more or less the same temperature as the platform or base.

All the heat to be dissipated will enter the heat sink through the thermal contacts of the heat sink to the base or platform. As the heat is dissipated from the fin the temperature of the fin will decrease progressively from a maximum at the root to a minimum at the tip. It can be shown by using fin theory (see the standard heat transfer texts listed in the reference list) that the temperature decrease is of exponential form.

Because of this decrease in temperature, the ΔT between the fin and air decreases along the length of the fin. The effectiveness of the fin to transfer heat is therefore less at the tip than at the root. Moreover, there is some thermal resistance to heat flow along the fin by conduction heat transfer.

Thus, to maximize the heat transferred from a heat sink, it is necessary to

1. Attach the heat sink to the base or platform in such a way as to minimize the thermal interface resistance to heat flow
2. Use material for the heat sink with a high coefficient of thermal conductivity so as to reduce the resistance to heat flow along the length of the fin
3. Limit the length of the fin because if it is too long then the area near the tip is

ineffective for heat transfer; increasing the thickness of the fin will decrease the resistance to heat flow
4. Have sufficient space between the fins to allow free passage of the cooling air

In heat sink design the optimum solution is therefore a complicated compromise affecting the cost, size, weight, volume, and effectiveness of the unit.

Cold Electronics

Many other advantages accrue from reducing the operating temperature of semiconductor devices: increased speed of operation, reduced "noise," and generally better operation.

Some devices, such as infrared detectors for night vision equipment or missile guidance systems require cooling to far lower temperatures; 80 K is a common temperature for operating infrared systems. This is because over the past 40 years of their development, liquid nitrogen (boiling point 77 K) was used in research on infrared devices. As a result, many of these are optimized to work best at the temperature of liquid nitrogen. Liquid nitrogen is a relatively safe, inert fluid to use that is readily available at low cost, even cheaper than milk in most places in North America.

Other semiconductor electronic devices have been shown to operate more effectively at lower temperatures, although in many cases it is rarely necessary or worthwhile going below 80 K.

The rapid and progressive development of electronic systems with continuous reduction in the physical size of component and the increasing power density poses urgent and difficult cooling problems. These problems are most acute in the field of high capacity computers. In one supercomputer, the main "board" is less than 1 m square but is so packed with large-scale integrated circuits or silicon chips that although each produces only milliwatts of resistance heat in aggregate the total heating load is nearly 1 kW.

This is cooled very effectively by plunging the whole unit in a bath of liquid nitrogen and boiling the nitrogen away, either releasing it to the atmosphere or recondensing the vapor with a cryogenic refrigerator.

Recent developments in high temperature superconducting materials have enormously elevated the interest and development potential of superconducting (as opposed to semiconducting) electronic devices.

The phenomenon of superconductivity (wherein a material loses all its electrical resistance) has already been demonstrated at temperatures greater than 90 K. This is a staggering advance over the previous 20 years when the upper limit of superconductivity appeared limited to about 20 K. This necessitated liquid helium cooling, a much more difficult and demanding process than the liquid nitrogen cooling, which appears viable with the new superconducting materials.

As a result there is considerable interest in whole new fields of technology generically called "cold electronics."

Although liquid nitrogen is cheap, readily available, safe, and easy to use, it would, in many situations, be very much more convenient if there were miniature cryogenic refrigerators available. These might be small compact reliable systems, electrically powered, integrated with the cryogenic chip or sensor so as to be simply plugged in to ambient temperature circuits. The present status and some future possibilities for these miniature units have recently been reviewed by Walker [1989].

HEAT SOURCES

Introduction

A heat source, in a literal sense, is any form of energy source providing energy in the form of heat, usually (but not always) at temperatures greater than ambient.

There are many forms and varieties of heat source:

1. The heat contained in any thermal effluent (i.e., hot water, hot gas, etc.) going to waste from an industrial or commercial plant.
2. Heat generated by combustion of coal, oil, gas, wood, or other biomass.
3. Heat produced by collecting and concentrating solar energy in a solar collector/concentrator.
4. Geothermal heat, using the hot water and steam available below the surface of the earth.
5. Radioisotope or nuclear reactor heat produced by the radioactive decay of certain materials to produce highly energetic particles that heat the dense material through which they attempt to pass.
6. Electric heat.

Obeying the injunction "brevity above all," we consider only this last source.

Electric powered heating is widely used in industry, commerce, and domestically because of its convenience, reliability, cleanliness, and, for some processes and applications, its unique characteristics.

Electricity is not cheap; indeed, it is among the most expensive ways to purchase heating. This high cost of electricity is not simply a price the market will withstand but is intrinsic to the whole energy situation.

Electricity is produced in large central power stations where fossil fuel, coal, oil, gas, or a nuclear reactor is used to provide heat to raise high pressure. High temperature steam that expands in a steam turbine to drive the electric generator produces the electric power.

The Second Law of Thermodynamics precludes the possibility that all of the heat produced by burning the fuel can be converted to electricity. Most of it must be rejected as low grade waste heat. In practice we are lucky to convert as much as

40 percent of the energy released in burning a ton of coal to electric power at the power station. Usually it is nearer to 30 percent.

Then, a good deal of that power is lost in distributing the power to factories, homes, offices, hotels, etc. Thus, the amount of electric energy supplied to the consumer is generally only about 20 percent of the amount of energy released in burning the fuel. When electricity costs two or three times as much as oil or coal, we should therefore be thankful the cost is not five times.

The reason it is not higher is that the costs of distribution of coal, gas, and oil in small amounts to consumers make the costs of these fuels very much higher than when they are supplied in vast quantities to a power station.

The price difference between electricity and fossil fuel is therefore reduced to only two or three times as much, a premium that consumers (industrial, commercial, and domestic) appear willing to pay so as to gain the advantage of convenience.

Resistance Heating

Resistance heating is by far the most common form of electric powered heating.

Resistance heating arises simply as a result of applying a voltage across the two ends of a resistive element causing a current to flow. The same I^2R loss we looked at in the previous section, causing so much trouble that it requires the provision of a heat sink for its dissipation, is now the friend and ally we utilize to provide local heating.

Resistance heating is used for a tremendous variety of purposes ranging from defroster elements in domestic refrigerators to industrial furnace heating elements operating at the highest temperatures (1600 °C and above).

Resistance heaters come in all shapes and sizes. By definition they include some resistance heating element contained within, or upon, some supporting structure that allows the resistance to be located safely and, when properly used, without hazard to life or property [see Watson, 1988].

The materials used for the resistance element may be broadly classified into two groups: (1) oxidation resistant materials and (2) nonoxidation resistant materials.

Oxidation resistant materials can be used in air without protection. They include metals, ceramic-metals (cermets), and ceramics.

Metal oxidation resistant elements are the most widely used for electric resistance heating. They include the nickel-chrome alloys, widely used in North America, and the iron-chrome-aluminum alloys, widely used in Europe and in many other parts of the world. For high temperature (1600 °C) heating, platinum and platinum alloys are sometimes used.

Ceramic metals, cermets, are materials that exhibit properties of both ceramics and metals. One that is sometimes used for electric resistance heating is molybdenum disilicide, widely valued for its high oxidation resistance. It can be used for element temperatures up 1800 °C.

Silicon carbide is a ceramic material sometimes used for resistance heating elements principally because of its high compressive strength at elevated temperatures. Lanthanum chromite and zirconia are other ceramic materials sometimes used for electric resistance heating.

Ceramic heating elements generally require the provision of a special power supply to regulate the voltage supplied as a function of the temperature and time. In their use, users should always be guided by the manufacturers and distributors.

Nonoxidation resistance materials used for resistance heating may be classified as metallic or nonmetallic materials. The metals include molybdenum, tungsten, and tantalum.

These metals have a high melting point and so may be used at high temperatures, as high as 2500 °C for tungsten, 2200 °C for tantalum, and 1900 °C for molybdenum. They must be protected from oxidation by operation in vacuum or a protective atmosphere. Tantalum is sensitive to gases containing oxygen, nitrogen, hydrogen, and carbon, but the other two may be used in almost any atmosphere that protects from oxidizing.

The ordinary filament electric light bulb is the most common form of electric resistance heating unit. These have a tungsten resistance element operating in a vacuum environment while carried on relatively fragile supports. The tungsten wire is heated to such a high temperature that the heat is used primarily for the bright light it emits rather than the 100 watts or so of heat produced by the bulb.

Graphite is a nonmetallic, nonoxidation-resistant material sometimes used for electric resistance elements. In a nitrogen, hydrogen, or rare gas atmosphere, carbon elements may operate at temperatures as high as 3000 °C and in a vacuum to temperatures up to 2200 °C.

All these materials find embodiment in a wide variety of electric resistance heaters [see Melly, 1988]. The simplest are the *open coil* or *ribbon heaters*. These usually take the form of an exposed resistance element of resistance wire coil or ribbon wound onto a ceramic rod or fitted into recesses in a flat ceramic plate. They are widely used for radiative and convective heating.

Cartridge heaters are generally cylindrical and consist of a resistance element wound on a ceramic core inside a protective metal tube filled with magnesium oxide to provide electrical insulation and a heat transfer medium. Cartridge heaters are used for local heating in all kinds of industrial equipment and in a wide variety of shapes and sizes.

Mineral insulated cable heaters are formed by encasing resistance heaters in a metal sheath with mineral insulation between the central heater and the sheath. In small diameters the cables may be bent or formed into virtually any shape and are widely used for many purposes.

Tubular heaters are somewhat similar to cartridge heaters and may be found with circular cross section or as flat tubular heaters containing as many as three separate parallel heating elements. Sometimes fins are swaged onto the outer sheath for the heater to be used as a perimeter heating for space heating purposes.

Tubular heaters are often used as immersion heaters in tanks containing water or other liquids.

Flexible heaters consist of a resistance element of wire coil or foil held between two covers of electrically insulating material, often silicone rubber, mica, or one of the flexible plastics materials. They are taped, glued, or clamped to surfaces to provide local surface heating generally at very moderate temperatures.

Electrical heating tape of similar construction is used for wrapping exposed piping and other equipment needing some local heating to prevent freeze up.

Ceramic-fiber heaters consist of a resistance wire embedded in the face of a moulded rigid ceramic fiber body. The element may be fully embedded or partially exposed. In some cases a surface coating, either for protection or decoration, is added.

In brief, electric resistance heaters are very widely used in a broad range of sizes, shapes, configurations, and applications.

Dielectric and Microwave Heating

When an electrically insulating (dielectric) material is exposed to a rapidly changing (1 MHz or higher frequency) electric field, dielectric or microwave heating occurs [see Moffit, 1988].

The heat is generated by molecular vibration directly within the work material. This allows very precise control of the amount of heat, and it tends to be very much faster than alternative processes that require the transfer of heat by conduction from the outside. This is particularly true for dielectric heating of materials that are poor thermal conductors.

Radio frequency (RF) dielectric heating has been used industrially since the Second World War but is best known nowadays in the microwave oven commonly found in home kitchens.

It is used industrially for plastics heating, welding, and processing; in food processing; in many drying or curing operations of adhesives, paper, textiles, etc.; in medical applications, diathermy, hyperthermia, and cauterizing; and in all kinds of material and chemical processing.

RF heaters are a possible source of harmful interference to communications equipment such as two-way radios, broadcast stations, and microwave repeaters. For this reason they are subject to government regulation and control.

Although any high frequency could be used for dielectric heating, certain frequencies for industrial, scientific, and medical use are specified by appropriate government agencies (in the United States, the Federal Communications Commission, and in many other countries the post office authority).

The frequencies of 915 and 2450 MHz are often called microwave frequencies and are favored for many drying, bulk, or volumetric heating processes.

At a low power the frequency 2450 MHz is attractive because of the availability of low cost magnetron generator tubes to service the consumer microwave market.

Radio frequency radiation may pose a hazard to living systems, particularly the people who operate RF heating equipment. The American National Standards Institute (ANSI) C95.1-1982 is a widely accepted guideline limiting personnel exposure to radio frequency radiation.

The U.S. Occupational Safety and Health Administration has regulations limiting the exposure of personnel to RF radiation.

Electric Induction Heating

When an electric current passes through a conductor, a magnetic field is generated around the conductor. When the conductor is wound in the form of a coil and alternating current is passed through the coil, a strong, alternating magnetic field is established within the coil.

If now an electrically conductive material, a metal bar for example, is placed within the coil, then it will become heated. This is because the alternating magnetic field establishes strong alternating current in the metal bar (called eddy currents) that heat the bar by resistance heating. This process is called induction heating [see Ross, 1988].

Induction heating can be used to heat any electrically conducting material but is primarily applied to heating metals for processing or forming. The largest units of 200 MW capacity are used in steel mills, and sizes range down to small induction heaters of 1kW or less used for heat treating small components.

The advantage of induction heating compared to combustion heating with fossil fuels is that it is quick, clean, and well suited to automatic control. Problems with surface scaling, uneven heating, hot and cold spots, and adverse impact on the environment are avoided. There are energy savings in the elimination of start-up, stand-by, and shut-down heat losses. Moreover, induction heating is very compact, highly flexible, and requires less labor to operate.

The advantage of induction heating over normal electric resistance heating is the ability to deliver heat to metal pieces without direct contact with the electric resistance heaters or any heating medium. This is an important advantage with composite assemblies involving both metallic and nonmetallic parts permitting selective heating of the electrically conductive parts only.

Plasma Heating

Plasma heating is akin to the more widely known electric arc welding. An electric "spark" is started and maintained through a gaseous conductor (the plasma arc column). The high electric resistance of the arc converts the electricity flowing through it to heat. The temperatures achieved are very high, up to 12,000 °C, well beyond those achievable by fossil fuel burners [Camacho, 1988].

Plasma heating is accomplished by using a plasma torch. There are different types of torches available to suit the particular application. The process is used to heat and melt solids, heat and vaporize solids and liquids, and to heat gases. The temperatures achieved by plasma heaters are the highest sustainable temperatures routinely achieved in practice.

The plasma arc column may be initiated and maintained in almost any gaseous medium and virtually any pressure from a modest vacuum to 20 atmospheres.

The advantages of plasma heating follow:

1. Since virtually no mass of fuel and air is consumed, the process effluent is very much less than with fossil fuel burners.
2. The plasma torch can be used in virtually any atmosphere whether or not it contains oxygen.
3. Very high temperatures result in high heat fluxes and rapid reaction rates.
4. Furnace size is reduced since the heating process does not involve the generation of large quantities of gas.

Applications of plasma heating include small hand-held plasma torches of 60 kW or less for metal cutting, welding, and deposition of metal on metal, metal on refractory, refractory on refractory and metal, etc. Larger units up to 6000 kW are used to melt steel and other metals, smelt ores, provide the principal or supplementary heating for blast furnaces and electric arc furnaces, and to pyrolyze hydrocarbons such as oil, coal, peat, plastic, and rubber.

Plasma torches are used in the production of powder metals, high temperature ceramics, and refractories to salvage precious metals from spent catalysts and in the pyrolitic destruction of hazardous materials.

Laser Heating

A laser is an electrically powered device that can produce light that is very highly concentrated and focused to a small-diameter beam of high power.

Lasers are used for a variety of purposes. For thermal work the material processing lasers are of most interest. They can be used for welding, hole drilling, cutting, and heat treatment of both metals and nonmetals as well as local surface heating of such power density as to cause surface melting of metals [Llewellyn, 1988].

Laser energy can be highly controlled and manipulated. Lasers can produce continuous power or energy pulses at frequencies up to 15 kHz. They can deliver small, accurate quantities of energy and, when combined with high-speed positioners directing the laser beam, heat can be produced on isolated areas of a larger part in such a way that other temperature-sensitive areas are not affected.

HEAT STORAGE

Introduction

A heat storage system allows energy that may be available in surplus abundance at one time to be collected and stored for use at a later time when the normal energy supply is restricted or depleted in some way.

Solar energy is a good example. Sunlight, available in abundance in the warm daylight part of the day, is collected, and usually concentrated, in the form of high temperature heat and is then stored in the thermal storage system to be used in the cool, dark hours for space heating or electric power production.

The thermal storage system may be nothing more sophisticated than a relatively large volume of water (an excellent heat transfer fluid) contained in a tank or pool, perhaps in the basement of the house or building and probably insulated with low cost insulation (polyurethane, styrofoam, etc.).

During the day the water is pumped through flat plate solar collectors on the roof of the house where it is heated. On its way to the roof the relatively cool water may pass through convective "coolers" in the house, maintaining the interior at a comfortable level.

The heated water returns to the storage pool and then during the dark hours may again be pumped around the house, this time heating the interior space rather than cooling during the day.

This dual function is possible because water has a very high specific heat compared with air (about four times on a mass basis and hundreds of times on a volumetric basis).

Sometimes rocks or large gravel are added to the water pool to act as the thermal storage media and so reduce the mass of water involved.

Instead of water, some solar heating systems involve a large thermal mass of masonary, rocks, or gravel with solar-heated air circulating to heat the mass during the day and withdrawing the heat during the night. This system functions much like a regenerative heat exchanger with a "hot" blow extending over the daylight hours and a cold blow extending over the dark hours.

Phase-Change Systems

The next level of sophistication involves a phase change of the heat storage medium, usually the transition from liquid to solid (freezing) or the reverse process (melting).

When a liquid is cooled, the temperature decreases progressively as heat is removed. Eventually, at some temperature characteristic of the liquid, called the *ice point* or *triple point,* the process of solidification begins.

Small quantities of the liquid form solid crystals. Further withdrawal of heat from the liquid-solid "slush" causes no further decrease in temperature, simply the formation of more solid crystals.

There is in fact no further reduction in temperature until all the liquid has been converted to the solid phase. Thereafter, further withdrawal of heat from the frozen solid will cause the temperature to decrease.

The natural process of freezing water to ice is, of course, familiar to all those who live outside the tropics. Many of those who live in the tropics are familiar with man-made ice makers.

The same process occurs at different temperatures to all liquids save one, a form of helium liquid (helium 3) that does not freeze even at the lowest achievable temperature (i.e., 0.001 K, virtually absolute zero degrees Kelvin).

The quantity of heat withdrawn from the liquid in the process of solidification is called the *latent heat of fusion*. It is a quantity that varies from one substance to another. For thermal storage systems it is useful for the heat transfer media to have a high latent heat of fusion so as to allow heat to be added and withdrawn at the same temperature. In thermal processing the highest thermodynamic efficiencies are achieved when the heat is added or subtracted at constant temperatures. It may be possible and justifiable to slightly superheat the liquid and subcool the solid so as to extend the total heat exchanged in the process of cyclic freezing and thawing of the heat exchanger media.

For domestic or commercial air conditioning and space heating, water is an excellent distribution medium to pump around the perimeter heaters or cooling systems. It is not a good fluid for a phase change thermal storage medium. This is because the process of solidification occurs at $0\,°C$, a temperature most people find uncomfortably low.

Other heat transfer media with higher fusion temperature exist naturally or have been devised by those interested in the field of energy conservation and air conditioning systems. Lane [1988] has given an excellent review of the available alternative phase change thermal storage media and the principal advantages and disadvantages of the main types.

As well as having a high latent heat of fusion at the proper temperature, the phase-change material should be a good heat transfer medium. Its desirable physical properties include a small volume change when changing phase, a low vapor pressure, relatively high density (to keep the volume small), and to have no storage or unusual phase behavior.

There should be little subcooling before crystallization starts, and the rate of crystallization should be sufficiently high to permit rapid withdrawal of the heat.

The material should have long-term chemical stability and be compatible with the materials of construction, that is, not cause corrosion of the enclosure.

It should not be toxic and present no fire hazard or any other nuisance factor. It should also be readily available at low cost.

Of course it is difficult to satisfy all these, sometimes conflicting, demands and so a compromise must often be accepted with priority given to factors the user perceives to be most important.

Lane [1988] lists a number of commercially available phase change solutions ranging from the Glaubers salt-based material, $Na_2SO_4.10\ H_o - NH_4Cl - KCl$, with melting point 8 °C, heat of fusion 29 cals/g, density 1.49 g/cm^2, gel form, to the high temperature material, magnesium chloride hexahydrate, $MgCl_2.6\ H_2O$, with a phase transition temperature of 114 °C, a heat of fusion 40 cal/g and a density 1.45 g/cm^2 with a dozen other fluids between.

High Temperature Thermal Storage

Thermal storage systems at very much higher temperatures than those discussed are also in use. These provide input to the heat engines of power systems to produce mechanical work or electric power.

One potential application under serious consideration is the provision of a thermal battery in a car. This would be charged overnight with the car stationary by using natural gas combustion or low-cost off-peak electricity.

When the car is driven during the day, heat is withdrawn from the thermal battery to energize a Stirling engine either by driving the car directly or driving a battery charger replenishing the battery charge. The attractive features of this system are the savings in fuel and the absence of environmental pollutants [see Walker, 1980].

A similar system is used for underwater power systems including unmanned, untethered roving vehicles. The high temperature thermal storage installed on board is sufficient to provide several hours or even days of submersible operation before the vehicle is raised to the surface or returned to the mother ship (perhaps a nuclear submarine) for recharging the thermal battery.

Figure 13.2 shows the energy density or heat capacity on a weight and volume basis, of possible energy-storage materials for a high temperature thermal storage system. This figure was prepared for the energy transfer per unit mass or volume to effect a change in temperature from 538 °C (1000 °F) to the maximum temperature shown alongside each material. The maximum temperatures were selected on considerations relating to the heat-storage materials and without regard to containment or insulation characteristics.

The first four materials are molten salts in which a high proportion of the heat transfer is latent heat associated with a phase change occurring at the maximum cycle temperature. Lithium hydride has an exceptionally high heat capacity on a mass basis, but this salt readily dissociates at temperatures only slightly above its melting point. A preferred material is lithium fluoride, which has a higher melting temperature and a better volumetric heat capacity. Unfortunately, the latent heat capacity on a mass basis is less than half that for the hydride. High latent heat capacity is favorable to maintain a constant temperature as heat is withdrawn to operate the engine.

Most work on thermal-energy storage has been done with lithium fluoride but other materials have been studied, particularly aluminum oxide. Percival [1967] describes one experimental unit using 60,000 hexagonal pellets of aluminum oxide in an insulated tank. The system operated between temperature limits of 1482.2 °C (2700 °F) and 815.6 °C (1500 °F) with a storage capacity of approxi-

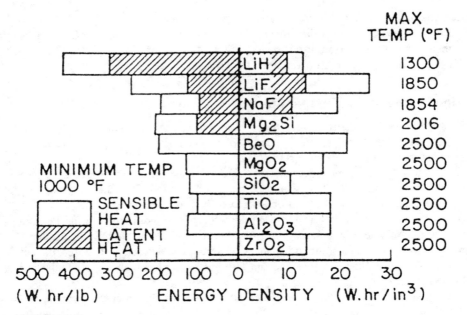

FIGURE 13.2
Energy density of thermal energy-storage materials on a weight and volume basis (minimum temperature 1000 °F). (After Mattavi et al. [1969].)

mately 73.6 kWh brake power (100 brake hp hours) when used in conjunction with a 22 kW (30 hp) Stirling engine. Mattavi et al. [1969] provided the data shown in Fig. 13.3 for the thermal energy storage/Stirling engine system weight and volume requirements as a function of the missing duration. Three different power levels, 7.3, 37, and 74 kW (10, 50, and 100 hp) were considered.

Low Temperature Thermal Storage

Low temperature thermal storage systems are also used extensively. In this case the temperatures are below ambient, and it is the useful refrigeration effect or "coolness" that is stored. There seems to be no generally accepted and elegant term to describe this activity. Cold storage is probably as close as we shall come.

The use of ice, frozen water, to store cold is well established. There are early accounts of ice cut from the Alps carried to Rome during the reign of the Caesars. Until quite recently, the practice was widespread of cutting ice from rivers and lakes and storing it in ice houses for sale in the summer. Now, of course, this has been replaced by mechanical vapor compression refrigerators.

These have permitted the widespread development of cold stores of frozen food and other items in an enormous range of capacities from huge warehouses to the tiny freezer in small domestic refrigerators.

Various proprietary salt-ice mixtures have been marketed as cold storage materials with phase change temperature at temperatures less than frozen water, 0 °C.

Very Low Temperature Storage: The Liquid and Solid Cryogens

There are many military, medical, scientific, and other civil applications for sensitive instruments, electronic devices, and intense, superconducting magnets that require cooling to very low temperatures frequently in the cryogenic range.

Many devices and instruments are optimized to work at the temperature of liquid nitrogen, 77 K. This is because liquid nitrogen is readily available in a

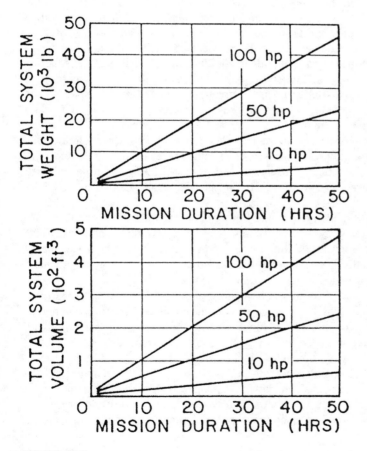

FIGURE 13.3
Weight and volume requirements as a function of power level and mission duration for thermal energy storage and stirling-engine underwater power systems. (After Mattavi et al. [1969].)

laboratory situation, is relatively safe to use and is low in cost, cheaper than milk in many parts of North America.

Other systems and instruments require cooling to lower temperatures. Liquid hydrogen at 20 K and liquid helium at 4 K fulfill most requirements, but there are also others with cooling needs at temperatures less than 4 K or intermediate temperatures between 4, 20, and 80 K. Other fluids are available at higher temperature levels so that cooling may be provided in the range 4–300 K.

Cooling by immersion in a bath of liquid nitrogen, hydrogen, helium, or other liquid is an effective method of cooling and maintaining temperatures. Alternatively, the device may be contained in an enclosure, itself immersed in the liquid bath.

Sometimes, mechanical refrigerators driven by electric motors or cooling systems energized from high pressure gas bottles are used as alternatives to the liquid bath. Nevertheless, cryogenic and noncryogenic liquid baths are widely used. The boil-off vapor may be reliquefied at the site, collected in large plastic bags for subsequent reliquefaction, or, more usually, vented to the atmosphere.

The fluids may be stored as liquids in equilibrium with their vapors (subcritical two phase system) or at higher pressures and temperatures as supercritical homogeneous fluids. This latter form is sometimes more useful in space-borne systems because the absence of gravity prevents precise location of the liquid fraction in a two-phase system.

In earthbound laboratories and airborne systems, the two-phase liquid plus vapor system is invariably used because of the advantages of simplicity and relatively low weight of the highly insulated storage vessel known as a dewar flask (after Sir James Dewar, a Scottish scientist of the 19th century who pioneered the study of cryogenics).

Sometimes the liquid cryogens are further cooled to the solid state. These are particularly useful for space-based systems. The solid cryogenic sublimes to gas directly without changing to the liquid stage and is vented to the high vacuum of space. The solidified mass is contained in a highly insulated container, and a conducting path is provided to the device to be cooled.

The advantages of a solid versus liquid system include independence from orientation requirement, higher storage of cooling effect per unit mass, a higher density for reduced volume, and a lower temperature, which is sometimes an appreciable advantage in the sensitivity of the cooled instruments.

Surprisingly, long lifetimes for liquid and solid cryogen cooling systems have been achieved in some spacecraft and satellite systems extending to 5 or 6 years. This obviously depends on the use of superbly insulated containers, the provision of relatively large masses of the cryogen, and the provision of a low thermal load, either very low continuous power or intermittent use of a higher power level.

SUMMARY

In this chapter we have looked, albeit superficially, at a number of diverse topics of current interest to thermal specialists.

As electronic systems are continuously developed to pack a higher density of integrated circuits onto smaller chips, the cooling problems intensify and in some cases already dictate the operating conditions and expected lifetimes of the systems.

In industry and commerce a wide spread of specialist heating requirements can be fulfilled with electric heating of one sort or another. We briefly considered a range of the different systems available.

There are many instances where heat or coolness collected or generated at one time may be used later with the energy stored until that time. Systems range from very high temperature thermal stores for power production, through moderate temperature systems for building space heating and cooling to very low temperature cryogenic storage for spacecraft instruments.

REFERENCES

Camacho, S. L.: Plasma Heating, Chap. 5, in E. C. Guyer (ed.), "Handbook of Applied Thermal Design," McGraw-Hill, New York, 1988.

Lane, G. A., Phase Change Thermal Storage Materials, Chap. 1, in E. C. Guyer (ed.), "Handbook of Applied Thermal Design," McGraw-Hill, New York, 1988.

Llewellyn, S. A., Laser Heating, Chap. 6, in E. C. Guyer (ed.), "Handbook of Applied Thermal Design," McGraw-Hill, New York, 1988.

Lorenzetti, V., Air and Liquid Cooled Heat Sinks, Chap. 5, in E. C. Guyer (ed.), "Handbook of Applied Thermal Design," McGraw-Hill, New York, 1988.

Mattavi, J. N., F. E. Heffner, and A. A., Miklos, The Stirling Engine for Underwater Vehicle Applications, pp. 2376–2400, S.A.E. Paper 690731, or General Motors Research Publ. GMR-936, 1969.

Melly, J. P., Characteristics and Applications of Electric Resistance Heaters, Chap. 2, in E. C. Guyer (ed.), "Handbook of Applied Thermal Design," McGraw-Hill, New York, 1988.

Moffit, P., Dielectric and Microwave Heating, Chap. 3, in E. C. Guyer (ed.), "Handbook of Applied Thermal Design," McGraw-Hill, New York, 1988.

Percival, W. H., The Stirling Engine for Naval Applications, pp. 196–209, Proceedings Conference on Energy Sources, NRC/CHW, 0340, Washington, D.C., April 1967.

Ross, N. V., Induction Heating, Chap. 4, in E. C. Guyer (ed.), "Handbook of Applied Thermal Design," McGraw-Hill, New York, 1988.

Walker, G., "Stirling Engines," Oxford University Press, New York, 1980.

Walker, G., "Miniature Refrigerators for Cryogenic Sensors and Cold Electronics," Oxford University Press, Oxford, U.K., 1989.

Watson, R., Electric Resistance Heating Element Materials, Chap. 1, in E. C. Guyer (ed.), "Handbook of Applied Thermal Design," McGraw-Hill, New York, 1988.

Appendix A   Glossary

Brittle fracture: The type of failure that is characteristic of brittle material.

Charpy impact strength: See *Izod impact strength.*

Creep: The phenomenon whereby the deformation of material stressed beyond the elastic limit continues indefinitely but at a progressively diminishing rate.

Ductile-brittle transition: The phenomenon whereby some materials that normally exhibit ductile failure become brittle at low temperatures. The impact strength declines drastically when the transition from ductile to brittle behavior occurs.

Ductile fracture: The type of fracture that is characteristic of ductile materials.

Fatigue: The phenomenon whereby a metal subject to pulsating, reversing, or vibrating loads eventually fails at a mean stress level that is far below the ultimate stress.

Heat exchanger: A device for enhancing the flow of heat.

Impact strength: See *Izod impact strength.*

Izod impact strength: The energy absorbed in fracturing a material specimen rigidly held in a vice in the path of a heavy pendulum bob. The specimen may be notched with standard-size notches or unnotched. The Izod or Charpy impact strength is a useful indicator of the ability of a material to withstand impact or shock loads and of the degree to which cracks will propagate in the material.

Proof stress: The stress at which a line drawn parallel to the straight elastic region crosses the stress-strain curve. The line is drawn for a specified strain, say 0.1 percent of the specimen gauge length (5 cm or 20 cm). To avoid confusion the stress must be identified as, for example, the "0.1 percent proof stress on a 2-in gauge length."

Recuperative heat exchanger: A heat exchanger provided with separate flow channels for the different fluid streams and through which the fluids usually flow continuously.

Regenerative heat exchanger: A heat exchanger consisting of a porous matrix or having a single set of flow channels through which the different fluids flow alternately and periodically, usually in opposite directions.

Strain: The extension of a specimen under load divided by the length of the specimen.

Stress: The load applied to a material divided by the cross-sectional area of the member supporting the load.

Ultimate stress: The minimum stress causing a material to fail (usually qualified to include the type of stress—tensile, bending, shear, compression, etc.).

Yield stress: The maximum stress at which a material behaves as an elastic solid.

Appendix B

Typical purchasing specification for tube-and-shell heat exchangers

INTRODUCTION

Scope

This specification, together with the purchase order, purchasing specification summary sheets, exchanger specification sheets, and setting plans, covers the requirements for the design, materials, and fabrication of tubular exchangers of the cylindrical shell-and-tube type.

All conflicts in the requirements of this specification, purchasing specification summary sheets, related specifications, purchase orders, standards, codes, setting plans, or exchanger specification sheets shall be referred to the Purchaser for clarification before the Seller proceeds with the manufacture or procurement of the affected item. Nothing contained in this specification, purchase order, purchasing specification summary sheets, or exchanger specification sheets shall be construed as relieving the Seller of the responsibility for designing and constructing the item to meet the specified operating conditions.

The issue of the codes and standards specified herein shall be as follows:

1. ASME Boiler and Pressure Vessel Code, Sections I, V, and VIII (1974), including addenda through Summer 1980
2. Standards of the Tubular Exchanger Manufacturers Association (TEMA), 6th ed., 1978
3. ASTM A193-74, A249-74, A263-74, A264-74, A265-74, A320-74, A388-75, A450-74, E23-72, E94-68, E142-72

This specification is based on the purchasing specification of the leading U.S. engineering contractor. Although described as "typical," it is in fact much more comprehensive than most purchasing specifications. It is recommended as a model to be emulated as far as possible at the discretion of the purchaser. (Reproduced from course notes on Tube and Shell Heat Exchangers by F. Rubin.)

4. API 660-73
5. ANSI B2.1-68, B16.5-68

Bids

The Bidder shal make one quotation in accordance with this specification and the specified terms and conditions. It shall be understood and clearly stated that the item offered is in complete accordance with the inquiry; otherwise the bid will be rejected. In addition,

1. Provided that all divergencies are indicated, the Bidder may also submit alternative bids or take exceptions or make deviations. No work shall be performed relative to the alternative bids unless written approval has been obtained from the Purchaser.
2. Optional auxiliaries, components, and controls not specified shall be quoted as separate costs.

Proposals shall include the cost of providing documentation and testing all exchangers as required by the applicable code, and according to any requirements listed on the purchasing specification summary sheets and exchanger specification sheets based on the use of the Bidder's testing equipment.

Proposals shall include unit prices for examinations and tests which may be specified by the Purchaser, including but not limited to examinations by radiographic, magnetic-particle, and liquid-penetrant methods, and tests for hardness.

It shall be the Bidder's responsibility to ascertain and comply with the mandatory requirements of all applicable rules (e.g., codes, regulations, ordinances, and statutes) of all bodies (e.g., city, provincial, state, and national) having jurisdiction in the location where the exchanger is to be installed. If this information is unavailable to the Bidder, the Purchaser will attempt to supply the available documents on request.

All exchanger items shall be built in compliance with the applicable code. Proposals on items to be built to the ASME code or other codes or governmental rules or regulations, including all those indicated in the following paragraph, shall include the cost of required inspection(s). With reference to items built in the United States and conforming to the ASME code, National Board standard numbers shall be assigned by the Fabricator to and stamped on items conforming to Section VIII but not on equipment conforming to Section I.

All pressure-containing equipment shall have qualified third-party inspection and certification as follows:

1. Exchangers constructed and stamped to the ASME code: The inspector shall be qualified as required by the code. For parts fabricated within the United States, the inspector shall hold a valid National Board commission.
2. Exchangers constructed wholly or partly to the ASME code without ASME code symbol stamping: For parts fabricated in the United States, the

inspector shall hold a valid National Board commission. For parts fabricated outside the United States, the inspector shall be regularly employed for such inspection by an insurance company authorized to write boiler and pressure-vessel insurance; or by an inspecting organization recognized by the governmental authority to make pressure-vessel inspections. The inspection agency selected is subject to approval.

3. Exchangers normally within the jurisdiction of the ASME code constructed to standards other than the ASME code: All such parts shall be constructed according to the terms of the referenced. The inspector shall be regularly employed for such inspection by an insurance company authorized to write boiler and pressure-vessel insurance. For parts fabricated outside the United States, the inspector shall be qualified as above or by an inspection organization recognized by the governmental authority to make pressure-vessel inspection. The inspection agency selected is subject to approval.

4. In all cases, the qualified inspector must be acceptable to the authority having jurisdiction at the plant site.

The Bidder shall furnish quotations in the number of copies specified in the inquiry document.

For fixed-tube-sheet units, the Manufacturer shall include in the price the cost of expansion joints, when required. If expansion joints are not required, the Manufacturer shall so state, and submit calculations to verify that conclusion.

The Bidder shall include the cost of all packing, crating, and cartage in the quotation. All packing, crating, and cartage shall conform to the requirements of this specification and those of the Purchaser's order.

Six copies of data sheets giving performance and construction information shall be supplied with the bids.

The Bidder, as the Manufacturer, shall submit both mechanical and thermal design calculations to the Purchaser, when requested, at no additional cost.

Drawings and Data

The Purchaser will furnish the Manufacturer with setting plans and exchanger specification sheets specifying the conditions of design and showing the shape, general dimensions, materials for all primary parts, and certain mandatory constructional details, including the location and projection of nozzles and supports.

In addition to the information to be furnished by the Seller as a Bidder, the Seller as the Supplier shall promptly furnish the following information for planning, layout, installation, maintenance, and record purposes:

1. Outline drawings showing all dimensions, the size, location and projection of every connection to the equipment, clearance for removing bundle and cover, shipping weight, bonnet weight, total weight of equipment when filled with water, and location and size of all foundation bolts

2. Drawings showing the locations of all parts that are to be stress-relieved, gasket sketches for all internal and external joints, the method of holding gaskets in place during assembly when joints are not of the confined type, fabrication details of any specially constructed parts and of parts constructed of alloy materials, and other fabrication details when required by the purchase order
3. ASME manufacturer's data reports and all ASME partial data reports

All drawings and data shall be identified with the job number and complete purchase-order number, Client's name, job location, item number, and service. Revision boxes shall be provided to describe the latest revisions in full detail.

The Manufacturer shall prepare a complete fabrication procedure and a schedule of all operations, and submit them for review together with the plans.

The following data must be supplied by the Seller for all items fabricated under the ASME code:

1. Three unsigned copies of ASME manufacturer's data report forms and stamping reproductions (pencil rubbings) well in advance of the submission of certified forms and shipment of components
2. Six signed original ASME manufacturer's data report forms and pencil-rubbing reproductions of the required code stamping

Unless otherwise specified on the purchase order, all drawings, lists, and instructions shall be sent

Attention: Vendor Data Expediting

Records

Records, as defined below and where required thereby, shall be fully identified with the specific materials they represent, made available for examination by the Purchaser or the Purchaser's representative at the time of inspection, and kept for a period of 5 years after the item is shipped.

The following definitions apply:

1. "Mill test reports" shall mean the original Manufacturer's mill test reports. Such reports shall state the specification to which the material complies, the heat, lot, or melt number, the heat treatment (if any), and the results of chemical analysis and both mechanical and nondestructive tests. When heat-treatment information is omitted from mill test reports, certificates will be acceptable only when the applicable code permits this omission.
2. "Certificate" shall mean a written declaration by the material Supplier (or to the extent that the Seller's work or records permit, by the Seller) referencing a particular material specification, such as ASTM or ASME, to which the mechanical and chemical properties and heat treatment conform.

Mill test reports are required for the following cases:

1. When required by the ASME code or by ASTM or ASME material specification
2. For all pressure parts
3 For all stainless-steel parts in urea or carbamate services

Certificates are required for the following cases:

1. When heat-treatment information is missing from mill test reports, if allowed by the applicable code
2. For bolting materials and for pressure and nonpressure part materials obtained from stock for which mill test reports are not available
3. Where the markings of materials are missing or are removed during fabrication operations

Other required records include:

1. Welding procedures specification procedure qualification record and operator qualification tests.
2. Radiographic film for required radiographic tests and certificates for the results of any required magnetic-particle, fluid-penetrant, ultrasonic, or other method of examination.
3. Charts or other records of required hydrostatic, pneumatic, and other tests. Test logs shall include the date, time, duration of the test, temperature of the test fluid, test pressure, and signature of the inspector witnessing the test.
4. Pyrometer charts or other detailed records of heat treatment (postweld, normalizing, heating for forming).
5. Other material records, whenever specified by the applicable code or the material standard, or required by the Purchaser.

No records are required, unless specified by code or the material standard, for pipe fittings, valves, and flanges (with the exception of stainless steel for urea or carbamate service), provided they are made and marked in accordance with an acceptable standard (such as ANSI), nor for pipe marked in accordance with the ASTM or ASME material standard.

Inspection

The Bidder shall have, as part of Bidder's usual business practice, an established, routine quality-control program that can ensure that all variables affecting the requirements for reliability at the end item have been considered, evaluated, and controlled. The said program shall, at the Purchaser's option, be subject to review.

All materials of construction shall be new and, unless otherwise specified, shall be the Seller's standard suitable to fulfill all operating conditions, if any, as specified in the purchase order.

All materials and methods of packaging shall conform to the specifications, if any, covering the same as included in the purchase order.

The Manufacturer shall also furnish inspection to ensure that all design and fabrication requirements of the Purchaser have been met.

In addition to any required code inspection or testing, all fabrication, materials, and packaging shall be subject to inspection by the Purchaser at all stages of fabrication. Shop drawings of the item shall be available to the Purchaser's inspector at the time of the inspection. Surfaces or parts shall not be painted until the inspection is completed.

Stacked units shall be shop assembled to check the location and alignment of mating nozzles, and the assembly shall be subject to inspection.

Rejection

Equipment or parts or materials indicating defects originating with the Seller's design, materials, workmanship, or operating characteristics of such materials, or that are not in accordance with the requirements of the Purchaser's specification or purchase order, shall be subject to rejection.

Discovery of such conditions, after acceptance of the items by the Purchaser, does not relieve the Seller of the responsibility to comply with the requirements of this specification and those of the purchase order.

Shipping and Marking

All items shall be suitably packed and protected from damage during shipment. Packages shall include lifting lugs or designated lifting points. Each item, crate, or bag shall, in addition to the address, be durably marked with the Purchaser's complete requisition number. It shall also be marked to show the exchanger name and the item number. Special tools shall be separately boxed, shipped, and identified as "special tools."

The completed exchanger shall be marked in accordance with the requirements of the applying code. Markings shall be placed on separate corrosion-resistant nameplates brazed or welded to brackets located on the channel cover. Brackets shall be wide enough to clear any required insulation.

All uninsulated ferritic-steel (i.e., carbon, low-alloy, and ferritic-stainless) exterior parts having an operating temperature of 150°F (65°C) or lower, except finished machine parts, and insulated ferritic-steel exterior parts having an operating temperature under 70°F (21°C), shall be painted with at least one coat of a primer that is compatible with the final coat as indicated in the painted pipe and equipment list.

All uninsulated ferritic-steel exterior parts having an operating temperature above 150°F (65°C) and all exposed machined surfaces shall be coated with one of the following protective coatings:

E. F. Houghton Co.: Rust-Veto, No. 342
Exxon Chemical Co.: Rust-Ban, No. 324 or 326

Small parts of the exchanger shall be shipped bagged, boxed, or otherwise protected from damage and loss. Spare gaskets shall be shipped separately and not bolted in place between the flange and cover plate.

Exchangers with carbon-steel tubes shall be dried with compressed air after draining. Sufficient desiccant shall be placed in these exchangers to provide protection for 6 months. The desiccant shall be placed in suitable containers to prevent dispersion of the desiccant in the exchangers during shipment.

Weld bevels shall be coated.

Beveled and plain openings shall be suitably closed with solid, metal, wedge-type protectors that fit either outside or inside the connectors or by means of covers welded to the outside of the nozzle body by means of a cylinder extending from the shipping cover. Shipping covers shall not be welded to the bevel end of the nozzle. Details of shipping covers for this type of nozzle shall be submitted for review prior to fabrication.

Threaded openings shall be plugged with caps or hexagonal-head threaded plugs of the same material as the connected part and sealed with tetrafluoro-ethylene-tape thread sealant.

All flanged openings shall be protected and made waterproof with bolted full-size 10-gauge [3.5-mm (0.138-in)] minimum-thickness steel covers and $\frac{1}{4}$-in (6-mm) minimum-diameter sponge-rubber gaskets. Bolting shall be equally spaced, $\frac{1}{2}$-in (13-mm) minimum-diameter machine bolts. Flanges shall be furnished with one bolt for every other bolt hole but with not less than four bolts, of the size specified in Table C.1.

When specified, external surfaces shall also be protected against corrosive dusts and vapors.

Exchangers shall not be floated to the job site nor transported by ship as deck cargo without prior written approval of the Purchaser.

A master shipping list shall be furnished in advance of the shipment, giving total quantities of each section, subassembly, or piece required for the complete shipment.

A packing list shall be included in each shipment giving the contents by assembly piece mark, individual piece mark, or by item number of each box, crate, bag, and skid, and indicating whether the contents are complete or partial.

Table C.1

Flange-Cover Bolt Sizes

Number of flange holes	Minimum bolt size, in (mm)
4–28	1/2 (13)
32–48	3/4 (19)

On all shipments, a copy of each instruction manual and all drawings required for assembly or erection shall be included, preferably attached inside the largest crate of the shipment.

One extra copy of all data required shall accompany the shipment and shall preferably be attached inside the largest crate in the shipment.

DESIGN

General

Tubular exchangers shall be of the type specified and shall be designed for the service and performance conditions given on the exchanger specification sheets. In addition, they shall comply with all requirements of TEMA standards, Class R, and with any other code or governmental requirements designated on the purchasing specifications summary sheet. The exchanger type nomenclature used on exchanger specification sheets conforms to the requirements of TEMA.

The definitions contained in Appendix III to Section VIII, Division 1, of the ASME code shall be used for any terms not defined in Section I of the ASME code.

The Manufacturer shall be responsible for the thermal design. The heat-transfer surface specified on the exchanger specification sheet shall be taken as the minimum required for the type of tube specified. The Manufacturer shall maximize the tube count, yielding the maximum surface for each shell size. If a Manufacturer's standard shell size is within 95 percent of the minimum specified surface, the Manufacturer may so state and submit a suitable alternative design for review and approval by the Purchaser.

Equipment conforming to Section VIII, Division 1, of the ASME code shall meet the following requirements in addition to welded joints in Category A:

1. Welded joints in Category B (paragraph UW-3) shall be in accordance with Type (1) of Table UW-12 when practicable and with Type (2) when Type (1) is not practicable.
2. Welded joints in Category C (paragraph UW-3) shall be in accordance with Type (1) or Type (2) of Table UW-12 when any of the following conditions are met:
   a. Design pressure is over 600 lb/in^2 gauge (4.14 MPa).
   b. Design temperature is below 32°F (0°C).
   c. Shell or chemical thickness is greater than 1 in (25 mm).

Parts subject to both shell- and tube-side fluids shall be designed for the pressure, on one side only, that requires the maximum material thickness for the part, unless the exchanger specification sheet explicitly requires some other design condition or when relative to fixed-tube-sheet design, where TEMA shall govern.

Basic Allowable Stresses

The allowable stresses at design temperature for nonpressure parts, except as modified below, shall be $33\frac{1}{3}$ percent of the ultimate tensile strength, $66\frac{2}{3}$ percent of the yield strength, or 100 percent of the average stress to produce a creep rate of 1 percent in 100,000 h, whichever is lowest. In addition,

1. Welds attaching nonpressure parts to the pressure shell shall be designed to the allowable stresses for pressure parts and shall be made with a procedure qualified for pressure parts.
2. Anchor bolting shall be designed to a maximum allowable stress of 15,000 lb/in^2 (103.4 MPa).

Shell, Channel, Heads, and Supports

Shell covers for kettle-type reboilers and U-tube units shall be welded directly to the shell. Covers of all other units shall be removable.

Channel and floating-head partitions shall be of plate welded in place; for cast parts they may be welded in place or cast integrally. All partitions shall have a $\frac{3}{8}$-in (10-mm) minimum width at the gasket surface and be machined to a common plane at the gasket face.

When exchangers have bonnet-type channels, tube sheets shall be full diameter (the same outside diameter as the girth flanges) and shall be provided with suitable means of maintaining a pressure seal between the tube sheet and shell while the bonnet is removed.

All welded multiple-pass tube-side units with design pressures of 1500 lb/in^2 gauge (10.3 MPa) and over shall be provided with a bolted, removable pass compartment in the channel for easy access to the tubes when the access opening is not the full diameter of the channel.

Each support saddle shall be one piece and seal welded to the shell with a weep hole. Welding shall be continuous and smooth. Bevel washers shall be furnished for exchanger supports with sloping flanges.

The base supports of stacked units shall allow approximately $\frac{1}{2}$ in (13 mm) for shims. Provision shall also be made in one of the base supports to allow for expansion.

Tubes, Tube Bundles, Tube Sheets, and Baffles

Bare tubes shall be of welded or seamless minimum wall specification may be used provided the nominal wall thickness less the tolerance on wall thickness is at least equal to the specified minimum wall thickness. Low finned tubes shall have the nominal diameter at the plain ends greater than the maximum diameter of the finned portion and may be of average wall specification.

The outside surface of the tubes, between the inner faces of the tube sheets, shall be used as the effective heat-transfer surface.

Baffles shall be of the crossflow type. Two-pass shell construction may be used, subject to the Purchaser's prior approval.

Except for reboilers, impingement baffles shall be provided at the inlets in all services to prevent direct flow against the tubes. Entrance areas shall be at least equal to nozzle areas.

When impingement baffles at inlets are welded to tie rods or spacers, they shall be supported by at least two tie rods or spacers. Baffles welded to only one rod or spacer are not acceptable.

Packed-type longitudinal baffles shall not be used without written consent. Perforated longitudinal baffles shall be provided in horizontal thermosyphon reboilers at the horizontal centerline.

When segment-cut baffles are to be used, the baffles shall be furnished with "ears" whenever the cut would otherwise include a portion of the inlet and outlet knockout areas.

When specified on the exchanger specification sheet, or at the Manufacturer's option, at least four longitudinal sealing strips, approximately $\frac{3}{8}$ in (10 mm) thick and extending from the first to the last segment baffle, shall be furnished to maintain crossflow within the tube nest. Notches shall be cut in each segment baffle, and the sealing strips shall be tack-welded in the notches.

Clad tube sheets that have grooved tube holes shall have one groove in the cladding, and at least one groove in the base plate. Shell-side cladding on tube sheets is not acceptable.

When austenitic tubes are not strength-welded to the tube sheets, special tube-to-tube hole clearances, specified in TEMA, shall be used.

When units are stacked, removable bundles shall be made rotatable and symmetric for interchangeability, unless otherwise stated in the exchanger specification sheet.

Flanges and Nozzles

Flanges for the exchanger proper shall be of the through-bolted type unless otherwise specified.

Flanges for external piping connections of 24 in nominal size and smaller shall be in accordance with the dimensions and ratings of ANSI B16.5. Facings shall be specified on the exchanger specification sheet. Larger sizes shall be specified on exchanger specification sheets and setting plans.

Raised and flat-face pressure flanges shall be furnished with a stock, smooth, spiral-serrated, or concentric-serrated gasket surface finish. Only smooth (125 AARH) surface finishes shall be supplied when the flange rating exceeds 300-lb ANSI or the temperature exceeds 700°F (371°C) in 150-lb ANSI.

Connections

Connections of 2 in nominal pipe size and larger shall be flanged; those of $1\frac{1}{2}$ in nominal pipe size and smaller may be threaded.

The minimum nominal pipe size of threaded connections shall be $\frac{3}{4}$ in. When couplings are used, they shall be 6000-lb standard couplings. Thread shall be NPT conforming to ANSI B2.1.

Two $\frac{3}{4}$-in nominal-pipe-size temperature and pressure gauge connections shall be provided on each nozzle. For intermediate nozzles, the lower nozzle temperature and pressure connections may be omitted. Hydrogen and vacuum services require instrument connections on initial inlet nozzle and final outlet nozzle only.

A vent connection and a drain connection shall be provided at the high and low points of the exchanger, respectively, when the exchanger is not vented or drained by the nozzles.

Bolting

Pressure bolting shall be of the stud-bolt type and not less than $\frac{5}{8}$ in (16 mm) in diameter.

For stacked units, the Manufacturer must supply bolts and gaskets of appropriate material to satisfy design conditions for mated nozzles.

Test Equipment

When specified, test equipment consisting of a test ring and gland conforming to Fig. E-4.13.2 of TEMA shall be provided.

Lifting Device

All vertical exchangers weighing less than 75 tons (68 Mg) shall be equipped with one lifting lug capable of lifting the entire weight of the exchanger. The lug should be located at the top head or elbow. The lug shall accommodate a shackle and pin of the following size:

Weight	Hole size	Pin size
0–50 tons (0–45.4 Mg)	$2\frac{7}{8} + \frac{1}{8}$ in (73 + 3 mm)	$2\frac{3}{4}$ in (70 mm)
61–75 tons (45.4–68 Mg)	$3\frac{3}{8} + \frac{1}{8}$ in (86 + 3 mm)	$3\frac{1}{4}$ in (83 mm)

The pin hole shall be a minimum of 6 in (152 mm) above the cover to allow the sling or shackles to clear the cover bolts and nuts.

For exchangers weighing more than 75 tons (68 Mg), two lifting lugs may be used.

Expansion Joints

When expansion joints are to be provided, the Manufacturer shall submit the design for review and approval before proceeding with the fabrication. Suitable drawings and calculations shall be furnished to describe the type of joint proposed.

The supplier of the expansion joints shall furnish partial data reports and supporting data to enable the final inspector to authorize application of the code stamp to the complete exchanger.

Flanged-and-flued-type expansion joints are to be used exclusively in exchanger shells, unless otherwise permitted by the Purchaser. The gas tungsten-arc welding process must be used for flanged-and-flued-type expansion joints.

Bellows-Type Expansion Joints

1. Bellows-type expansion joints for the shell are not acceptable, unless specified otherwise on the exchanger specification sheet.
2. Bellows may be either rolled or hydraulically formed. When reinforcing rings are used to reduce internal pressure stresses, the bellows shall be formed with rings in place.
3. Bellows must be free of girth welds except where attached to nipples, unless other designs are approved in writing.
4. Tangent bands shall be machined and shrink-fitted on the bellows.
5. A cover or other device shall be furnished to protect the exterior surface of the bellows from foreign objects or mechanical damage.
6. Internal sleeves for streamlining are to be provided when the expansion joint is located in a flowing pipe connection.
7. Exchangers with internal expansion joints are to be designed and fabricated with the cold tube-side inlet or outlet piped at the bonnet end, with the exception of units with service conditions requiring condensing in the tubes.

MATERIALS

General

The nominal composition of the materials of exchanger parts will be specified on the exchanger specification sheets. The materials shall conform to any appropriate specification listed in the applicable code for the specified nominal composition. Materials subject to welding, including backing rings or strips to be left in place, shall be of weldable quality. Backing-ring material shall conform to the nominal composition requirements of the base-metal material standard.

Baffles may be of commercial-quality materials of the specified nominal analysis. Clad tube sheets may be of furnace silver-brazed copper alloy subject to the Purchaser's approval. Support plates, tie rods, and spacer spool materials shall be of a nominal analysis similar to baffle material.

When an austenitic-steel facing is required, clad plate or a weld-deposited overlay may be used. The deposit shall have a minimum thickness of $\frac{1}{8}$ in (3 mm) after machining. Loose-type liners are not permitted without prior approval of the Purchaser. Solid plates may not be substituted without prior approval.

On weld overlays, the outermost $\frac{1}{16}$ in (1.6 mm) layer shall meet the alloy composition required. The first layer may be made with electrodes of higher alloy content to compensate for dilution effects. If two or more layers are used, the procedure qualification test results shall include a chemical analysis to prove that the required analysis is met at the outermost layer. The finished surface shall be examined by the liquid-penetrant method of examination.

All welded tubes shall be subjected to a nondestructive electric test at the mill as described in ASTM A450. This test shall be in addition to the pressure test.

When welded stainless-steel tubes are used, they shall be in accordance with ASME SA-249.

Plate

Carbon-steel plate materials shall be of pressure-vessel quality when used as a pressure-containing part, and made to fine-grain practice.

The minimum thickness of cladding or overlay shall be $\frac{5}{64}$ in (2 mm) for shells, and $\frac{5}{16}$ in (8 mm) for tube sheets.

The heat treatment required for clad plate conforming to ASME SA-263 and SA-264 shall be performed at the mill.

All clad plate conforming to ASME SA-263 and SA-264 shall be cold-flattened, if required, only after final mill heat treatment and descaling.

Explosion clad plate not conforming to ASME SA-263 or SA-264 shall be stress-relieved in accordance with the times and temperatures required by the applicable code before commencing fabrication operations. If required, cold flattening shall be performed before stress-relieving.

Clad plate conforming to SA-265, nickel, and nickel-copper alloy (monel) shall be furnished in the as-rolled condition, except that normalizing is required if the thickness of the as-rolled "pack" exceeds 2 in (50 mm).

Tube sheets that are 2 in (50 mm) and over shall be ultrasonic tested according to ASTM A-435.

Couplings

All couplings shall meet code requirements in addition to the following.

Pressure couplings in unlined carbon-steel exchangers or sections shall be of weldable-quality, open-hearth or electric-furnace mild carbon steel, with the carbon content limited to 0.35 percent. In unlined exchangers or sections other than carbon steel, they shall be of the same material as the exchanger of section.

Couplings and threaded fittings in clad exchangers or sections shall be wrought and of the same materials as the cladding.

Bolting and Gaskets

Materials for bolts and nuts shall be selected from Table C.2 by reference to a "type of bolting." The manufacture and heat treatment of the materials shall be in accordance with the specifications listed in Table C.3 for the respective types.

Unless otherwise specified, stud bolts shall be threaded full length.

The manufacture, dimensions, and tolerances of gaskets shall be in accordance with the Manufacturer's standards unless otherwise specified herein.

Asbestos gaskets shall be cut from homogeneous compressed asbestos sheets having a minimum weight of 4.8 lb/yd^2 (2.6 kg/m^2) per $\frac{1}{16}$ in (1.6 mm) thickness. Compressed asbestos sheet shall contain a minimum of 70 percent of the whole by weight of chrysotile asbestos fiber, a maximum of 10 percent of the whole by weight of natural or synthetic rubber or a mixture of the two, and suitable mineral fillers. The tolerance in thickness shall be 10 percent.

The metal jackets of metal-jacketed asbestos gaskets (except brass-jacketed gaskets) shall be annealed.

Solid-metal gaskets shall have the following maximum Brinell hardnesses:

120 for low-carbon iron, low-carbon steel, and nickel
125 for 67 percent Ni–30 percent Cu (monel)
140 for 5 per cent Cr–$\frac{1}{2}$ percent Mo steel
160 for 18 percent Cr–8 percent Ni steel (AISI Type 304) and 16 percent Cr–13 percent Ni–3 percent Mo steel (AISI Type 316)

Except for ring-joint gaskets, a spare gasket shall be furnished for each one required.

Only metal-jacketed or solid-metal gaskets shall be used in confined joints.

Welding Electrodes

Welding electrodes and welding-rod materials shall be selected. Welding materials used for welding the base metal of clad material shall conform to the requirements of the base material.

FABRICATION

Design of Welded Joints

Girth, longitudinal, and nozzle attachment welds in exchanger shells shall be full-penetration welds.

TABLE C. 2

Selection of Type of Bolting

Use	Materials joined	Metal temperature, °F (°C)						
		−425 to −151 (−256 to −101)	−150 to −21 (−101 to −30)	−20 to 450 (−29 to 232)	451 to 900 (233 to 482)	901 to 1000 (483 to 538)	1001 to 1100 (539 to 594)	1101 to 1500 (595 to 762)
External bolting	Ferritic		L7	B7	B7*	B7*	B5†	
	Austenitic	B8	L7	B7	B8CT	B8CT	B8CT	B8M
Internal bolting	C, C-$\frac{1}{2}$% Mo, $\frac{1}{2}$ to 3% Cr-Mo, 2 to 6% Ni steel		L7	B7	B7	B7	B5	
	4–6 Cr-Mo steel			B5	B5	B5	B5	
	11–13% Cr steel			B6	B6			
	AISI Type 304	B8	B8	B8	B8	B8	B8	B8
	AISI Type 321 or 347	B8	B8	B8CT	B8CT	B8CT	B8CT	B8CT
	AISI Type 316	B8M	B8M	B8M	B8M	B8M	B8M	B8M

*B5 bolting may be used in floating heads when higher strengths are required in the temperature range of 800 to 1040°F (427 to 560°C) inclusive.

†When type B5 bolting is used for ANSI flanges, bolt sizes shall be calculated.

379

TABLE C.3

Material specifications

Type of bolting	Bolts		Nuts	
	ASME Material specification	Mark	ASME Material specification	Mark
B5	SA-193, gr B5	B5	SA-194, gr 3	3
B6	SA-193, gr B6	B6	SA-194, gr 6	6
B7	SA-193, gr B7	B7	SA-194, gr 2H	2H
B8	SA-193, gr B8	B8	SA-194, gr 8	8
B8CT	SA-193, gr B8C or B8T	B8C, B8T	SA-194, gr 8, 8C, or 8T	8, 8C, 8T
B8M	SA-193, gr B8M	B8M	SA-194, gr 8M	8M
L7	SA-320, gr L7	L7	SA-194, gr 4	4

Tube joints shall be in accordance with **TEMA** standards. For units requiring the strength-welding of tubes to tube sheets, the proposed method of fabrication, including complete sequential joint preparation, cleaning, assembly, welding, testing, and other inspection procedures, together with a representative sample of the tube-to-tube-sheet weld, shall be furnished for review and comments before commencing fabrication. In addition,

1. A strength weld of tubes to tube sheet must have a minimum shear dimension of 1.4 times the actual tube wall thickness. Tubes may be rolled in the tube sheet.
2. If the shear length of the weld is less than 1.4 times the actual tube wall thickness, it shall be defined as a seal weld, and the tubes must be rolled into grooves per **TEMA** "R."
3. Some combinations of tube and tube-sheet materials, when welded, result in welded joints having lower ductility than required in the material specifications. Appropriate tube-to-tube-sheet joint geometry, welded method, and/or heat treatment shall be used with these materials to minimize this effect.
4. Particular attention shall be given to air-hardening materials, and, as part of the weld procedure, the Manufacturer must post-heat-treat or justify the omission of post-heat-treatment by submitting appropriate qualification test results as part of the weld procedure.
5. Post-heat-treatment of the strength welds of all material for ASME Section I application is mandatory unless written waiver is requested by the Manufacturer and written approval is given.

Preparation for Welding

On alloy-clad plates, if flame-cutting is used for preparing plate edges, any flame-gouged areas at the junction of the alloy and the base metal shall be repaired by stripping the cladding back from the gouged area and building up the base metal with base-metal weld deposits, and the clad surface with alloy weld metal.

Weld metal shall not be used to build up the edges of plates that are too short or contain large cavities without the approval of the Purchaser.

The Manufacturer shall prepare the ends of nozzles that are to be welded by others. The profiles of such welding ends shall be in accordance with the purchasing specification summary sheets for the size, schedule, and style of bevel specified on the setting plans. Temporary closures for testing purposes shall be designed so that they can be removed without damaging the prepared nozzle ends or shall be repaired after closures are removed.

Welding

Welding-procedure qualification tests for the electroslag, single-pass electrogas, and metal-arc inert-gas-shielded processes shall include side-bend tests.

Fabrication involving welding shall not be sublet to others without prior approval of the Purchaser.

All weld slag shall be removed from austenitic stainless steels and nickel-iron-chromium alloy material welds and parts by grinding or sand blasting when the design metal temperature is above 750° (399°C), to avoid the possibility of corrosion at elevated operating temperatures.

Temporary attachment welds on pressure shells shall be removed. The surface under such welds and under backing rings that have been removed shall be properly conditioned to eliminate surface stress raisers. Such surfaces shall be examined by either the magnetic-particle or liquid-penetrant method of examination.

Arc strikes on pressure parts shall be avoided. If they occur, the surface shall be properly conditioned to eliminate surface stress raisers and shall be examined by either the magnetic-particle or liquid-penetrant method of examination.

Postweld Heat Treatment

Exchangers shall be postweld heat-treated in accordance with the applicable code, the exchanger specification sheets, the purchasing specification summary sheets, and the following additional requirements:

1. For P3 materials, postweld heat treatment shall be at a temperature of 1100°F (593°C) minimum. Note 1 to Table UCS-56 of the ASME code, Section VIII, Division 1, shall not apply.
2. Postweld heat treatment for welds between dissimilar metals shall conform to the requirements of the material having the more stringent requirements and shall be verified by the procedure qualification tests. The proposed postweld heat treatment for welds between ferritic and austenitic steels shall be reviewed by the Purchaser.

3. Local postweld heat treatment shall not be performed without prior written approval of the Purchaser.
4. The maximum stress-relief or postweld heat-treatment temperature shall not exceed the lower of the following values:
   a. The maximum temperature listed in the applicable code.
   b. The tempering temperature at which mill test-report test pieces and the components were tempered if tempering was performed.
   c. The following temperatures for the listed materials:
      (1) Carbon steel: 1200°F (649°C)
      (2) $\frac{1}{2}$ percent Mo and Mn-Mo steels: 1275°F (691°C)
      (3) $\frac{1}{2}$ percent Cr-$\frac{1}{2}$ percent Mo steel: 1275°F (691°C) except that for design metal temperatures of 900°F (482°C) and over the postweld heat treatment shall be in the temperature range of 1250 to 1350°F (677 to 732°C) for a period of 4 h minimum
      (4) $\frac{3}{4}$ to 9 percent Cr-Mo steels: 1375°F (746°C)
      (5) $2\frac{1}{2}$ to $3\frac{1}{2}$ percent Ni steels: 1100°F (593°C)

Plate, seamless heads, parts of built-up heads, and similar pressure-carrying parts subjected to cold or hot bending or forming or forging shall be heat-treated as required by Table C.4. Stress relief shall be performed in accordance with the postweld heat-treatment requirements contained in the applicable code and as modified herein. Annealing, normalizing, and tempering required by Table C.3 shall be performed in accordance with Table C.5.

Welds with a thickness of $\frac{3}{4}$ in (19 mm) or greater made by the electroslag and single-pass electrogas processes shall be normalized.

Austenitic alloy-steel exchanger U-tube bundles shall be heat-treated to relieve the stress caused by cold bending of the tubes. When tubes are of type 304L, 316L, 321, or 347 material, heat treatment shall consist of heating to 1600 to 1650°F (871 to 899°C) for 10 min minimum, followed by air cooling. This heat treatment for low-carbon or stabilized grades may be applied to the whole length of the bent tubes before installation in the tube sheet or to a portion of the assembled bundle that includes the tube bends and a minimum of 1 ft (305 mm) of the straight tangents. Alternative solution annealing for 10 to 15 min at 1950 ± 25°F (1066 ± 14°C) may be applied.

When stress-relieving of fixed-tube-sheet exchanger shells is required, the Manufacturer shall submit a description of the proposed fabrication and stress-relieving procedure, including thermocouple locations, for review before commencing fabrication.

When postweld heat treatment is required, no welding (other than diaphragm closures) to pressure parts shall be performed after such heat treatment. After final inspection, a notice stating WELDING OTHER THAN DIAPHRAGM CLOSURES ON THIS EXCHANGER IS PROHIBITED should be attached to the exchanger.

TABLE C.4

Bending, Forming, and Forging Heat-Treatment Requirements

Type of material	Cold bending or forming[*]	Hot bending or forming	Forging
C steel	None[†]	None	Normalize if heated above 1800°F (982°C)
C-$\frac{1}{2}$ % Mo steel	None[†]	None	Normalize
$\frac{1}{2}$ % Cr–$\frac{1}{2}$ % Mo steel	None[†]	None	Normalize and temper
1 to 9% Cr-Mo steel	None[†]	Normalize and temper to conform to mill-test-report heat treatment	Normalize and temper to conform to mill-test-report heat treatment
AISI Types 304, 316, 321, 347	None	Anneal	Anneal
2 to 6% Ni steel	None	Normalize and temper	Normalize and temper
Inconel	None	None	None
Incoloy (ASME code Case 1325, gr 2), AISI Type 310	Solution anneal	Solution anneal	Solution anneal

[*]Bending and forming is defined for ferritic materials as cold when the temperature is below the maximum temperature permitted, and hot when above 1600°F (871°C) or under and as hot when over 1100°F (593°C). Bending and forming shall not be performed on ferritic materials at temperatures below stress-relieving temperature.

[†]Pressure-vessel formed parts having a ratio of thickness to local radius greater than 5 percent shall be stress relieved prior to subsequent operations if formed below stress-relieving temperature.

[‡]For U bends, see applicable paragraph.

Tolerances

Unless otherwise specified, tolerances shall conform to the applicable code and TEMA standard.

The tolerances on the specified thickness of the alloy on integrally clad plate shall be as given in Table C.6.

The dimensions and tolerances of U-bend steel heat-exchanger tubes shall conform to the requirements of Paragraphs 12a through 12h inclusive of ASME SB-395.

Painting

The exchanger shall be shop-painted in accordance with the painted pipe and equipment list.

Machined gasket surfaces shall not be painted.

Weld bevels made by the Fabricator on the ends of ferritic-steel components that are to be welded by others shall be coated on the inside and

TABLE C. 5
Heat-Treating Requirements

Heat treatment	Type of material	Soaking temperature °F (°C)	Holding time, h	Method of cooling
Anneal	AISI Types 304, 316, 321, 347	1900–2000 (1038–1093)	1 per inch (25 mm) of thickness, but not less than $\frac{1}{2}$	Water quench or air blast
	Incoloy (ASME code Case 1325, grade 2)	2100* (1149)		
	AISI Type 310	2100 (1149)		
Normalize	C steel, C–$\frac{1}{2}$ % Mo steel, $\frac{1}{2}$ to 9% Cr-Mo steel	1650–1750 (899–954)	1 per inch (25 mm) of thickness, but not less than $\frac{1}{2}$	Still air
	2 to 6% Ni steel	1500–1550 (816–843)		
Temper	$\frac{1}{2}$ to 9% Cr-Mo steel	1300–1400 (704–760)	1 per inch (25 mm) of thickness, but not less than 1	†
	2 to 6% Ni steel	1100–1200 (593–649)		

*Two thermocouples shall be attached, one at the expected coldest and one at the expected hottest portion of the Incoloy part or batch, to obtain a thermocouple chart. The thermocouples shall be calibrated before and after use. The temperature range during heat treatment shall be plus 0, minus 50°F (28°C).
†After the soaking temperature is attained and maintained for the required holding time, the cooling time to 800°F (427°C) shall not be less than 1 h.

TABLE C.6

Tolerances

Sections	Composite plate thickness	Tolerances for the clad thickness specified	
		Nominal thickness	Minimum thickness
Flat plate as rolled and cylindrical and conical sections after rolling	$\frac{3}{16}$–2 in (5–50 mm) inclusive	−2% but not less than − $\frac{1}{64}$ in (0.4 mm); +5%	−0%; +7%
	Over 2–10 in (50–255 mm) inclusive	− $\frac{1}{64}$ in (0.4 mm); + $\frac{1}{16}$ in (1.6 mm)	−0 in (0 mm); + $\frac{1}{16}$ in (0.6 mm)
Other formed sections	$\frac{3}{16}$–2 in (5–50 mm) inclusive	−2% but not less than − $\frac{1}{64}$ in (0.4 mm); +8%	−0%; +10%
	Over 2–10 in (50–255 mm) inclusive	− $\frac{1}{64}$ in (0.4 mm); subject to Purchaser's approval, Vendor shall determine plus tolerance	

outside for a distance of 3 in (75 mm) from the end of the component with Deoxaluminite (manufactured by Special Chemicals Corp., 100 South Water St., Ossining, N.Y. 10562).

When austenitic and high-nickel-alloy (Inconel, Hastelloy, etc.) steels are to be marked with fabrication or shipping information, the marking shall be done with water-insoluble ink that contains no metallic pigments and no sulfur. An acceptable ink is made by The Garvey Corp., St. Louis, Mo. 63132. Other inks may be used with prior approval.

Austenitic-alloy steel and high-nickel-alloy steels with design metal temperatures above 750°F (399°C), except as required as mentioned above, shall be kept free of paint to avoid the possibility of corrosion at elevated operating temperatures. All traces of paint that accidentally spatters such parts shall be removed.

TESTS, WELD INSPECTION, AND REPAIRS

Pressure Testing

Before hydrostatic testing, all internal surfaces of the exchanger shall be cleaned by sweeping, vacuum cleaning, or other methods so that equipment will be free of welding slag and flux, welding rod stubs, loose scale, dirt, and debris. The hydrostatic testing of exchangers with austenitic stainless-steel-clad material or containing austenitic stainless-steel components shall be done with boiler condensate or with demineralized water with 0 to 5 ppm maximum chloride content or other chloridefree fluid. The Vendor shall obtain prior approval if fluids other than water are to be used for the test. All insulation in contact with austenitic stainless steel shall be kept dry at all times. Insulation that may contact test water shall be removed before the test, or an alternative test medium shall be used and so stated to Purchaser. After completion of the test, the water shall be drained, the interior dried, and all openings closed, sealed, or covered to prevent access to water, dirt, or corrosive dusts and vapors during shipment and site storage.

It is the responsibility of the Manufacturer to select a hydrostatic test temperature that is high enough so that all parts of the exchanger remain above their transition (ductile-brittle fracture) temperatures. The minimum metal temperatures during the hydrostatic test shall be no lower than those listed in Table C.7.

Service bolting and gaskets to be furnished with the exchanger may be used

TABLE C.7

Minimum Metal Temperature at Time of Testing[*]

Item	Material	Nominal wall thickness, in (mm)	Minimum metal temperature	
			°F	°C
Weld metal	Carbon and ferritic alloy steels	All	−20	−29
	Austenitic steels	All	−20	−29
Pipe, tubing, fittings	SA-53 seamless, SA-106	$1\frac{1}{2}$ (38) and less	−20	−29
	Not post-heat-treated ferritic	$\frac{3}{4}$ (19) and less	60	16
	Post-heat-treated ferritic alloy	1 (25) and less	25	−4
	steels	Over $1-1\frac{1}{2}$ (25–28)	40	4
ANSI forgings and castings[†]	SA-105, SA-181, SA-216	$1\frac{1}{2}$ (38) and less	−20	−29
	Not post-heat-treated ferritic alloys	$\frac{3}{4}$ (19) and less	60	16
	Post-heat-treated ferritic alloy	1 (25) and less	25	−4
	steels	Over $1-1\frac{1}{2}$ (25–38)	40	4
	Austenitic steels	All	−20	−29
Nonstandard forgings and castings	Not post-heat-treated SA-181, SA-105, SA-216	$1\frac{1}{2}$ (38) and less	90	32
	Post-heat-treated SA-181,	1 (25) and less	60	16
	SA-105, SA-216	Over 1–2 (25–50)	80	27
		Over 2–3 (50–75)	100	38
		Over 3–4 (75–100)	120	49
	SA-541 class I with supplementary requirements S3 and S5 at 32°F (0°C), with or without post-heat-treatment	3 (75) and less	32	0
	Post-heat-treated ferritic alloy	$\frac{1}{2}$ (13) and less	10	−12
	steels (normalized and	Over $\frac{1}{2}-1$ (13–25)	25	−4
	tempered forgings and	Over 1–2 (25–50)	50	10
	castings)	Over 2–3 (50–75)	70	21
	Austenitic steels	All		

TABLE C.7

Minimum Metal Temperature at Time of Testing (*Continued*)

Item	Material	Nominal wall thickness, in (mm)	Minimum metal temperature	
			°F	°C
Plate, not post-heat-treated	SA-204	$\frac{1}{2}$ (13) and less	32	0
		Over $\frac{1}{2}$–1 (13–25)	80	27
		Over 1–1$\frac{1}{2}$ (25–38)	100	38
	SA-442, grade 60	$\frac{3}{4}$ (19) and less	5	−15
		Over $\frac{3}{4}$–1$\frac{1}{2}$ (19–38)	32	0
	SA-516 (all grades)	1 (25) and less	−20	−29
		Over 1–1$\frac{1}{2}$ (25–38)	5	−15
	Austenitic steels	Any	−20	−29
Plate, post-heat-treated	SA-204, SA-285, SA-515, grade 65, 70	$\frac{3}{4}$ (19) and less	32	0
		Over 1–1$\frac{1}{2}$ (25–38)	70	21
		Over 1$\frac{1}{2}$–4 (38–100)	90	32
	SA-442, grade 60	Over $\frac{3}{4}$–1$\frac{1}{2}$ (19–38)	5	−15
	SA-516 (all grades)	1 (25) and less	−20	−29
		Over 1–1$\frac{1}{2}$ (25–38)	5	−15
		Over 1$\frac{1}{2}$–3 (38–75)	5	−15
		Over 3–4 (75–100)	32	0
	SA-387 (normalized and tempered plate)	$\frac{1}{2}$ (13) and less	10	−12
		Over $\frac{1}{2}$–1 (13–25)	25	−4
		Over 1–2 (25–50)	50	10
		Over 2–3 (50–75)	70	21
	Austenitic steels	Any	−20	−29

*Materials and thicknesses not listed will be considered individually.

†Flanges, fittings, and valves conforming to ANSI or MSS standards only.

‡The thicknesses indicated for flanges, fittings, forgings, and castings are at the ends at which butt welds are made.

for shop tests, provided that any such material damaged during the test shall be replaced by the Fabricator with new material.

If any pressure parts of an exchanger are broken subsequent to a successful test, the exchanger shall be retested after repairs are made.

Weld Inspection

All butt welds, except stud welds, in exchangers fabricated from chromium-molybdenum steels with over $\frac{1}{2}$ percent (nominal) chromium shall be given a radiographic examination.

Welds placed in any part, regardless of the material, thickness, or service, before the part is subjected to severe working (ratio of thickness to local radius greater than 5 percent) by any means including spinning, pressing, and rolling shall be given a complete radiographic and either a magnetic-particle or liquid-penetrant examination after the completion of the severe working and before further fabrication is performed.

Welds made by the electroslag and single-pass electrogas processes shall be subjected to an ultrasonic examination after heat treatment is completed, in accordance with Paragraph N625 of Section III of the ASME code.

Weld surfaces of ferritic materials shall be examined by either the magnetic-particle or liquid-penetrant method, and those of austenitic materials by the liquid-penetrant method, in accordance with Table C.8 after any required postweld heat treatment is completed. When the vessel wall thickness is greater than 2 in (50 mm), such examination shall be performed after any required postweld heat treatment and hydrostatic testing are completed.

The welds attaching welding pins or studs shall be tested by striking with a light hammer.

In addition to ASME code requirements, the image-quality indicator of the radiographic examination shall be in accordance with ASTM E142, except that the penetrameter design shall conform to the ASME code. Radiographic film

TABLE C.8

Examination of Weld Surfaces

Material	Condition of vessel	Radiographed	Longitudinal and girth welds	Nozzle welds[*]	Attachment welds[†]
Ferritic steels with maximum of $\frac{1}{2}$ % (nominal) chromium	Not postwelded	Not 100%	No	No	No
	Heat treated	100%[‡]	No	Yes	No
	Postwelded	Not 100%	No	No	No
	Heat treated	100%[‡]	No	Yes	Yes
Ferritic steels with over $\frac{1}{2}$ % (nominal) chromium		100% or less	No	Yes	Yes
Austenitic steels with design metal temperature of 750°F (399°C) or less and thickness of $\frac{3}{4}$ in (19 mm) or less		Not 100%	No	No	No
		100%[‡]	No	Yes	No
Austenitic steels with design metal temperature above 750°F (399°C) or thickness over $\frac{3}{4}$ in (19 mm)		100% or less	Yes	Yes	Yes

[*]Nozzle welds include the welds between the nozzle and the reinforcing pad, the vessel and the reinforcing pad, and the nozzle and vessel under the reinforcing pad. Welds under the reinforcing pad shall be examined before the pad is attached.

[†]Attachment welds include structural, tray-support, and skirt-attachment welds.

[‡]Surface examination is necessary only when the radiography is required by the applicable code or by the purchasing specification summary sheets.

shall be of the fine-grain nonscreen type, possessing high contrast and definition (ASTM E94, types I or II). For any lot of film examined, the film density over the weld shall average 2, and no individual film shall show a density of less than 1.5. For film densities greater than 2, the adequacy of the viewing equipment shall be approved by the Purchaser's inspector.

When liquid-penetrant examination is required, it shall be performed on all accessible weld surfaces. Any surface irregularities that interfere with the examination shall be removed by grinding, machining, or wash blending with a heliarc torch. When the surface examined is of austenitic alloy steel or nickel-iron-chromium alloy materials having a design metal temperature above 750°F (399°C), the liquid-penetrant material shall be tested for sulfur content or, after completion of testing, the surface shall be cleaned with Alconox (manufactured by Alconox Inc., 215 Park Ave., New York, N.Y. 10003) or an approved alternate.

Repairs

Defects shall be removed and rewelded using qualified welding procedures, and the repair weld shall again be postweld heat-treated (if originally required) and reexamined by the original method for freedom from defects. If a correction involves serious alterations, the approval of the Purchaser shall be obtained before such correction is made.

LOW-TEMPERATURE OPERATION

General

Unless otherwise indicated on the exchanger specification sheets, the special material and fabrication requirements of this section shall be applicable for exchangers having design metal temperatures below 60°F (16°C). These requirements are in addition to the requirements of the ASME code and the preceding sections of this specification.

Materials and metal temperatures shall be selected in accordance with the engineering standards listed on the purchasing specification summary sheets.

Materials

Materials required to be purchased to impact-test specifications shall be tested at a temperature at least as low as the minimum temperature listed in the ASME specification for the material.

Unless a part is specifically noted as "non-impact-tested," all parts welded to impact-tested parts, except backing rings (which are to be removed) and insulation studs, shall be of impact-tested materials.

Backing rings that are to remain in place and are welded to plates conforming to SA-516 shall be made of SA-516 plate material or of impact-tested material.

Forging shall be in accordance with SA-350.

All welds joining impact-tested materials and the heat-affected zone (HAZ) shall be impact-tested to the same standards and at the same temperature.

Fabrication

For design metal temperatures below 32°F (0°C), the welding employed on pressure and nonpressure parts except welding studs in the construction of an exchanger shall be performed with the make and brand of electrode (and flux, if the automatic submerged-arc welding process is being used) that has been used in the qualification of the welding procedure.

For design metal temperatures between 32°F (0°C) and −20°F (−29°C), the welding qualification tests shall include impact tests made in the weld metal only, at or below the minimum design metal temperature. Impact tests shall be Charpy V-notch tests. The average value for three standard specimens shall be 15 ft·lb (20.3 J), and the minimum value for any one specimen shall be 12 ft·lb (16.3 J).

When postweld heat treatment is required by the applicable code or this specification on materials with design metal temperatures below 60°F (16°C), no welding to pressure parts shall be performed after such heat treatment.

After shop postweld heat treating, the note WELDING NOT PERMITTED shall be stencilled in 3-in (75-mm) high letters on the exchanger, by the Manufacturer.

HIGH–PRESSURE OPERATION

General

Unless otherwise indicated on the exchanger specification sheets, the special material and fabrication requirements of this section shall be applicable for exchangers having design operating pressures of 1500 lb/in² gauge (10.3 MPa) and above. These requirements are in addition to the requirements of the ASME code and the first five sections of this specification.

Seamless tubes in exchangers with design pressures of 1500 lb/in² gauge (10.3 MPa) and above shall be eddy-current tested at the mill according to ASTM A450. This test shall be in addition to the hydrostatic test.

Nozzles, branch connections, and piping connections of exchangers shall be designed without reinforcing pads, except where noted on the exchanger specification sheet.

Fabrication

Before any welding is done at any point on the exchanger, the surfaces must be clean and scalefree.

Single-pass fillet welds are not permitted for connecting attachments to pressure parts.

All piping and nozzle attachment welds shall be ground smooth on the inside surface, or the root pass shall be made by inert-gas-shielded methods.

Where nozzles consist of heavy-walled forgings and pipe extensions, welding of the pipe to the forging shall be done before the exchanger shell (or channel) is stress-relieved. Local stress relief of this weld is not permitted.

Nozzles S_1, S_2, T_1, and T_2 are to be extended by welding pipe extensions and/or elbows to the nozzle forgings in accordance with the applicable drawing. Such welding shall be done before the heat exchanger is stress-relieved.

Butt welding end of the nozzle shall be designed in accordance with Engineering Standard 4-6S and Drawing 314A1.

Preheat temperatures found necessary during welding procedure tests shall be reported on the procedure test report. During fabrication, preheating shall be done in accordance with these temperatures.

Heat Treatment

All carbon-steel and low-alloy steel materials forming pressure parts of the exchanger or welded directly to pressure parts shall be normalized. All material originally normalized must be stress-relieved if cold-formed during fabrication.

Heat treatment (normalizing, postweld heat treatment, and preheating) shall conform to the requirements of ASME Section VIII, Paragraph UCS-56. Recording thermocouples, to prove this compliance, shall be placed at the high and low points at each end of the piece and on the heaviest and thinnest parts being heated. The difference between the maximum and minimum temperature of the section being heat-treated shall not exceed $144°F$ ($80°C$) during the heat-treating cycle. Sufficient evidence shall be presented to prove that the thermocouples are reliable. The furnace shall be such that the uniform temperature can be maintained within $-25°F$ ($-4°C$) for a sufficient period of time to assure the soaking temperature. The furnace atmosphere shall be neutral or slightly decarburized.

Normalizing shall be done in a manner that ensures rapid and uniform air cooling.

Preheating, when necessary, shall be done in a manner that achieves as uniform a temperature as possible and minimizes thermal stress. Inspection shall review proposed preheating procedures in advance.

Impact Test

All plate materials over 2 in (50 mm), except exchanger tubes, forming pressure parts of the exchanger shall be impact tested.

All forgings such as tube sheets and nozzle flanges (except vents and drains) shall be impact tested.

All welding procedures shall be qualified with impact testing.

Carbon and low-alloy attachments that are $\frac{1}{2}$-in (13 mm) thick or more and are welded directly to pressure parts shall be impact tested (except supports and davit brackets).

Impact testing is not required on carbon and low-alloy steel attachments that are less than $\frac{1}{2}$-in (13 mm) thick and welded directly to pressure parts, provided that the attachment material is fine-grain steel.

Samples selected for impact testing shall be subject to a thermal history similar to that of the item each sample represents. Impact tests shall be done on Charpy V-notch specimens of the form required by ASTM E23-72 unless otherwise approved in advance.

Impact tests shall be performed at $+30°F$ ($0°C$) unless otherwise required by the appropriate design specifications. Required values of impact tests are as required in Table UG-84.1, ASME Section VIII, Division 1.

Inspection Testing

Inspection will be carried out in the normal manner, using specifications and standards in accordance with the requirements of the purchase order.

Following hydrostatic testing, all welds shall be inspected by magnetic-particle or penetrant-oil examination on all accessible surfaces. Welds joining tubes to tube sheets shall be inspected by penetrant-oil examination.

Magnetic-particle testing shall be done before heat treatment, if heat treatment is required. The magnetic field shall be aligned to obtain maximum sensitivity in the search for cracks at toes of fillet welds, parallel to the axis of the weld (Section V, Article 7 of ASME code).

All nozzle welds in exchangers with design pressures of 1500 lb/in² gauge (10.3 MN/m²) and above shall be given a radiographic examination. When the welds cannot be radiographed, the weld surfaces of ferritic materials shall be examined by the magnetic-particle method on the first pass, on the cover weld of the first side, after backchipping from the second side, and on the cover pass of the second side. If welded from one side only, magnetic-particle inspection shall be performed on the first pass, an intermediate pass, and the cover pass. Austenitic materials shall be examined by the liquid-penetrant method both inside and outside after hydrostatic testing is completed. However, when a weld is not accessible after testing, it shall be examined before testing. The Purchaser or Purchaser's designated representative shall witness these tests.

All forgings such as tube sheets, nozzles, and flanges shall be ultrasonically inspected in accordance with ASTM A–388 or Vendor's equivalent. Vendor must submit quality control specifications, including acceptance and rejection standards, for approval.

All butt welds and nozzle welds 2 in (50 mm) thick and over shall be subjected to ultrasonic examination in accordance with Specification UTS-CT in addition to the required radiography and magnetic-particle or penetrant-oil testing.

Directory of heat exchanger manufacturers

The following list is reprinted with permission from U.S. Heat Exchanger Market: Summer 1989, Report number A-2156, published by Frost & Sullivan, Inc. It may not be copied, given, lent, or sold to third parties without written permission of Frost & Sullivan, Inc.

For additional information and sales, contact Frost & Sullivan, Inc., 106 Fulton Street, New York, New York, New York 10038. Telephone (212) 233-1080; telex 235986; fax (212) 619-0831.

United States

Adams (R. P.) Company, 237 East Park Drive, Buffalo, NY 14240; 716-877-2608

Aerco International, Inc., 159 Paris Avenue, Northvale, NJ 07647; 201-768-2400; 800-526-0288

Aerofin Corporation, 4623 Murray Place, Lynchburg, VA 24506; 804-845-7081

Ahlstorm Machinery, Inc., Pruyns Island, P.O. Box 74, Glens Falls, NY 12801; 518-797-9541

Air Products & Chemicals, Inc., P.O. Box 538, Allentown, PA 18195; 215-481-4911; 800-633-7227

Air-X-Changers (HARSCO), P.O. Box 1804, Tulsa, OK 74101; 918-266-1850

Alaskan Copper Works, P.O. Box 3546, Seattle, WA 98124; 206-623-5800; 800-547-8533

Alfa-Laval, Inc., 8115 Linwood Ave., Ft. Lee, NJ 07024; 201-592-7800

Allegheny Bradford Corporation(Top-Line Corp.), Box 200, Bradford, PA 16701; 814-362-4626, 800-458-6095

Alpha United, Inc., Affiliate of Lyron, Inc., 1301 El Segundo Blvd., El Segundo, CA 90245; 213-322-9570

American Air Filter, P.O. Box 35530, Louisville, KY 40232; 502-637-0011

American Heat Exchanger Co., Inc., 9 Old Dock Road, Yaphank, NY 11980; 516-345-3590

Ametek, Haveg Division, 902 Green Bank Rd., Wilmington, DE 19808; 302-995-0560

Ametek Heat Transfer Division, 2300 West Marshall Drive, Grand Prairie, TX 75051; 214-647-2626

Amtrol, Inc., 1400 Division Road, West Warwick, RI 02893; 401-884-6300

Aqua-Chem, Inc., Box 421, Milwaukee, WI 53201; 414-962-0100

Armstrong Engineer Assocs., Inc., Box 566, West Chester, PA 19380; 215-436-6080

Astro Metallurgical Div., Harsco, 3225 Lincoln Way West, Wooster, OH 44691; 216-264-8639; 800-543-5810

Atlas Industrial Mfg., Co., 80 Somerset Pl., Clifton, NJ 07012; 201-779-3970

Automated Controls & Systems, Inc., 395 Lively, Wood Dale, IL 60191; 312-860-6860

Babcock & Wilcox, 1010 Common St., New
    Orleans, LA 70161; 504-587-4411
Baltimore Aircoil Co., Box 7322, Baltimore,
    MD 21227; 301-799-1300
Bas-Tex Corporation, A Brown Fin Tube Co.,
    P.O. Box 40082, Houston, TX 77240;
    713-466-3535
Beltran Associates, Inc., 1133 East 35th
    Avenue, Brooklyn, NY 11210;
    718-338-3311
Bepex Corp., 3 Crossroads of Commerce,
    Rolling Meadows, IL 60008;
    312-506-0100
Berdell Industries, Inc., 8 43rd Avenue, Long
    Island City, NY 11101; 800-221-1522
Blackstone Corporation, 1001 Allen St.,
    Jamestown, NY 14701; 716-665-2620
Blaw-Knox Equipment Co., P.O. Box 1041,
    Buffalo, NY 14240; 716-895-2100
Blissfield Manufacturing Co., 626 Depot St.,
    Blissfield, MI 49228; 517-486-2121
Bohn Heat Transfer, A Division of Gulf &
    Western, 1625 East Voorhees St., Danville,
    IL 61832; 217-446-3710
Bos-Hatten, Inc., A subsidiary of Nitram
    Energy, Inc., 10 French & Old Union
    Roads., Buffalo, NY 14224; 716-668-8111
Brighton Corp., A Division of Trinity
    Industries, 11862 Mosteller Rd.,
    Cincinnati, OH 45241; 513-771-2300
Brown Fintube Co. (Identified as Bas-Tex
    Corp.), P.O. Box 40082, Houston, TX
    77240; 713-466-3535
Bry Air, Inc., P.O. Box 239, Route 37W,
    Sunbury, OH 43074; 614-965-2974
Bryan Steam Corporation, Route 19 North,
    Peru, IN 46970; 317-473-6651
Buffalo Forge Co., 465 Broadway, Buffalo,
    NY 14240; 716-847-5121
Burn-Zol, Box 109, Dover, NJ 07801;
    201-361-5900

C-E Air Preheater, Amdover Road, Wellsville,
    NY 14895; 716-593-2700
C-E Lummus Crest, 1515 Broad Street,
    Bloomfield, NJ 07003; 201-893-1515
C-E Lumus Heat Transfer Systems Co., 425
    Market St., Elmwood Park, NJ 07407;
    201-796-5800
C-E Power Systems, 1000 Prospect Hill Road,
    Windsor, CT 06093; 203-688-1911

Camac Industries, 27 Dwight Pl., Fairfield,
    NJ 07006; 201-575-1831
Capitol Temptrol Corp., Colton Road, Old
    Lyme, CT 06371; 203-739-4421
Carbone Corp., 410 Myrtle Ave., Broonton,
    NJ 07005; 201-334-0700; 800-526-0877
Carrier Corporation, Carrier Parkway,
    Syracuse, NY 13221; 315-451-2600
Ceramic Cooling Tower Co., 1100 Northway
    Drive, Fort Worth, TX 76131;
    817-232-4661
Champ Products, Inc., 951 Second St.,
    Ronkonkoma, NY 11779; 800-645-0200
Chattanooga Boiler and Tank Co., 1010 East
    Main St., Chattanooga, TN 37401;
    615-266-7118
Cherry-Burrell, Inc., 2400 Sixth St. S. W.,
    Cedar Rapids, IA 52406; 319-399-3200
Chester-Jenson Co., 345 Tilghman, Chester,
    PA 19013; 215-876-6276
Christian Engineering, Hunters Point, San
    Francisco, CA 94124; 415-822-1080
Coil Company, Front & Walnut Sts., Colwyn,
    PA 19023; 215-461-6100; 800-523-7590
Colmac Coil Mfg., P.O. Box 72, Colville, WA
    99114; 509-684-2595
Compressed Air Components, 757 East Ferry
    St., Buffalo, NY 14211; 716-892-6111
Condensing Heat Exchanger Corp., 678
    Troy-Schenectady Road, Latham, NY
    12110; 518-785-1234
Continental Products, Inc., P.O. Box 418228,
    Indianapolis, IN 46241; 317-241-4748
Coolentheat, Inc., 1 Stercho Rd., Lindon, NJ
    07036; 201-925-4473
Corning Glass Works, Corning, NY 14831;
    607-974-8991
Cryochem, Inc., P.O. Box 32, Boyertown, PA
    19512; 215-689-9531
CVM Corp., 402 Vanderer Ave., Wilmington,
    DE 19802; 302-654-7070
Cyclotherm Divison, 157 East First St.,
    Oswego, NY 13126; 315-343-0660

De Dietrich (USA), Inc., Highway 22, Union,
    NJ 07083; 201-686-4900
Dean Products, Inc., 985 Dean St., Brooklyn,
    NY 11238; 718-789-4444
Delta Southern Co., Box 3034, Baton Rouge,
    LA 70821; 504-356-4431
Deltak Corp., Box 9496, Minneapolis, MN
    55440; 612-544-3371

Denver Equipment Co., 621 South Series Madre, Colorado Springs, CO 80903; 719-471-3443

Des Champs Laboratories, Inc., P.O. Box 440, East Hanover, NJ 07936; 201-884-1460

Doucette Industries, Inc., P.O. Box 1641, York, PA 17405; 717-845-8746

Dow Corning Company, P.O. Box 1767, Midland, MI 48640; 800-248-2345

Doyle & Roth Mfg. Co., 26 Broadway, New York, NY 10004; 212-269-7840

Dunham-Bush, Inc., 101 Burgess Rd., Harrisburg, VA 22801; 703-434-0711

Dynatherm Corp., One Beaver Court, Cockeysville,MD 21030; 301-666-9151

Ecodyne Corp., Cooling Products Division, P.O. Box 1267, Santa Rosa, CA 95403; 707-544-5833

Ecodyne Corp., MRM Division, 608 First St., SW, Massilon, OH 44646; 216-832-5091

Ecolaire Process Management Co., P.O. Box 390, 1500 Lehigh Drive, West Easton, PA 18042; 215-250-1000

Edwards Engineering Corp., 101 Alexander Avenue, Pompton Plains, NJ 07444; 201-835-2808; 800-526-5201

EG&G Wakefield, 60 Audobon Rd., Wakefield, MA 01880; 617-245-5900

Electro-Impulse Laboratory, Inc., 116 Chestnut Street, Red Bank, 07701; 201-741-0404

Energizer Corporation, Bach-Buxton Road, Amelia, OH 45102; 513-753-4817

Energy Exchanger Company, 1844 North Garnett Rd., Tulsa, OK 74116; 918-437-3000

Enerquip, Inc., 611 North Rd., Medford, WI 54451; 715-748-5888

Engineers & Fabricators, Box 7395, Houston, TX 77248; 713-869-3461

Essex Cryogenics, 8007 Chivvis, St. Louis, MO 63123; 314-832-8077

Fansteel, Inc., 1 Tantalum Pl., North Chicago, IL 60064; 312-689-4900

First Thermal Systems Int'l., 200 Compress St., Chattanooga, TN 37405; 615-265-3441

Foster Wheeler Energy Corp., 110 S. Orange Ave., Livingston, NJ 07039; 201-533-1100

Frick Co., Subsidiary of York Int'l., 345 West Main St., Waynesboro, PA 17268; 717-762-2121

Gamewell Manufacturing, Inc., Box 2309, Salisbury, NC 28144; 704-637-6770

Garrett Thermal Systems, Division of Allied-Sugnae, Inc., 9225 Aviation Boulevard, Los Angeles, CA 90045; 714-891-1640

Gaston County Fabrication, P.O. Box 308, Stanely, NC 29164; 704-263-4765

GEA Power Cooling System, Inc., 10240 Sorrento Valley Road, San Diego, CA 42121; 619-457-0086

General Resource Corp., 203 S. 3rd St., Hopkins, MN 55343; 612-933-7474

Graham Manufacturing Co., 20 Florence Ave., Batavia, NY 14020; 716-343-2216

Hague International, 3 Adams St., South Portland, ME 04106; 207-799-7346

Harris Thermal Transfer Products, Inc., P.O. Box 339, Tualarin, OR 97062; 503-692-1260

Harrison Radiator Division, Upper Mountain Rd., Lockport, NY 14094; 716-439-2011

Haydon, Inc., Box 848, Corona, CA 91720; 714-735-4900; 800-854-4757

Heat Exchangers, Inc., 8100 North Monticello Avenue, Skokie, IL 60076; 312-679-0300

Heat Process Equipment Corp., 34252 Mills Road, Avon, OH 44011; 216-327-6051

Heat Transfer Equipment Group, American Precision Industries, Inc., 2777 Walden Ave., Buffalo, NY 14225; 716-0 = 684-6700

Heat Transfer Technology, Inc., Fox Pavilion 1432, Jenkintown, PA 19046; 215-884-2488; 800-223-COIL

Hex Industries, Inc., 15001 South FIguerao St., Gardena, CA 90248; 213-770-2130

High Performance Tube, Inc., 1640 Morris Ave., Union, NJ 07083; 201-964-8520

Hiross, Inc., Box 290 L.P.O., Niagara Falls, NY 14304; 716-283-1911

Hirt Combustion Engineers, 931 S. Maple Ave., Montebello, CA 90640; 213-728-9164

Hitachi America, Ltd., 6 Pearl Court, Allendale, NJ 07401; 201-825-8000

Holland (R.W.) Company, Inc., P.O. Box 1106, Indianapolis, IN 46206; 317-636-4321

Hudson Products Corp., 6855 Harwin Drive, Houston, TX 77036; 713-785-4000

Hughes-Anderson Engineering, 1001 N. Fulton Avenue, Tulsa, OK 74115; 918-836-5967

Industrial Energy Corporation, P.O. Box 775, Windsor Locks, CT 06096; 800-982-0030

Ionics,Inc., P.O. Box 99, Bridgeville, PA 15017; 412-343-1040

ITT Bell & Gossett, 820 North Austin Avenue, Morton Grove, IL 60053; 800-243-8160

ITT Standard, Inc., 175 Standard Pkwy., Buffalo, NY 14227; 716-897-2800

Janitrol Aero Division, Midland-Ross, 4200 Surface Rd., Columbus, OH 43228; 614-276-3561

Karmazin Industries, Inc., 3776 Eleventh St., Wyandotte, MI 48192; 313-282-3776

Kennedy Tank & Mfg. Co., Inc., 835 East Sumner Avenue, Indianapolis, IN 46227; 317-787-1311; 800-247-1355

Kentube Division, Tranter, Inc., 4150 So. Elwood, Tulsa, OK 74107; 918-446-4561

Krueger Engineering & Mfg. Co., 12001 Hirch Rd., Houston, TX 77016; 713-442-2537

Kusel Equipment Co., 820 West Street, Watertown, WI 53094; 414-261-4112

Larkin Coils, Inc., 519 Memorial Drive, Atlanta, GA 30371; 404-688-3171

Lee Industries, Inc., P.O. Box 668, Philipburg, PA 16866; 814-342-0461

Limco Manufacturing Corp., 1 Gravies Point Road, Glen Cove, NY 11542; 516-671-7400

LORI, 510 South Lansing, Tulsa, OK 74120; 918-587-4105

Lundell Manufacturing Co., 5200 W. State St., Milwaukee, WI 53208; 414-476-9934

Lytron, Inc., 2 Dragon Ct., Woburn, MA 01801; 617-933-7300

Mammoth Division, Lear Siegler, 13120-B County Rd. 6, Minneapolis, MN 55441; 612-559-2711

Manning & Lewis Engineering Co., 675 Rahway Avenue, Union, NJ 07083; 201-687-2400

Marine Coolers, Inc. (Champ Products, Inc.), 68 Newton Plaza, Plainview, NY 11803; 800645-0200

Marley Cooling Tower Co., 5800 Foxridge Drive, Mission Woods, KS 66202; 913-362-1818

Marlo Coil/Nuclear Cooling, Box 171, High Ridge, MO 63049; 314-677-6600

McLean Midwest, 4000 83rd Ave., Brooklyn Park, MN 55443; 612-561-9400

McQuay, Inc., Box 27906, Milwaukee, Wi 53227; 414-276-9229

Midland-Ross, Janitrol Aero Division, Box 28503, Columbus, OH 43228; 614-276-3561

Midwesco Energy Systems, Subsidiary of Midwesco, Inc., 7720 Lehigh Ave., Niles, IL 60648; 312-966-2150

Modine Manufacturing Co., 1500 De Koven Ave., Racine, WI 53401; 414-636-1200

Moorhead Machinery & Boiler Co., 35th & University Ave., Minneapolis, MN 55418; 612-789-3541

Mueller (Paul) Co., Box 828, Springfield, MO 65801; 417-831-3000; 800-641-2830

Niagara Blower Co., 673 Ontario St., Buffalo, NY 14207; 716-876-2000; 800-426-5169

Nickell (E.L.) Co., Box 97N, Constantine, MI 49042; 616-878-2475

Nooter Corp., 1414 So. 3rd St., P.O. Box 451, St. Louis, MO 63166; 314-621-6000

North Atlantic Technologies, Inc., 7801 Bush Lake Rd., Bloomington, MN 55435; 612-835-9146

NRC, Inc., 45 Industrial Pl., Newton, MA 02164; 617-969-7690

O-I Schott Process Systems, Inc., 1640 Southwest Blvd., Vineland, NJ 08360; 609-692-4700

Oat (Joseph) Corporation, 2500 Broadway, P.O. Box 10, Camden, NJ 08104; 609-541-2900

Ohmstede Machine Works, Box 2431, Beaumont, TX 77704; 409-833-6375

Old Dominion Fabricators, 13200
Ramblewood Dr., Chester, VA 23831;
804-748-6453

Packless Industries, P.O. Box 20668, Waco,
TX 76702; 817-666-7700
Pasilac Inc., 666 Taft St. NE, Minneapolis,
MN 55413; 612-331-7710
Patterson-Kelley Co., 101 Burson Street, East
Stroudsburg, PA 18301; 717-421-7500
Perry Products Corp., Mt. Laurel Rd.,
Hainesport, NJ 08036; 609-267-1600
Pfaudler Inc., P.O. Box 1600, Rochester, NY
14692; 716-235-1000
Pick Heaters, Inc., P.O. Box 516, West Bend,
WI 53095; 800-233-9030
Proctor & Schwartz, Inc., 251 Gibraltar Rd.,
Horsham, PA 19044; 215-443-5200
Product Information Center, 17325 Euclid
Avenue, Cleveland, OH 44112;
216-531-3000

Q-dot Corp. (Combustion Engineering, Inc.),
701 North First Street, Garland, TX
75040; 214-487-1130

Refrigeration Research, 525 North Fifth St.,
Brighton, MI 48116; 313-227-1151
Repco Engineering, Inc., 1721 So. Bluff Rd.,
Montebello, CA 90648; 213-723-1106
Rigidbilt, Inc., 9240 Belmont, Franklin Park,
IL 60131; 312-671-1512
Rosenblad, Inc., P.O. Box 2325, Princeton,
NJ 08540; 609-452-2626

Schmidt-Brettenn, Inc., 1612 Locust Avenue,
Bohemia, NY 11716; 516-589-2112
Slagle Manufacturing Corp., Subsidiary of
Lamson and Sessions, 909 N. Wheeling,
Tulsa, OK 74110; 918-584-2434
Smith Engineering & Environmental Corp.,
1718 Highland, Duarte, CA 91010;
818-357-1186
Smithco Engineering, P.O. Box 571330,
Tulsa,OK 74157; 918-446-4406
Southern Heat Exchanger Corp., P.O. Box
2788, Tuscaloosa, AL 35403;
205-345-5335
Southwestern Engineering Co., Box 54940,
Terminal Annex, Los Angeles, CA 90054;
213-726-0641
Southwind Division Stewart-Warner, 1514

Drover St., Indianapolis, IN 46221;
317-632-8411
Spirec N.A., 141 Lanza Avenue, Garfield, NJ
07026; 201-478-1701
Standard Refrigeration, 2050 North Ruby St.,
Melrose Park, IL 60160; 312-345-5400
Struthers Wells Corp., Box 8, Warren, PA
16365; 814-726-1000
Sundstrand Heat Transfer, Inc., 415 E. Prairie
Ronde St., Dowagiac, MI 49047;
616-782-2141
Superior Welding Co., 900 E. Division St.,
Decatur, IL 62525; 217-422-4333

Taco, Inc., 1160 Cranston St., Cranston, RI
02920; 401-942-8000
TFC Technologies, Inc., 3401 8th Avenue
North, Birmingham, AL 35201;
205-324-7511
Thermacore, Inc., 780 Eden Rd., Lancaster,
PA 17601; 717-569-6551
Thermon Manufacturing Co., 100 Thermon
Dr., San Marcos, TX 78666;
512-396-5801
Thrush Products, Inc., (Division of Amtrol,
Inc.), Box 228, Peru, IN 46970;
317-472-3351
Tico Metal Co., 24579 Crestview Court,
Farmington Hills, MI 48018;
313-478-4704
Titanium Industries, 110 Lehigh Drive,
Fairfield, NJ 07006; 201-227-5300
Toshiba America, Inc., 2900 MacArthur
Blvd., Northbrook, IL 60062;
312-564-5631
Trane Co. (Division of American Standard),
3600 Pammel Creek Rd., La Crosse, WI
54601; 608-787-2000
Tranter, Inc., 1054 Claussen Rd., Suite 314,
Augusta, GA 30907
Turbotec Products, Inc., 651 Day Hill Road,
Windsor, CT 06095; 203-683-2005

Uniflux/Exotherm (Bryson Tank Co.), 8303
Southwest Frwy., Houston, TX 77074;
713-981-9100
United Aircraft Products, Inc., P.O. Box
90007, Dayton, OH 45490; 513-898-1811
USA Coil & Air, Inc., Box 161, Landsdown,
PA 19050; 215-622-4414

Vapor Corporation, Exchange Products Group, 6240 West Howard St., Chicago, IL 60648; 312-967-8300

Vilter Mfg. Co., 2217 S. First St., Milwaukee, WI 53207; 414-744-0111

Vogt (Henry) Machine Co., 10th & Ormsby Sts., Louisville, KY 40210; 502-634-1511

Votator Division, Cherry Burrell, 10300 Brunsen Way, Louisville, KY 40299; 502-491-4310

Vulcanium Corp., 3045 Commercial Ave., Northbrook, IL 60062; 312-498-3111

Wiegmann & Rose Int'l (RSI Corp.), 2801 Giant Road, Richmond, CA 94804; 415-529-2121

Wing Industries, Inc., 125 Moen Ave., Cranford, NJ 07016; 201-272-3600

Xchanger, Inc., 617 14th Ave. S., Hopkins, MN 55343; 612-933-2559

York International Corp., P.O. Box 1592, York, PA 17405; 717-771-7890

York-Shipley, Inc., P.O. Box 349, York, PA 17405; 717-755 = 1081

Young Radiator Co., 2825 Four Mile Rd., Racine, WI 53404; 414-639-1010

Yuba Heat Transfer, 2121 North 161st St. E., Tulsa, OK 74101; 918-234-6000

Yuba Corp., 332 Bryant Ave., Bronx, NY 10474; 212-991-0900

Zimpro, Inc., Military Rd., Rothschild, WI 54474; 715-359-7211

Zink (John) Co., 4401 So. Peoria St., Tulsa, OK 74170; 918-747-1371

Zurn Industries General Air Division (Also Energy Division) One Zurn Place, Erie, PA 16505; 814-452-2111

Zurn Constructors Inc. (Cooling Towers), P.O. Box 24718, Tampa, FL 33623; 813-870-0040

Europe

ALENA GmbH, Puchstrasse 18, 8055 Graz

AL-KO Kober KG, Hauptstrasse, 8742 Obdach

Maschinenfabrik Andritz AG, Reichsstrasse 66, 8045 Graz-Andritz

Alfred A. Arnold KG, Sportklubstrasse 6, 1020 Wien

J. L. Bacon KG, Linzer Strasse, 140, 1140 Wien

Kesselfabrik Otto Berger, KG, Brigittenauer Lände 234, 1205 Wien

Joseph Bertsch GmbH, Herrengasse 23, 6700 Bludenz

Josef Biedermann u. Söhne, Mollardgasse 73, 1060 Wien

Bohr-u. Rohrtechnik GmbH, Gonzagagasse 1, 1010 Wien

Ferd. Brunnbauer KG, AKaziengasse 36, 1234 Wien

Dumag OHG, Schwarzenbergplatz 7, 1037 Wien

ECO Wärmeaustascher GmbH, PF 18, Industriestrasse 450, 9640 Kötschach-Mauthen

Eisbär Kälte-u.Klimatechnik GmbH, Rheinhofstrasse 3, 6845 Hohenems

FBK Apparatebau-u.Handelsges.mbH, PF27, Karl Tornay-Gasse 2-4, 1234 Wien

Dipl.-Ing. Ernst P. Fischer Maschinen-u.Apparatebau GmbH Schleiergasse 17, 1100 Wien

Fröling Heizkessel-u.Behälterbau GmbH, Industriestrasse 30, 4170 Grieskirchen

Franz Greiner, Murmühlweg 10, 8112 Gratwein

Ing. Grill & Grossman, Industriestrasse 21, 4800 Attnang-Puchheim

Josef Mayr Kühl-u.Wärmetechnische Apparate GmbH & Co. KG Atzgersdorf Ziedlergasse 6-8, 1230 Wien

ÖKG Österreichische Klima-Technik GmbH, Feuerwerksanstalt, 2700 Neustadt

Metallwerk Plansee, GmbH, 6600 Reute

Obstverwertung Rauch OHG, 6830 Rankweil-Vorarlberg

Schiff & Stern KG, Erste Haidequerstrasse 3, 1110 Wien

Simmering-Graz-Pauker AG, Mariahilfer Strasse 32, 1071 Wien

Erich Strobl GmbH, Lastenstrasse 14, 5020 Salzburg

Hugo Thalhammer KG, Griesplatz 19-20, 8020 Graz

Unitherm (Öst. G. für Universelle

Wärmtechnik mbH), Nemelkegasse 9,
1110 Wien
VEW Vereinigte Edelstahlworker AG,
Elisabethstrasse 12, 1010 Wien
Vöest-Alpine AG, PF 2, 4010 Linz
Waagner-Birö AG, PF11, Stadlauer Strasse 54,
1221 Wien

BelgiumLuxembourg

Ateliers de Construction de la Meuse SA,
Sclessin-Ougrée
SA Carnoy-Vandensteen BV, Koopvaardijlaan
56, 9000 Ghent
A. De Bruyn, Hertog Van Brabantlei 9, 2710
Hoboken
Francois d'Hondt SA, Chaussée de Namur 66,
Parc Industriel, 1400 Nivelles
C. Douinet Fils Sprl, Ave. Fr. Roosevelt 27,
4540 Visé
Engetil SA, Bld. Auguste Reyers 41, BP9,
1040 Bruxelles
Ets. de Fays Sprl, Ave. des Dauphins 40,
BP17 1410 Waterloo
SA Gränges Graver NV, Molenweg 107, 2660
Willebroek
Hamon-Sobelco SA, Rue Capouillet 50/60,
1060 Bruxelles
Lauffer SA, 36 rue de la Résistance, 4530
Hermalle-sous-Argenteau Ateliers Lebrun,
Rue Maouzin 37, 7450 Mons (Nimy)
Mecan Arbed Sarl, Ave. de la Liberté 19, BP
1802, Luxembourg
Nichols Engineering SA (NESA), 51 rue du
Moulin à Papier, 1160 Bruxelles Tôleries
Gantoises SA, 9810 Ghent (Drongen)
Wanson SA, Ave. de la Woluwe 30, 1130
Bruxelles
NV Welders SA, Wijngaardveld 5, 9300 Aalst

Denmark

A/S Atlas, Baltorpvej 154, 2750 Ballerup
Burmeister & Wain Energi A/S, Teknikerbyen
23, 2830 Virum
Elektrogeno A/S, Fabriksvej 12/14, 600
Kolding
Gjettermann & Nielsen A/S, Kulholmsvej 24,
Postbox 38, 8900 Randers
Helsingö Vaerft A/S, 3000 Helsingör
Joseph Levin & Co. A/S, Vadgårdsvej 42,
2860 Söborg-Köbenhavn
Markussen & Kristiansen A/S, Sandager 14,
2600 Glostrup-Köbenhavn

Möller & Jochumsen A/S, Vejlevej 3-5, 8700
Horsens
Pasilac Therm A/S, Olaf Ryes Gade 7, 6000
Kolding
Quitzau Industri Sönderborg A/S, Grundtvigs
Alle 170/172, 6400 Söndeborg
Randers Rustfri Staal-Industrie A/S,
Bogensevej 40, 8900 Randers
Rustfri Stalmontage, Sandtoften 10, 2820
Gentofte, Köbenhavn
F. L. Smidth & Co A/S, 77 Vigerslev Allé,
2500 Valby
Thomsen Tempcold Köling og Luftkond. A/S,
Rovsingsgade 82, 2200 Köbenhavn N
Völund A/S, Abildager 11, 2600
Glostrup-Köbenhavn

Finland

A Ahlström Osakeyhtio, Boiler Works, POB
184, 78201 Varkaus
A Ahlström, POB 329, 00101 Helsinki
Oy Ja-Ro AB, PO Box 15, Rautatiekatu 24,
68601 Pietarsaari
Oumet Oy, Lumijoentie 3, 90120 Oulu 12
Rauma-Repola Oy, Snellmaninkatu, 13,
PL203, 00171 Helsinki 17
Oy W. Rosenlew AB, Antinkatu 2, PL69,
28101 Pori 10

France

Air-Industrie, 19-21 ave. Dubonnet, 92411
Courbevoie
Soc. L'Air Liquide—DCVM, 57 ave. Carnot,
94500 Champigny-sur-Marne
Armand Interchauffage SA, 63 rue de
Gerland, Lyon 7
SA Arpin, 24 q.des Gresillons, 92230
Gennevilliers
A.S.E.T., 2 rue de Bourgogne, BP 25, 69800
St. Priest
Ateliers Chantiers de Bretagne (ACB), Prairie
au Duc, 44040 Nantes
Athen SA, route de Harslzirchen, 67260
Sarre-Union
Auchère et Blavier SA, Z.I., route de la
Charité, BP 9 18390 Saint-German-de-Puy
Barbier, Benard & Turenne, 82 rue Curial,
75940 Paris 19
Barriquand Sàrl, 9-13 rue St. Claude, 42300
Roanne
Sté.B.B.M., 39 rue du 8 Mai 1945, 69320
Feyzin

Sté. J. Berthier, 309 ave. de Reims,
Villeneuve St. Germain, BP 97, 02203
Soissons

Sté. des Fabrications Biraghi-Entrepose, 75
rue Tocqueville, 75017 Paris 17

Soc. Nouvelles des Ets. Bracq-Laurent, 71 rue
Marcel Delis, BP 1, Achicourt, 62000
Arras

Ets. Brangoleau SA, 7 rue de la Traquette,
49000 Angers

Ets. Gérard Brouillon, 67 rue des Isserts,
47200 Marmande

Brown Fintube France Sàrl, Z. I. de Vongy,
74200 Thonon-les-Bains

BSL (Bignier Schmid-Laurent), 25 quai
Marcel-Boyer, BP 205, 94201 Ivry sur
Seine

Le Carbone-Lorraine, 45 rue des Acadias,
75821 Paris Cedex 17

Castor SA—S.I.R.A.C., 52 rue Ste. Héléne,
59350 Saint-André

C.E.F.A. Chaudronnerie et Forges d'Alsace,
Z.I., BP 11, 67250 Soultz-sous-Forêts

C.F.E.M. SA, BP 318, 16, 57 boul.de
Montmorency, 75781 Paris 16

C.G.E. (Cie. Générale d'Electricité) SA, 54
rue la Boétie, 75008 Paris 8

Chaudronnerie de l'Atlantique SA, 5 rue
Buffon, BP 348, 44010 Nantes

Chaudronnerie F. Groux SA, Z.I., Chemin
Corps de Garde, Chelles

La Grande Chaudronnerie de Lorraine SA,
21/23 rue du Crosne, 54000 Nancy

Chaudronnerie des Roches SA, St.
Clair-du-Rhône, 38370 Les Roches de
Condrieu

Chaudronnerie de Saint-Priest, 98 rue du
Dauphiné, 69800 Saint-Priest

CIAT, 30 ave. Jean-Falconmeir, 01350 Culoz

C.I.R.P. SA, 42 rue de Montigny, BP 77,
95101 Argenteuil

Constructions Chaudronnées du Centre SA,
Z.I. de Felet, 63300 Thiers

Constructions Mécaniques Mota SA, 38 ave.
de la Timone, 13010 Marseille 10

Les Constructions soudées du Coteau, 4 boul.
des Etines, 42120 Le Coteau

Damois Frères SA, 1-27 boul. de Graville,
76600 Le Havre

de Dietrich et Cie SA, Reichshoffen, 67110
Niederbronn-les-Bains

Delas Weir, 12-14 rue d'Alsace, 92532
Levallois

Sté. Delaunay et Fils, 6 rue de Valmy, BP
429, 76057 Le Havre

Didier–S.I.P.C., 55 rue de Chateaudun, 75009
Paris

E.C.A.N. d'Indret, 44620 La Montagne

Sté. des Echangeurs Trépaud, 44 rue la
Boétie, 75008 Paris 8

Ergé-Spirale SA, 2 rue de l'Electrolyse, BP 6,
62410 Wingles

Ets. E. Fischer et Fils SA, 62-68 ave.
Voltaire, 54303 Luneville

Fives-Cail Babcock, 7 rue Montalivet, 75383,
Paris Codex 08

Foure-Lagadec SA, 2 rue de la Vallée, 76600
Le Havre

Friedlander SA, 39 boul. Sainte-Lucie, 13262
Marseille 7

Fryer et Cie SA, 29 rue Marquis, 76100
Rouen

SA Pierre Guérin, 179 Gr4ande Rue, BP 12,
79210 Mauze-sur-le-Mignon

Ets. Eugéne Halard Sárl, 17 rue
Richard-Lenoir, 75011 Paris 11.

Heurtey Industries SA, 30-32 rue Guersant,
BP 323, 75823 Paris Cedex 17

Jeaumont Schneider, 31/32 Quai National,
92806 Puteaux

Sté. Julin, 78 rue d'Elbeuf, 76040 Rouen St.
Clement

Kestner SA, 7 rue de Toul, BP 44, 59003
Lille

Lozai Sàrl, 20 rue Etienne-Dolet, 76140 Le
Petit-Quevilly

Ets. Madelaine, 56 rue de Lille, BP 8, 59940
Estaires

Sté. Nouvelle des Ets. A. Maguin SA, BP 1,
Charmes, 92800 La Fère

Metraflu Sàrl, 24 bis Chemin des Mouilles,
69130 Ecully

Michon Frères SA, 80-88 ave. de Canéjan, BP
35, 33602 Pessac

F. Mock SA, 45 route du Général de Gaulle,
67300 Schiltigheim

Munch Frères SA, route de Champignelles,
54930 Frouard

NAT Nouvelles Applications Technologiques,
370 ave. Napoléon Bonaparte, 92500 Reuil
Malmaison

Ets. Neu, 47 rue Fourier, 59000 Lille

Neyrtec, 4 ave. de Général du Gaulle, Pont de
  Claix, BP 61X, 38041 Grenoble
Obringer, SA, 2 Place du Marché, BP 104,
  57503 St. Avold
J. Parmilleux SA, 3 bis rue de Montbrillant,
  69003 Lyon 3
Sté. Industrielle pecquet Tesson, 183 ave. du
  Général Leclerc, 78220 Virflay
Pichon SA, 24 rue Wilson, BP 104, 69150
  Decines-Charpieu
Entrepose Générale de Chauffage Industrial
  Pillard, 13 rue Raymond Teissere, BP 56,
  13268 Marseille Cedex
Ponticelli Fréres SA, 5 place des Alpes, 75013
  Paris 13
Quiri et Cie SA, 56 route de Bischwiller,
  Schiltigheim, BP 40, 67042 Strasbourg
Sand et Cie Sárl, Sainte-Agathe, 57140
  Woippy
Secathen Sárl, route de Harskirchen, BP 51,
  67260 Sarre = Union
Ets. Sereys Sárl, 11-13 rue Mondet, 33130
  Bégles
S.E.T.R.E.M., 3-5 rue de Metz, 75010 Paris
  10
SGN Saint-Gobain Techniques Nouvelles, 23
  boul. Georges Clémenceau, 92400
  Courbevoie
S.E.C.E.R. Sárl, 22 rue Etienne Dolet, 94140
  Alfortville
Soudox SA, 8 quai Mesnil-Chatelain, 60700
  Point-Saint-Maxence
S.P.E.R.I. Sárl, 9 rue Ernest Psichari, BP
  12207, 75326 Paris Cedex 7
Spiro-Gills SA, Route Départementale 50, BP
  36,59970 Fresnes-sur-Escaut
S.P.P.D. SA, 64 rue du Dauphiné, 69800
  Saint-Priest
Ets. Sprunck et Cie., 5 rue Maréchal-Joffre,
  57250 Moyeuvre-Grande
S.T.E.F.I. Procédés A. Terrier,
  Saint-Quen-l'Aumone, BP 407, 95005
  Cergy
Teal, 1 rue Isabey, 92500 Rueil Malmaison
Thimon SA, 7 rue Clement Ader, BP 175,
  73104 Aix les Bains
Tournaire SA, Route de la Paoute, Le Plan de
  Grasse, BP 4, 06338 Grasse
Sté. Trane SA, Z.I., rue de Fort, Golbey, BP
  127, 88004 Epinal
U.I.E. (Union Industrielle et d'Entreprises),
  49 bis ave Hoche, 75008 Paris 8

(SVCM) Sté. Valenciennoise de Chaudronnerie
  et ade Mécanique SA, 80 rue Jean Jaurés,
  BP 51, 59920 Quiévrechain
Vicarb SA, 24 ave. Marcel-Cachin, 38400 St.
  Martin d'Héres

Germany

Ahrens & Bode GmbH & Co. Alversdorfer
  Weg 1, 3338 Schöningen
Friedrich Ambs GmbH & Co. KG,
  Karl-Friedrich Srasse 87b, PF 1560, 7830
  Emmendingen
Anderson GmbH & Co. KG, Hesslingsweg
  71, 4600 Dortmund
Arge Jasper-Eikmeier Rohrleitungsbau,
  Josef-Baumann-Strasse, 4630 Bochum
Atlantik Gerätebau GmbH, Scharnhorststrasse
  7, PF 3570 Stadtallendorf 1
Balcke-Dürr AG, Homberger Strasse 2, PF
  1240, 4030 Ratingen 1
W. Bälz & Sohn GmbH & Co., PF 1346,
  7100 Heilbronn
Baumco G. für Anlagentechnik mbH,
  Müller-Breslau-Strasse 30a, 4300 Essen
Bavaria Anlagenbau GmbH, Hasenheide 7,
  8080 Furstenfeldbruck
Fr.u.K.Bay GmbH & Co. KG, Zeppelinstrasse
  35, 7120 Bietigheim-Bissingen
Gebr. Becker Apparatebau, Zementstrasse
  112, 4720 Beckum
Behncke, Energie-Spar-Technik GmbH,
  Wernher von Braun Strasse 1, PF 600,
  8011 Putzbrunn
Bepex GmbH, Daimlerstrasse, PF 9, 7105
  Leingarten-Heilbronn a.N.
Bergfeld u. Heider, PF 1120, 5093 Burscheid
  1
Bermann u. Sigmond GmbH & Co., PF 175,
  6660 Zweibrücken
Gottfried Bischoff GmbH & Co. KG,
  Gärtnerstrasse 44, 4300 Essen 1
BKW-Kälte-Wärme Versorgungstechnik
  GmbH, Benzstrasse 12, 7441 Wolfschlugen
Bleiwerk Goslar KG, Besserer u.Ernst, Im
  Schleeke 8, 3380 Goslar 1
Bleiwerk Gebr. Röhr KG, Bruchfeld 52, 4150
  Krefeld-Linn
Blohm = Voss AG, PF 100720, 2000
  Hamburg 1
BMA Braunschweigische Maschinenbauanstalt,
  Am Alten Bahnhof 5, 3300 Braunschweig

R. A. Böhling GmbH & Co. KG, 118
Grossmannstrasse, 2000 Hamburg 28
Bolin GmbH, Berner Strasse 42, 6000
Frankfurt 56
Bomag GmbH & Co. KG, Ströherstrasse 1-3,
3100 Celle
Joh. Heinr. Bornemann & Co. KG,
Bornemannstrasse, 3063 Obernkirchen
Borsig GmbH, Berliner Strasse 27-33, 1000
Berlin 27
Robert Bosch GmbH, Geschäftsbereich
Junkers, PF 1309, 7314 Wernau
Braukmann Kessel GmbH, 6966 Seckach
Brown Boveri-York Kälte-u. Klimatechnik
GmbH, Gottlieb-Daimler-Strasse 6, 6800
Mannheim
Buckau-Walther AG, Lindenstrasse 43, PF
100460, 4048 Grevenbroich 1
Buderus AG, PF 1220, 6330 Wetzlar
Carl Canzler Apparate-u.Maschinenbau,
Kölner Landstrasse 332-350, 5160 Düren
CASS International GmbH, Gänseberg 5, 2070
Ahrensburg
C. Conradty Nürnberg GmbH & Co. KG, PF
1752, 8500 Nürnberg 1
Deggendorfer Werft u.Eisenbau GmbH,
Werftstrasse 17, PF 1209, 8360
Deggendorf
Wilhelm Deller KG, PF 223160, 5900 Siegen
DET Dräger-Energie-Technik GmbH,
Marktstrasse 300, 4150 Krefeld
Deutsche Babcock AG, Duisberger Strasse
375, PF 100347, 4200 Oberhausen 10
Deutsche Gerätebau GmbH, PF 1140, 4796
Salzkotten
Diessel GmbH & Co., Steven 1, PF 470,
3200 Hildesheim-Bavenstedt
Artur Dietz Wärme-u.Lufttechnische Apparate,
Neumannstrasse 81, 1000 Berlin 2
DSD Dillinger-Stahlbau GmbH, Henry
Fordstrasse, PF 1340, 6630 Saarlouis
DRAKA-PLAST GmbH, Nibelungenstrasse
40, PF 210369, 5600 Wuppertal 21
Druna Heizung GmbH, Girardetstrasse 64,
4300 Essen
Eisenwerke Fried. Wilh. Düker GmbH & Co.,
Hauptstrasse 39, 8752 Laufach
Dynamit Nobel AG, Verkauf Trovidur, 5210
Troisdorf
Ebner & Co. KG, Industriestrasse 8, 6419
Eiterfeld (Rhön)

Ludwig Edel GmbH & Co,
Ravensburgerstrasse 71, PF 135, 7988
Wangen/Allgäu
effitherm Wärmetechnik GmbH, Berliner Ring
55, 4834 Harsewinkel
Edwin Eikmeier Apparate-u.Rohrleitungsbau,
Josef-Baumann-Strasse, 4630 Bochum
Eisenwerke Kaiserslautern Göppner GmbH,
Barbarossastrasse 30, 6750 Kaiserslautern
Anton Ellinghause Masschinenfabrik
u.Apparatebauanstalt KG, Oelder Strasse
4-6, 4720 Beckum (Bz. Münster)
Heinrich Engert GmbH & Co. KG, PF 1249,
5910 Kreuztal 1
Entec GmbH, Königsberger Strasse 48, 5830
Schwelm
Otto Estner GmbH, PF 110197, 4600
Dortmund 1
EVT Energie-u.Verfahrenstechnik GmbH, PF
395, 7000 Stuttgart 1
Fickel Maschinen u.Kessel GmbH,
Nürnberger Strasse, 8820
Gunzenhausen/Bay.
Karl Fischer Industrieanlagen GmbH,
Holzhauser Strasse 159-165, 1010 Berlin
27
Freier Grunder Eisen-u.Metallwerke GmbH,
Fritz-Schäfer-Strasse 26, 5908
Neunkirchen
Friedrichsfeld GmbH, Steinzeugstrasse 50,
6800 Mannheim 71
C. Fuhrmann Apparatebau, Braunschweiger
Strasse 38, 3307 Schöppenstadt
FUNKE Wärmeaustauscher Apparatebau KG,
PF 10, 3212 Gronau (Leine)
GEA Ahlborn GmbH & Co. KG, Voss-Strasse
11-13, 3203 Starstedt
GEA Kühlturmbau u.Systemtechnik GmbH,
Königsallee 43-47, 4630 Bochum 1
GEA Luftkühlergesellschaft Happel GmbH &
Co. KG, Dorstener Strasse 18, 4690 Herne
2
Geka-Wärmetechnik, Gottfried Kneifel GmbH
& Co. KG, Dieselstrasse 8-10, 7500
Karlsruhe 41
GEMOC Verfahrenstechnik GmbH & Co.,
Wolfshofstrasse 21, 4600 Dortmund 76
Gerberich GmbH & Cie. Maschinenfabrik,
Zielstrasse 6, 6800 Mannheim 1
Gustav F. Gerdts KG, Hemmstrasse 130, 2800
Bremen 1

GFG Ges. f. Freizeit-u.Gesundheitstechnik
mbH, Munchner Strasse 1, 8084 Inning
GHH Gutehoffnungshütte Sterkrade AG,
Bahnhofstrasse 66, 4200 Oberhausen 11
Graphit Apparatebau Dipl.-Ing. Hans
Neumann, Feldbergerstrasse 18, 7867
Maulburg
GST Ges. f. Systemtechnik mbH, Am
Westbahnhoff 2, 4300 Essen 1
Hans Güntner GmbH, Industriestrasse 14,
8080 Furstenfeldbruck
W. Ernst Haas & Sohn GmbH & Co., PF 46,
6349 Sinn-Hess 1
Gebr. Hagemann KG Apparatebau,
Sudhoferweg 55, 4720 Beckum
Gottfried Hagen AG, Rolshoverstrasse 95/101,
500 Köln 91
Halberg Maschinenbau GmbH, Halbergstrasse
1, 6700 Ludwigshafen am Rhein
HAW Harzer Apparatewerke KG, 3205
Bockenum 2 (OT Bornum)
W. C. Heraeus GmbH, Heraeusstrasse 12-14,
6450 Hanau
A. Hering AG, Herrnhüttestrasse 33-35, 8500
Nürnberg
Herwi-Solar GmbH, Rollfelder Strasse 17/18,
8761 Röllbach
Hiros Denco GmbH, Kaiserstrasse 109, 4050
Mönchengladback 1
Hock-Temperatur-Technik GmbH, PF 1922,
4900 Herford
Holstein u.Kappert GmbH, Zechenstrasse 49,
4750 Unna-Köningsborn
HDW Howaldtswerke-Deutsche Werft AG, PF
111480, 2000 Hamburg 11
Hydac Ges. f. Hydraulik-Zubehör mbH,
Industriestrasse, 6603 Sulzbach
Janetschek & Scheuchl
Verfahrenstechnik-Anlagenbau, J.-F.
Kennedy-Strasse 9, PF 220, 8038
Gröbenzell
Deutsche Richard kablitz Ges., Bahnhofstrasse
72-78, 6970 Lauda
Paul Kahle GmbH Rohrleitungsbau, Kölner
Strasse 170, PF 44620, 4000 Düsseldorf
Fr. Kammerer GmbH, Goethestrasse 2-8,
7530 Pforzheim
Kermi GmbH & Co., KG, 8350
Plattling/Pantofen 54
KHD Kumboldt Wedag AG, Wiersbergstrasse,
5000 Köln 91 (Kalk)

Walter Kidde GmbH, PF 1909, 2120
Lüneburg
KKW Kulmbacher Klimageräte-Werk GmbH,
PF 1523, 8650 Kulmbach
Alb. Klein GmbH & Co. KG,
Konrad-Adenauer-Strasse 108, 5241
Niederfischbach/ Sieg
Otto Klein GmbH, Im Wiesengrund 11, 3503
Lohfelden 2
Theordor Klein, Knollstrasse 20-28, 6700
Ludwigshafen
Kleinewefers Energie-u.Umwelttechnik
GmbH, Kleinewefers-Kalanderstrasse,
4150 Krefeld
Eugen Klöpper GmbH & Co., PF 6868, 4600
Dortmund 1
A. Knoevenagel GmbH & Co. KG,
Emil-Meyer-Strasse 20, 3000 Hanover 1
Kölsch-Fölzer-Werke AG, Hohler Weg 75,
5900 Siegen
Konus-Kessel Ges. f. Wärmetechnik mbH &
Co., Robert-Bosch-Strasse 3-5, 6830
Schwetzingen
Körting Hannover AG, PF 911363, 3000
Hannover 91
fried. Krupp GmbH, Altendorfer Strasse 103,
4300 Essen 1
Krupp Industrie-u.Stahlbau,
Franz-Schubert-Strasse 1-3, 4100 Duisburg
14
KÜBA Kühlerfabrik Heinrich W. Schmitz
GmbH, PF 1126, 8021 Baierbrunn
Kübler Industrieheizung GmbH, Neckarauer
Strasse 106-116, 6800 Mannheim 1
Kühlturmbau Ernst Kirchner GmbH,
Königsallee 47, PF 500941, 4630 Bochum
1
Kühlturm GmbH, Köllestrasse 17, 750
Karlsruhe 21
Kühlturmtechnik Emmerich Rehsler, Grubach
Weg 27, 8990 (Lindau (Bodensee)
Lahn-Technik GmbH, PF 240, 5408
Nassau/Lahn
Langbein & Engelbrecht GmbH, Hattinger
951, PF 500175, 4630 Bochum 5
Alois Lauer Stahl-u.Rohrleitungsbau GmbH,
Industriestrasse 1, 6638 Dillinger
Ferdinand Lentjes
Dampfkessel-u.Maschinenbau, Hansa-Allee
305, 4000 Düsseldorf 11
Linde AG, Werksgruppe TVT München, Carl

von Lindestrasse 6-12, 8023
Höllriegelskreuth

Linde AG Werksgruppe
Kälte-u.Einrichtungstechnik, Sürther
Hauptstrasse 178, 5000 Köln 50 (Surth)

Eisenwerk Theodor Loos GmbH, Nürnberger
Strasse 73, 8820 Gunzenhausen

Lüco Wärmetechnik GmbH, Fröndenberger
Strasse 42, PF 620, 5750 Menden 1

Regelstechnische Geräte P. Lüthge GmbH,
Handelshof 26, 2805 Stuhr 1

Luwa-SMS GmbH Verfahrenstechnik
Butzbach, Kaiserstrasse 13-15 6308
Butzbach Wilhelm Mass GmbH & Co.
KG, Zeche Ernestine 18, 4300
Essen 1

MANN Maschinen-u. Stahlbau,
Frankenstrasse 150, PF 440100, 8500
Nürnberg 44

Mannesmann-Anlagenbau AG, Theodorstrasse
90, 4000 Düsseldorf 30

Molenda KG Mess-u.Regeltechnik, Breslauer
Strasse 75, 5630 Remscheid

Julius Montz GmbH, PF 530, 4010 Hilden

Hubert Münch GmbH, Steinstrasse 22, 5000
Köln 1

Fr. August Neidig Söhne Maschinenfabrik
KG, Friesenheimer Strasse 3-8, 6800
Mannheim 1

Neue Wärmetechnik GmbH, Warnstedtstrasse
28/32, 2000 Hamburg

F. A. Neuman GmbH & Co. KG, PF 1260,
5180 Eschweiler

Gebr. Neunert GmbH, PF 840, 2200
Elmshorn

NIEROS-Fördertechnik GmbH,
Industriestrasse 38, 7253 Renningen

Ges.f.Oeltechnik mbH, Lessingstrasse 32,
6833 Waghäusel

Ohl-Industrietechnik, Theodor Ohl AG,
Blumenröderstrasse 3, 6250 Limburg
(Lahn)

Olsberg Ges. f. Verwalting u.Vertrieb mbH,
PF 27, 5787 Olsberg

Josef van Opbergen KG, Rampenstrasse 2, PF
807, 4040 Neuss 22

Oschatz GmbH, Westendhof 10-12, PF
102843, 4300 Essen 1

Pampus-Fluorplast, Am Nordkanal 37, 4156
Willich 3

Pass + Co. GmbH + Co. KG, PF 210405,
5900 Siegen 21

Dipl.-Ing. Penkert, Keferloherstrasse 43, 8000
München 40

Claudius Peters AG, Kapstadtring 1, 2000
Hamburg 60

Gebr. Plersch Spezialfabrik f. Kältemaschinen
GmbH & Co., Dietenheimer Strasse
12-16, PF 228, 7918 Illertissen

Pneumatex GmbH, Bosenheimer Strasse 85,
PF 127, 6550 Bad Kreuznach

Paul Pollrich GmbH & Co., PF 609, 4050
Monchengladbach 1

Rekuperator KG Dr. Ing. Schack & Co.,
Sternstrasse 9-11, 4000 Düsseldorf 30

Renzmann & Grünewald KG, 6551
Monzingen/Nahe

Rhein-Bleiapparatebau Hagen & Röhr GmbH,
Bruchfeld 52, PF9007, 4150 Krefeld-Linn

Rohleder Kessel-u.Apparatebau GmbH, PF
300829, 7000 Stuttgart 30

Rosenthal Technik AG, Wittelsbacherstrasse
49, 8672 Selb

Kurt Rosskamp Apparate-u.Behälterbau,
Wittauer Strasse, 7180 Crailsheim

Apparatebau Rothemühle Brandt & Kritzler
GmbH, Wildenburgerstrasse 1, PF 5140,
5963 Wenden (Biggetal) 5

Sabroe Kältetechnik GmbH, PF 787, 2390
Flensburg Weiche

S. & K. Graphit Apparatebau GmbH, PF
1241, 5276 Wiehl 1

SAPRO Ingenieurbüro von Stosch KG,
Fichtenweg 2, 6270 Idstein

Schäfer Werke GmbH, PF 1120, 5908
Neunkirchen

C. F. Scheer & Cie, Ludwigsburger Strasse
13, 7000 Stuttgart 30

Schmidding-Werke Wilhelm Schmidding
GmbH & Co., Emdener Strasse, 10, 5000
Köln 60 (Niehl)

W. Schmidt KG, Pforzheimer Strasse 46, 7518
Bretten (Baden)

(SHG) Schmidt'sche Heissdampf GmbH,
Ellenbacher Strasse 10, 3500
Kassel-Bettenhausen

R & G Schmöle Metallwerke GmbH & Co
KG, Fröndenberger Strasse 29, PF 620,
5750 Menden 1

Aug. Schnakenberg & Co. GmbH,
Beyenburger Strasse 146-168, 5600
Wuppertal 2

Peter Scholz Ingenieurges.mbH,
Mauerkircherstrasse 4, 8000 Munchen 80

Harald Schönstein Apparatetechnik PF
    520403, 2000 Hamburg 52
Schott Glaswerk Geschäftsbereich
    Chemie-Apparatebau, Hattenbergstrasse
    10, 6500 Mainz
Wilh. Schulz GmbH, Kuhleshütte 85, 4150
    Krefeld 13
Schwarz Apparate-Behälterbau GmbH,
    Gottlieb-Daimler-Strasse 9, 7297
    Dornstetten 4
Schwelmer Eisenwerk Müller & Co.
    GmbH,Loher Strasse 1-3, 5830 Schwelm
Dipl.-Ing. Paul Schwingel GmbH,
    Overfeldweg 80, 5090 Leverkusen 1
SEH-Energietechnik GmbH, Matzenhofer Weg
    10, 7919 Unterroth
Seico Industrie-Elektrowärme GmbH, An der
    Autobahn 29, PF 1107, 3012
    Langehagen/Hann.
Selas-Kirchner GmbH, Schillerstrasse 44,
    2000 Hamburg 50
Siegener AG Geisweid, Birlenbacher Strasse
    17, 5930 Hüttental-Geisweid
Sigri Elektrographit GmbH,
    Werner-von-Seimens-Strasse 18, 8901
    Meitingen b.Augsburg
Solar-Energie-Technik GmbH, Industriestrasse
    1-3, PF 6831, Altlussheim 1
Spillingwerk GmbH, Werftstrasse 5, PF
    110529, 2000 Hamburg 1
Stahl-Apparate-u.Gerätebau GmbH, Grosser
    Stellweg 23, PF 1127, 6806 Viernheim
Johann Stahl Kessel-Apparatebau GmbH &
    Co. KG, Rheinaniastrasse 58-60, 6800
    Mannheim-Neckarau
J. Stahlhaacke, Bäckerweg 16, 3012
    Langenhagen 4/Engelbostel
Standard Filterbau, Dr. E. Andreas GmbH &
    Co. Rosnerstrasse 6/8, 4400 Münster
Standardkessel GmbH, Baldusstrasse 13, PF
    120403, 4100 Duisburg 2
Paul Stehning GmbH, Frankfurter Strasse,
    6394 Grävenwiesbach
Anton Steinecker Maschinenfabrik GmbH, PF
    1960, 8050 Freising
L. & C. Steinmüller GmbH, PF 100855, 5270
    Gummersbach 1
W. Strikfeldt & Koch GmbH, PF 1164, 5276
    Wiehl 1
Sümak Maschinenfabrik GmbH & Co.,
    Ditzinger Strasse 75, 7250 Leonberg 6

Sumitomo Metal Industries Ltd., Königsallee
    48, 4000 Düsseldorf
Thermal Werke Wärme-Kälte-Klimatechnik
    GmbH, PF 1680, 6832 Hockenheim
Thermotech GmbH, Bergrund 11, 7554
    Kuppenheim
Thies GmbH & Co., Borkener Strasse 155,
    4420 Coesfeld/Westfalen
Thyssen Maschinenbau GmbH Witten-Annen,
    Stockumer Strasse 28, 5810 Witten-Annen
Friedrich Uhde GmbH Werk-Hagen,
    Buschmühlenstrasse 20, 5800 Hagen 1
Union Carbide Deutschland GmbH,
    Mörsenbroicher Weg 200, 4000 Düsseldorf
    30
VAD Beschichtungstechnik GmbH & Co. KG,
    Mittelweg, 8752 Kleinostheim
Venus Wärmetechnik AG, Industriestrasse 1,
    4802 Halle/Westf.
Viessman Werke AG, Schäferstrasse, 3559
    Allendorf (Eder)
Voith Turbo GmbH & Co. KG, Voithstrasse 1,
    PF 460, 7180 Crailsheim
Fritz Voltz Sohn GmbH & Co., Solmsstrasse
    56-68, PF 900960, 6000 Frankfurt/Main
    90
Ges. Wärme Kältetechnik mbH, PF 2128,
    5883 Kierspe 2
Werner & Pfleiderer, PF 301220,
    Theodorstrasse 10, 7000 Stuttgart 30
A G Weser Seebeckwerft, Riedermannstrasse
    12, 2850 Bremerhaven 1
WETAG Wärmetechnische Apparateges.
    Müller & Co., PF 1345, 3203 Sarstedt
Wibau-Maschinenfabrik Hartmann AG, PF
    1520, 6460 Gelnhausen 1
Wichmann & Söhne, PF 2969, 5880
    Lüdenscheid
F. Widmann & Sohn GmbH & Co. KG,
    Wittenerg Strasse 3, PF 810429, 6800
    Mannheim 81-Rheinau
Apparatebau Wiesloch GmbH, Baiertaler
    Strasse 92-98, PF 1130, 6908 Wiesloch b.
    Heidelberg
Weissner GmbH, PF 3120, 8580 Bayreuth
Theodor Winkel KG, PF 2107, 4190
    Kleve 1
H. Wolff Apparatebau KG, PF 2067, 4937
    Lage, Lippe
P. J. Wolff & Söhne, Elberfelder Strasse 6-12,
    5160 Düren

Ireland

Heat Recovery Ireland Ltd., Mill Street Road, Macroom, Co. Cork

Italy

Aeroto Srl, Via Privata Santa Maria 17, 20047 Brugherio (MI)

Aertermica Srl, Viale Teodorico 19/2, 20149 Milano

A.T.I. Snc, Via Procaccini 41, 20154 Milano

Antonio Badoni SpA, Corso Matteotti 7, 22053 Lecco (CO)

Gino Battaglini SaS, Via G. P. Orsini 10/12, 50126 Firenze

Fratelli Berengo SaS, Via Elettricità 2, 30175 Venezia-Marghera

Boldrocchi SpA, Via Trento e Trieste 93, 200046 Biassano (MI)

Ing. Bono SpA, Via Resistenza 12, 20068 Peschiera Borromeo (MI)

Bosco Industrie Meccaniche SpA Terni, Piazzale A. Bosco 3, 05100 Terni

Breda Standard SpA, Z. I., Via R. De Blasio, 70010 Bari

Breda Termomeccanica, Viale Sarca 336, 20126 Milano

C.A.M.P.I. SpA, Via Diamantina 15, Mizzana, 44100 Ferrara

(CNR) Cantieri Navali Ruiniti SpA, Via Cipro 11, 16129 Genova

C.M.C. SpA, Via Asiago 244, 21042 Caronno Pertusella (VA)

(COMBER) Costr. Meccaniche Bergamasche SpA, Nansen 15, 20156 Milano

COMTEA SpA, Via Tagliamento 15, 20048 Carate Brianza (MI)

Contardo SpA, CP 13, 21040 Uboldo (VA)

COREMA Impianti SaS, Via de Amicis 59, 20092 Cinisello, Balsamo (MI)

(CMB) Costr. Meccaniche Bernardini SpA, Via della Petronella 2, 00040 Pomezia (Roma)

(CMP) Costr. Meccaniche Pesaro SpA, Via Taramelli 26, 20124 Milano

(CMR) Costr. Metallurgische Riuniti Srl, Strada Nazionale 42-50, 31050 Villorba (Treviso)

Fratelli Delfino SpA, Via Anguissola 50/B, 20146 Milano

del Monego SpA, Piazza della Repubblica 8, 20121 Milano

Elettrografite Meridionale SpA, Via dell'Industria, 81100 Caserta

FA.S.T. Srl. Località Gerlotti, 15040 Castalletto Monferrato (AL)

F.A.T.T. SpA, Z.I., CP 75, 90044 Carini (PA)

Favra International SpA, Via Guerrazzi 9, 20145 Milano

FBM Costruzione Meccaniche SpA, Via Lambruschini 15, Milano

FERGAL Stabilimenti SpA, Via Europa 35, 20010 Pogliano Milanese (Ml)

Fillipi Fochi SpA, Via Portanova 3, 40123 Bologna

Fonderie e Officine di Saronno SpA, Via Legnano 6, 20121 Milano

Forain Srl, Via G. B. Boeri 11, 20141 Milano

Off. Mecc. Fradelloni SpA, Via Bandello 19, 09100 Cagliari

Nuova Frau SpA, 36010 Carrè (Vicenza)

Frigotecnica Industriale Chiavenna SpA, 23022 Chiavenna

Fratelli Gianazza SpA, Viale Cadorne 78, 20025 Legnano (MI)

Umberto Girola Impresa, Via Solferino 7, 20121 Milano

Heat Exchangers/Burioli Srl, Via Bellarmino 13, 20141 Milano

(ICM) Industrie Costr. Metalliche, Via Marconi 42, 30020 Quarto d'Altino

Industria Meccaniche di Scardellato, Via XV Juglio 14, 31100 Treviso

Industrie Meccaniche di Bagnolo SpA, Strada Paullese, 26010 Bagnolo Cremasco

Italimpianti (Soc. Ital. Impianti pA), Piazza Piccapietra 9, 16121 Genova

Jucker SpA, Via Campanini 6, 20124 Milano

Tito Manzini e Figli SpA, Via Tonale 11, 43100 Parma

G. Mazzoni SpA, Costr. Impianti Chimici, CP 421, 21052 Busto Arsizio

Nicola & albia Srl, Villa Pompea 1, 20060 Cassina de Pecchi (MI)

Nuovo Pignone Industria Meccaniche SpA, Via F. Matteucci 2, 50127 Firenze

O.C.S. SpA, Strada Battaglia, 35020 Albignasego (PD)

OLMI, Viale Europa, 24040 Suisio (Bergamo)

OLSA SpA, Via Porta d'Arnolfo 35, 20046 Biassono (MI)

Pressindustria SpA, Via Porta d'Arnolfo 35, 20046 Biassono (MI)

Ruths SpA, Via Palmaria, 16121 Genova

Guido Sartori SpA, CP 314, San Marco 2991,
30100 Venezia
Sasso, Lungo Torrente Secca 54, 16163
Genova
S.C.A.M. SpA, Via Arcivescovado 7, 10121
Torino
S.I.A.D. Macchine Impianti SpA, Via
Canovine 2/4, 24100 Bergamo
S.I.A.T. SpA, Via E. Cantoni 24, 20156
Milano
Stà. per Impianti Generali Srl, Viale Lunigiana
23, 20125 Milano
S.M.I.M. SpA, Piana de Signore, 93012 Gela
(CL)
SO.CO.ME, Strada di Serrabella 11, Sesto di
Rastignano (BO)
Taurus Srl, Z.I., Strada Monte d'Oro 11,
34147 Trieste
Termotecnica Industriale SpA, Via Cristina
Belgioioso 70, 20157 Milano
Franco Tosi SpA, Via Brisa 3, 20123 Milano
Tycon SpA, Via Kenedy, 30027 S. Dona di
Piave (VE)
Villa & Bonaldi, Via Stazione 9, 26013 Crema
(CR)
Worthington SpA, CP 3474, 20100 Milano

Netherlands

AER BV, Noordestraat 17, 9611 AA
Sappemeer
Machinefabriek Arnhem BV, Industriestraat 9,
PB 161, 6800 AD Arnhem
v/h E. H. Begemann NV, Kanaaldijk NW 31,
5700 AA Helmond
Breda Boiler, PB 3260, Breda
Bronswerk Heat Transfer BV, Stationsweg 22,
3860 AB Nijerk
Chemtac BV, Belvedereweg 7, PB 52, 3762
EE Soestdijk
Eskla BV, De Boelelaan, 7 PB 7811, 1008 AA
Amsterdam
F.I.B. Industriele Bedrijven BV, Einsteinweg
23, PB 314, 8901 BC Leeuwarden
Gems Metaalwerken BV, Zutphenseweg 7, PB
2, 7250 AA Vorden
Hudson Luftkuhler, 12 Amaliastraat, den Haag
Hydrochemie-Conhag BV, PB 177, 2280 AD
Ryswijk
Kasteel BV, Hoge Rijndijk 205, 2832 AL
Zouterwoede
Klima BV, Rogier van der Weijdenstraat 45,
5600 AG Eindhoven

Korpershoek BV, Brielselaan 125, 3081 AB
Rotterdam
Lalesse BV, P. Calandweg 11, 6827 BJ
Arnhem
Mueller Europa BV, Lievelderweg 68, PB 20,
7130 AA Lichtenvoorde
N. E. M. Boilers Division, Hollandse
Constructie Groep BV, PB 56,
Zoeterwoudsweg 1, 2300 AB Leiden
Noord Nederlandse Machinefabriek BV, Sint
Vitusstraat 81, 9673 AM Wiunschoten
Novum BV, Landweerstraat 91, 5340 AB Oss
Pijttersen BV, Pampuskade 13, PB 192, 8600
AD Sneek
Sif Holland, Mijnheerkensweg 33, 6041 TA
Roermond
Sombroek Zaandem BV, Aris Van Broekweg
9, PB 180, 1500 ED Zaandam
Stork Ketel-en Apparatenbouw, PB 20, 7550
GB Hengelo
Tankfabriek-Kooiman NV, Noordhoek 23, PB
10, 3350 AA Papendrecht

Norway

A/S Akers mek Verksted, Munkedamsveien
45, Oslo 2
Ali Installation A/S, PB 89, 3136
Tónsberg-Melsomvik
A. M. Anderssen Mek Verksted A/S,
Tevlingveien 4, Oslo 10
Busch Maskin & Montering A/S, Tunejordet,
1701 Sarpsborg
Fredriksstad Mek Verksted A/S, PB 96, 1601
Fredrikstad
Anders Halvorsen A/S, PB 173, 4401
Flekkefjord
Brödr. Hetland A/S, PB 235, 4341 Bryne
Kvaerner Varme A/S, Kvaernerveien 10, Oslo
1
Myrens Verksted A/S, PB 4200, Oslo 4
Stord Bartz Industri A/S, C. Sundtsgate 29,
5000 Bergen

Portugal

Equimetal, R. Rosa Araùjo, 2-5, 1200 Lisboa

Spain

Adaibra SA, Carratera de Calafell Km 9.3
Apdo. De. Correos 11, San Baudillo de
Llobregat, Barcelona
Astilleros Españoles SA, Apdo. 815, Padilla
17, Madrid 6

Auxiliares de Galvonatecnia SL, Basilio
Armendàriz 1, Burlada (Pamplona)
Averly SA, Paseo María Agustín 57-59, Ato.
36, Zaragoza 4
Beotibar SL, Belaunza, Tolosa (Guipuzcoa)
Comaq SA, Apdo. 2, Barrio Soravilla,
Andoain (Guipúzcoa)
Construcciones Navales P. Freire S.A., Avda,
Orillamar, Apdo. 2001, Vigo (Pontevedra)
Gosag SA, Menéndez Pelayo 6, Apdo. 14.
486, Madrid 9
Inoxa SA, Alcalde Martín Cobos, Apdo. 248
Burgos
Enrique Lorenzo y Cía. SA, Espiñeiro, Apdo.
1507, Vigo (Pontevedra)
La Maquinista Terrestre y Maritima SA,
Fernando Junoy 2-64, Apdo. 94,
Barcelona 30
Metalúrgica del Tormes SA, Plaza de la
Justicia 1, Apdo. 50, Salamanca
Talleres Orva SL, Barrio Careaga 107, Apdo.
109, Baracaldo (Vizcaya)
Andrés Suría, Napoles 49, Santa Maria de
Barbara (Barcelona)

Sweden

AGA-CTC Industri AB, Box 60, 372 01
Ronneby
Alfa Laval AB, Munkhäder AB, Främby, Box
342, 791 28 Falun
Asarums Industri AB, S. Industriv., 292 02
Asarum
Broby Rostfria AB, Industrig., 280 60 Brosby
Bulthen-Kanthal AB, 734 01 Hallstahammar
J. Edholms Verkstäder AB, Industrivej 6, Box
9, 592 00 Vadstena
AB K. A. Ekström & Son, Industriomradet
Näsby, Box 3007, 291 03 Kristianstad
Fläkt Evaporator AB, 614 00 Söderköping
Generator Industri AB, Box 95, 43301 Partille
Götaverken Ångteknik AB, Stjärng. 9, Box
8734, 402 75 Göteberg
Götaverken Motor AB, Box 8843, 402 71
Göteborg
Gränges Engineering AB, Birger Jarlsg. 52,
103 26 Stockholm
Hässleholms Verkstäder AB, N. Industrig.,
Box 194, 281 01 Hässleholm
Holmstrands Plåtindustri AB, Box 63, 660 50
Valberg
Karlskronavarvet AB, Box 1008, 371 24
Karlskrona

Martin Larsson Tryckkärl AB, S. Bang., Box
73, 690 70 Pålsboda
Luftkonditionering AB, Framnäsbacken 2, Box
110, 171 22 Solna
AB Carl Munters, Industriv. 1, 191 20
Sollentuna
Örnsköldsviks Mekaniska Verkstad AB,
Ångermanlandsg. 17, 891 00 Örnsköldsvik
Parca Norrahammer AB, 562 00
Norrahammer
PM-Luft AB, Frejgatan 14, Box 300, 535 00
Kvanum
AB Rosenblads Patenter, Box 29012, 100 52
Stockholm
Stal-Laval Apparat AB, Köpetorps
Industriområde, 581 01 Linköping
Stamo Maskin AB, Navigatorg. 6, 724 10
Västerås
Studsvik Energiteknik AB, 611 82 Nyköping
Sunrod International AB, Box 2501, 175 02
Järfälla
J. W. Torell AB, Tornbyv., Box 3105, 580 03
Linköping
Uddcomb Sweden AB, Box 1040, 371 24
Karlskrona

Switzerland

Aluminium AG Menziken, 5737 Menziken
Angst & Pfister AG, Thurgauerstrasse 66,
8052 Zürich
Apaco AG für Apparatebau, 4203 Grellingen
Bertrams AG, Eptingerstrasse 41, 4132
Muttenz
BBC AG Brown, Boveri & Cie, Haselstrasse,
5401 Baden
BTA-Beheizungstechnik AG, Hofackerstrasse
12, 4132 Muttenz
Gebr. Bühler AG Maschinenfabrik, 9240
Uzwil
Brunner AG, Bachtelstrasse 34, 8636 Wald
Buss AG, Hohenrainstrasse, 4133 Pratteln
Du Pont de Nemours International SA, 1211
Geneve 24
Elstet AG, Sternenfeldstrasse 40, 4127
Birsfelden
Escher Wyss AG, Escher-Wyss-Platz, 8023
Zürich (Stammhaus)
Estrella AG Apparatebau u. Emaillierwerk,
Lohweg 19, 4107 Ettingen
Fryma Maschinen AG, Theodorshofweg 20,
4310 Rheinfelden
Inox AG, Industriestrasse 180, 4600 Olten

Inter-Hoval AG, neugat, 9490 Vaduz,
Liechtenstein

Jäggi AG, Wangenstrasse 102, 3018 Bern

Koehler Bosshardt AG, Hochbergerstrasse 15,
4016 Basel

Koenig AG, St. Gallerstrasse 2, 9320 Arbon

Kühni AG, Mühlebachweg 9-15, 4123
Allschwil 2

Dipl.Ing. H. List Industrielle
Verfahrenstechnik, Vogelmattstrasse 15,
4133

Pratteln 2

Lugat AG Luft-u. Gastechnik, Isteinerstrasse
70, 4058 Basel

Luwa AG Verfahrenstechnik Zürich,
Anemonenstrasse 40, 8047 Zürich

Relutherm AG, Burgfelderstrasse 24, 4002
Basel

Gebr. Sulzer AG, Zürcherstrasse 9, 8401
Winterthur

J. M. Voith AG, Linzer Strasse 55, 3100 St.
Pölten

von Roll AG, 4702 Oensingen

Wotag AG, Kronleinstrasse 8, 8044 Zürich

United Kingdom

AAF Ltd, Bassington Industrial Estate,
Cramlington, Northumberland NE23 8AF

Acalor International Ltd, Crompton Way,
Crawley, W. Sussex RH10 2QR

Accumatic Engineering Group, Llay Hall,
Cefny-y-Bedd, Wrexham, Clwyd LL12
9YH

Acoustics & Envirometrics Ltd, Winchester
Road, Walton-on-Thames, Surrey KT12
2RP

Actric Ltd, Bull Lane Works, Brandon Way,
West Bromwich, W. Midlands, B70 9PQ

Advances Heat Pumps Ltd, 19 Priory Road,
Wrentham, Beccles, Suffolk N34 7LR

Aerogenerators Ltd, Anstey Mill Lane Works,
Newman Lane, Alton, Hants. GU34 2QW

Air Pollution Control Ltd, 413 Sydenham
Road, Croydon CR9 2XQ

Allbook & Hashfield Ltd, 153 Huntington
Street, Nottingham NG1 3NG

Anex (UK) Ltd, 39 Frogmoor, High
Wycombe, Bucks. HP13 5DG

Anstee & Ware Ltd, Avonmouth Way,
Avonmouth, Bristol BS11 9HE

APV Baker Plc, PO Box 4, Manor Royal,
Crawley, West Sussex RH10 2QB

APV Hall Products Ltd, Home Gardens,
Dartford, Kent DA1 1EP

Aquafan Cooling Towers Ltd, 183 Hampton
Road, Twickenham, Middx. TW2 5NG

Arcleth Fabrications Ltd, Bridgefield Works,
Elland Bridge, Elland, W. Yorks.

Armca Specialities Ltd, Armca House, 102
Beehive Lane, Ilford, Essex IG4 5EQ

Armstrong Patents Co. Ltd, GibsonLane,
Melton, North Ferriby, Humberside HU14
3HY

Associated Lead Manufacturers Ltd, Lead
Works Lane, Chester CH1 3BS

Babcock Energy Ltd, Maypole House,
128/132 Borough High Street, London SE1

Baltimore Aircoil International, 3 Old Walsall
Road, Great Barr, Birmingham B42 1NN

W. B. Bawn (Bickford) Ltd, Baron Works,
Russell Gardens, Wickford, Essex SS11
8BX

Beltran Ltd, Sunderland House, Sunderland
Street, Macclesfield, Cheshire SK11 6JF

Charles Benn & Sons Ltd, Fulcan Foundry,
Stourton, Leeds LS10 1RX

Beta-Plus Ltd, 177 Haydons Road, London
SW19 1AW

F. H. Biddle Ltd, Newtown Road, Nuneaton,
Warwicks.

Birwelco Ltd, Plant & Construction Division
Mucklow House, Mucklow Hill,
Halesowen, West Midlands, B62 8DG

Biwater Heat Exchangers Johnson-Hunt Ltd.,
Hall Works, Astley Street, Dunkinfield,
Cheshire SK16 4QT

R. Blackett Charlton (Vessels) Ltd, Jubilee
Estate, Ashington, Northumberland, NE63
8UD

Blairs Ltd, 143 Woodville Street, Glasgow
GS1 2RH

Blundell and Crompton Ltd, West India Dock
Road, London E14 8HA

BMB Industries Ltd, 4 High Street, Billericay,
Essex

E. J. Bowman (Birmingham) Ltd, Aston
Brook Street East, Birmingham B6
4AP

Bramah Process Plant Ltd, Holbrook Works,
Halfway, Sheffield S19 5GZ

Brierley, Collier and Hartley Equipment Ltd,
Bridgefield Road, Rochdale, Lancs. OL11
4EX

British Appliances Manufacturing Co. Ltd,

Longclose Works, 52 Dolly Lane, Leeds LS9 8NJ

Peter Brotherhood Ltd, Lincoln Road, Peterborough PE4 6AB

BTR Silvertown Ltd, Chemical Plant Division, Horninglow Road, Burton on Trent, Derbyshire DE13 0SN

Buckley and Taylor Ltd, Castle Iron Works, Owerens Street, Oldham, Lancs OL4 1HJ

Bunting Titanium Ltd, 139 Middlemore Industrial Estate, Handsworth, Birmingham B21 0AY

Burke Thermal Engineergin Ltd, Mill Lane, Alton, Hants.

Burnett & Rolfe Ltd, Commissioners Road, Strood, Rochester, Kent ME2 4EJ

W. P. Butterfield (Engineers) Ltd, PO Box 38, Otley Road, Shipley, W. Yorks.

Cabot Alloys Europe Ltd, Earlstress Road, Corby, Northants, NN17 2AZ

Caird & Rayner-Bravac Ltd, Otterspool Way, Watford By Pass, Watford, Hertz. WD2 8HL

Campbell Hardy Ltd, PO Box 25, 6-8 Nevill Terrace, Tunbridge Wells, Kent TN4 8YD

C & P Tube Fabrications Ltd, Bishopsgate Works, Union Road, Oldbury, Warley, W. Midlands B69 3ES

Carter Industrial Products Ltd, Bedford Road, Birmingham B11 1AY

Castletree Engineering Ltd, 26 Hadham Road, Bishops Stortford, Herts CM23 2QS

Charlton, Weddle & Co. Ltd, White Street, Walker, Newcastle upon tyne NE6 3QH

Charnock & Co. Ltd, Unit 3, Hall Street, Dudley, W. Midlands DY2 7DJ

Chem-Plant Stainless Ltd, Coppermill Lane, West Hyde, Rickmansworth, Herts. WD3 2XS

Clark Hawthorn Ltd, Northumberland Engine Works, POB 8, Wallsend, Tyne and Wear NE2 6QH

George Clark and Sons (Hull) Ltd, Hawthorn Avenue, Hull HU3 5LZ

Clayton Dewandre Co. Ltd, PO Box 9, Titanic Works, Lincoln

Clayton, Son & Co. Ltd, Moor End Works, Hunslet, Leeds LS10 2BH

Cleveland Fabricators Ltd, Norton Road, Stockton-on-Tees, Cleveland TS20 2AQ

Climate Equipment Ltd, Highlands Road, Shirley, Solihull, W. Midlands B90 4NL

Ralph Coidan Equipment Ltd, Eaglescliffe Industrial Estate, Stockton-on-Tees, Clevelland TS16 0PN

Concentric (Fabrications) Ltd, Hawksworth Industrial Estate, Swindon, Wilts, SN2 1DZ

Contrapol Ltd, Norton Priory, Selsey, Chichester, W. Sussex PO20 9DT

Cool Technology Ltd, Third Avenue, Pensnett Estate, Brierley Hill, W. Midlands DY6 7PP

Cooper Merseyside Ltd, Price Street, Birkenhead, Merseyside L41 3PT

Corning Process Systems, Corning Ltd, Stone, Staffs. ST15 OBG

Covrad Ltd, Sir Henry Parkes Road, Canley, Coventry CV5 6BN

A. F. Craig & Co. Ltd, Caledonia Engineering Works, McDowell Street, Paisley PA3 2NA

Crosse Engineering Ltd, Herriot House, North Place, Cheltenham, Glos. GL50 4DS

Curwen and Newbery Ltd, Redhills Road, Stoke-on-Trent, Staffs. ST2 7ER

Custom Coils Ltd, Woodside Avenue, Eastleigh, Hants. SO5 4JQ

Dantherm Ltd, Hither Green, Clevedon, Avon BS21 6XT

The Davenport Engineering Co. Ltd, 72 Harris Street, Bradford, W. Yorks. BD1 5JD

DAW Heat Transfer Ltd, Unit 9, Hill Top Industrial Estate, Shaw Street, West Bromwich, W. Midlands, B70 0TX

Delaney Gallay Dynamics Ltd, Egware Road, Cricklewood, London NW2 6LD

Delta R. A. Ltd, Hollands Road, Haverhill, Suffolk

Deltech Engineering Ltd, Albert Drive, Sheerwater, Woking, Surrey GU21 5R7

Denco Miller Ltd, Denco Water Systems, PO Box 11, Holmer Road, Hereford HR4 9SJ

John Dore & Co. Ltd, 51-55 Fowler Road, Hainault, Ilford, Essex IG6 3XF

Ductwork Engineering Systems Ltd, Airport Trading Estate, Biggin Hill, Kent

Durametallic UK, 10A Northenden Road, Sale, Cheshire M33 3BR

Environmental & Thermal Engineering Ltd, Victoria House, Walker Street, Macclesfield, Cheshire SK10 1BH

Equimet Ltd, Tyler Street, Sheffield, S.
Yorks. S9 1GL

Eurex Energy Ltd, Sittingbourne, Kent

Eurovent Environmental Control Ltd,
Middlemore Lane, Aldridge, W. Midlands
WS9 8SP

Evapco (UK) Ltd, 27 Causeway Road, Corby,
Northants, NN17 2DU

Extended Surface Tube Co. Ltd, Corby
Works, Weldon Road, Corby, Northants,
NN17 1UA

Fallon Engineering Co, 44A Packhorse Road,
Gerrards Cross, Bucks. SL9 8EF

F. & R. Cooling Ltd, Trading Estate,
Wellington, Somerset TA21 8ST

Fawcet Engineering Ltd, Dock Road South,
Bromborough, Wirral, Merseyside L62
6SW

Ferranti Resin Ltd, GRP Division, South West
Industrial Estate, Peterlee, Co. Durham
SR8 2HZ

Film Cooling Towers (1925) Ltd, Chancery
House, Parkshot, Richmond, Surrey TW9
2RH

Filtration and Transfer Ltd, Dawkins Road,
Poole, Dorset BH15 4JY

Flowline Water Cooling Ltd, Unit 13, Ashford
Industrial Estate, Shield Road, Ashford,
Middx.

Food and Beverage Development (UK) Ltd,
Cooper House, Dam Street, Lichfield,
Staffs. WS13 6AA

J. S. Forster and Co. Ltd, Locarno Works,
Powis Avenue, Tipton, W. Midlands DY4
0NA

Foster Wheeler Power Products Ltd, Greater
London House, Hampstead Road, London
NW1 7QN

Freeman, Taylor Ltd, Necton Street, Syston,
Leics. LE7 8HG

GEA Spiro-Gills Ltd, London Road,
Pulborough, W. Sussex RH20 1AR

GEC Diesels Ltd, Boiler Division, Vulcan
Works, Newton-le-Willows, Lancs. WA12
8RU

GEC Power Engineering Ltd, Electrical
Materials Division, Trafford Park,
Manchester M17 1PR

General Combustion Ltd, Brookers Road,
Billingshurst, W. Sussex RH14 9SA

Gibson Wells Products, The Old Vicarage, 2
Town Gate, Calverly, Leeds LS28 5NF

T. Giusti and Son Ltd, 202-228 York Ways,
Kings Cross, London N7 9AW

Glass of Mark Ltd, Woodford Park Estate,
Winsford, Cheshire CW9 2RA

Graham Manufacturing Ltd, Hadley House,
Bayshill Road, Cheltenham, Glos. GL50
3SP

Hamworthy Engineering Ltd, Fleets Corner,
Poole, Dorset BH17 7LA

William Hare Ltd, Weston Road, Bolton,
Lancs. BL3 2AT

F. & R. M. Harris (Birmingham) Ltd, Chemix
Building, Dudley Road, Halesowen, W.
Midlands B63 3LL

Graham Hart (Process Technology) Ltd,
Euroway, South Bradford Trading Estate,
Bradford, W. Yorks. BD4 6SG

Head Wrightson Process Engineering Ltd,
16-22 Baltic Street, London EC1Y OTD

Heap Economiser Ltd, Oldham Road,
Rochdale, Manchester OL11 1BZ

Heatsure Engineering (Kent) Ltd, 32 New
Street, Ashford, Kent TN24 8TS

Heat Transfer Ltd, 3/4 Bath Street,
Cheltenham, Glos. GL50 1YE

Heenan Marley Cooling Towers Ltd, PO Box
20, Pheasant Street, Worcester WR1 2DX

Hego Engineers Ltd, 12 Garthland Drive,
Arkley, Barnet, Herts.

Herman Smith Ltd, Cinderbank Works,
Dudley, W. Midlands DY2 9AH

Hesler Heat Exchangers Ltd, 6 Union Street,
Luton, Beds.

Hick, Hargreaves & Co. Ltd, Soho Iron
Works, Crook Street, Bolton, Lancs BL3
6DP

Hirt Combustion Engineers Ltd, Dane Works,
Water Street, Northwich, Cheshire CW9
5HP

Hitachi Zosen International SA, Winchester
House, 77 London Wall, London EC1

Holden and Brooke Ltd, Sirius Works,
Haverford Street, West Gorton,
Manchester M12 5JL

Holyhead Engineering Co. Ltd, Meadow
Lane, Coseley, Bilston, W. Midlands
WV14 9NJ

James Howden & Co. Ltd, 195 Scotland
Street, Glasgow G5 8PJ

H.P.C. Engineering Ltd, Victoria Gardens
Estate, Burgess Hill, W. Sussex RH15
9RQ

HRS Heat Exchangers Ltd, PO Box 230, Watford, Herts. WD1 2DW

Hubbard Heat Recovery Division, Hubbard Commercial Products Ltd, Unit 26, Perivale Industrial Park, Horsenden Lane South, Perivale, Greenford, Middx. UB6 7RJ

Hygrotherm Engineering Ltd, Botanical House, 1 Botanical Avenue, Talbot Road, Manchester M16 0HL

IMI Marston Radiator Services Ltd, Heat Exchanger Division, Ashville Trading Estate, Cambridge Road, Whetstone, Leics.

IMI Rycroft Ltd, Duncombe Road, Bradford, W. Yorks. BD8 9TB

J. K. Innes & Co. Ltd, Kingmoor Works, Kingmoor Road, Carlisle, Cumbria CA3 90L

Interchange, Invincible Road, Farnborough, Hants. GU14 7QU

Interweld Ltd, Harfreys Road, Harfreys Industrial Estate, Great Yarmouth NR31 OLS

IRD (International Research & Development Ltd), Fossway, Newcastle-upon-Tyne, Wallsend NE6 2YD

Isoterix Ltd, Mill Works, Mill Crescent, Tonbridge, Kent TN9 1PE

ITM Head Wrightson Teesdale Ltd, PO Box 10, Stockton-on-Tees, Cleveland TS17 6AZ

ITT Reznor, Park Farm Road, Folkestone, Kent CT19 5DR

W. E. James and Sons (Precision Engineers) Ltd, Kingston Road, Leatherhead, Surrey

Robert Jenkins Systems Ltd, Wortley Road, Rotherham, S. Yorks. S61 1LT

A. Johnson & Co. Ltd, 448 Basingstoke Road, Reading, Berks. RG2 0LP

Johnson Industrial Ltd, Alliance House, 9 Leopold Street, Sheffield, S. Yorks. S1 2GY

Jord Engineers Ltd, Unit 4, Loomer Road Ind. Est., Loomer Road, Chesterton, Newcastle under Lyme, Staffs. ST5 7LB

Joy Process Equipment Ltd, Capitol House, 2 Church Street, Epsom, Surrey KT17 4NY

John J. Kincaid & Co. Ltd, East Hamilton Street, Greenock, Strathclyde PA15 2AE

Kingswinford Engeering Co. Ltd, Shaw Road, Dudley, W. Midlands DY2 8TS

Kobe Steel Ltd, Machinery Division, Hanover House, 73/74 High Holborn, London WC1

Laidlaw, Drew & Co. Ltd, Bankhead Avenue, Sighthill Industrial Estate, Edinburgh EH11 4HG

Lamanco Ltd, Progress Drive, Bridgetown, Cannock, Staffs, WS11 3JE

Laporte Fluorides Ltd, PO Box 2, Moorfield Road, Widnes, Cheshire WA8 0JU

Largo-Lintec Ltd, Station Road, Buckhaven, Methil, Fife KY8 1JH

Ledward & Beckett Ltd, Rosemary Lane, Halstead, Essex CO9 1HR

Lenco Fabrications Ltd, Mossfield Road, Adderley Green, Longton, Stoke-on-Trent ST3 5BW

Lennox Industries Ltd, PO Box 43, Lister Road, Basingstoke, Hants. RG22 4AR

R. Lord and Sons Ltd, Heat Exchanger Division, Barnbrook Boiler Works, Bury, Lancs. BL9 6AF

Lyon and Pye Ltd, 543 Prescot Road, St. Helens, Merseyside, WA10 3BZ

Marine and Industrial Heat Ltd, Gazelda Works, Lower High Street, Watford, Herts WD1 2JL

Marston Palmer Ltd, Wobaston Road, Fordhouses, Wolverhampton WV10 6QJ

Matthews and Yates Ltd, Cyclone Works, Swinton, Manchester M27 2AP

M. L. Refrigeration & Air Conditioning Ltd, 286 Aberdeen Avenue, Trading Estate, Slough SL1 4BD

Morris, Warden & Co. Ltd, Woodhead Road, Glasgow G53, 7NS

Robert Morton (DG) Ltd, Trent Works, Derby Street, Burton-on-Trent, Derbys, DE14 2LH

Motherwell Bridge Engineering ltd, PO Box 4, Logans Road, Motherwell, Lanark ML1 3NP

MSM Group, Spring Vale Works, Middleton, Manchester M24 2HS

Myson Process Equipment Ltd, Nuffield Industrial Estate, Poole, Dorset BH17 7RA

NCC Engineers Ltd, Station Road, Lowdham, Notts. NG14 7DU

N.E.I.-Clarke Chapman Ltd, Victoria Works, Gateshead, Tyne and Wear NE8 3HS

New Metals & Chemicals Ltd, Chancery House, Chancery Lane, London WC2A 1RD

Newton Chambers Engineering Ltd, Process
   Engineering Division, Thorncliffe,
   Chapeltown, Sheffield S30 4PY
Nobel's Explosives Co. Ltd, Nobel House,
   Stevenston, Strathclyde, Ayrshire.
Northvale Engineering Ltd, Uxbridge Road,
   Melton Road, Leicester LE4 7ST
Norton Pampus Fluorplast Ltd, Chesterton
   Works, Loomer Road, Newcastle under
   Lyme ST5 7HR
Nutter Engineering, Oxford Street, Bilston, W.
   Midlands WV14 7EG
Nu-Way Eclipse Ltd, PO Box 14, Berry Hill
   Estate, Droitwich, Worcs. WR9 9BJ
Oakwood Thermo Products Ltd, Oak House,
   Market Place, Macclesfield, Cheshire
   SK10 1ER
Oil Coolers & Compressors Ltd, Heath Hill
   Industrial Area, Dawley, Telford TF4 2RH
Pathe Engineers Ltd, Shepley Street,
   Chesterfield, Derbyshire
Paul Fabrications Ltd, PO Box 72, Ipswich
   IP4 5TW
Peabody Encomech, Brenchley House,
   123/135 Week Street, Maidstone, Kent
   ME14 1RF
Peabody Holmes Ltd, Turnbridge,
   Huddersfield HD1 6RB
F. Pedley & Sons Ltd, Superbright Works,
   Crompton Street, Chadderton, Oldham,
   Lancs. OL9 9AA
Pentagon Radiator (Stafford) Ltd, Hixon
   Industrial Estate, Hixon, Staffs. ST18 OPY
Plastic Constructions Ltd, Tyseley Industrial
   Estate, Seeleys Road, Greet, Birmingham
   B11 2LP
Portobello Fabrications Ltd, Coleford Road,
   Sheffield S9 5PE
Potteries Engineering Ltd, Chemical
   Engineering Division, St. Mary's
   Engineering Works, Tunstall, Stoke on
   Trent, Staffs. ST6 4PS
P.P.E. Fittings (Birstall) Ltd, Industrial
   Trading Estate, Gelderd Road, Birstall,
   Batley, W. Yorks. WF17 9NE
Pullen Product Developments Ltd, 34 Bridge
   Road, Cowes, Isle of Wight PO31 7DB
Quietflo Engineering Ltd, Bentinck Road,
   West Drayton, Middx. UB7 7SJ
Redpoint Associates, Cheney Manor,
   Swindon, Wilts SN2 2PS

Refrigeration Appliances Ltd, 42 Hollands
   Road, Haverhill, Suffolk CB9 8PT
Rileys of Stockton Ltd, Ferry House, Ferry
   Road, Middlesbrough TS2 1PJ
Royles Ltd, Irlam, Manchester M30 5AH
S. & P. Coil Products Ltd, Evington Valley
   Road, Leicester LE5 5LU
Sasakura Engineering Co, 2 Grove House,
   Foundry Lane, Horsham, W. Sussex RH13
   5PL
Scheibler Filters Ltd, PO Box 5, Churchgate
   House, Retford, Notts. DN22 6PG
Scurrah Hytech Products Ltd, 6 Market Street,
   Soham, Ely, Cambs CB7 5JG
SDI Water Technology Ltd, 28 Birmingham
   Street, Oldbury, Warley, W. Midlands B69
   4DS
Searle Manufacturing Co. Ltd, Newgate Lane,
   Fareham, Hants. PO14 1AR
Senior Green Ltd, Economiser Works, Calder
   Vale Road, Wakefield WF1 5PF
Senior Platecoil Ltd, Otterspool Way, Watford,
   Herts. WD2 8HX
Serck Heat Transfer, PO Box 598B, Warwick
   Road, Birmingham B11 2QY
Serck Visco, 161 Stafford Road, Croydon
   CR9 4DT
Simmons and Hawker Ltd, North Feltham
   Trading Estate, Central Way, Feltham,
   Middx. TW14 OUQ
Sims Fabrications Ltd, Ranalagh Works,
   Fishbourne, Isle of Wight PO33 4EY
A & W Smith and Co. Ltd, 1 Cosmos House,
   Bromley Common, Bromley, Kent
Solar Fabricating and Engineering Ltd,
   Willenhall Trading Estate, Rosehill,
   Willenhall, W. Midlands
Solek Ltd, 16 Hollybush Lane, Sevenoaks,
   Kent TN13 3TH
S.O.S. Cooling Ltd, Unit 16, Saddington
   Road Estate, Churchill Way, Fleckney,
   Leics. LE8 0UD
Souter Shipyard Ltd, Thetis Road, Cowes, Isle
   of Wight PO31 7DJ
Specialist Heat Exchangers Ltd, Freemand
   Road, North Hykeham, Lincoln LN6 9AP
Special Metals (Fabrication) Ltd, Units 19/21,
   Horndon Industrial Park, West Horndon,
   Brentwood, Essex.
Spiral Tube (Heat Transfer) Ltd, Osmaston
   Park Road, Derby DE2 8BU

Spur Engineering Co. Ltd, 138 Kenley Road, Merton Park, London SW19

Steel Brothers Process Plant Ltd, Bredgar Road, Gillingham, Kent ME8 6PW

Steels Engineering Ltd, Leechmere Works, Leechmere Industrial Estate, Sunderland SR2 9TG

Stone International Ltd, PO Box 5, Gatwick Road, Crawley, W. Sussex RH10 2RN

Struthers Wells (UK) Ltd, 123a High Street, Crawley, Sussex RH10 1DQ

Thermal Developments Ltd, The Whins, North End, Sedgefield, Cleveland TS21 2AZ

Thermal Syndicate Ltd, PO Box 6, Neptune Road, Wallsend, Tyne and Wear NE28 6DG

Thermal Technology Ltd, Thermal House, 46 Hilperton Road, Trowbridge, Wilts. BA4 7JH

Thermex Ltd, Howard Road, Park Barn Ind. Estate, Redditch, Worcs. B78 7SE

Thermo Engineers Ltd, Chamberlain Road, Aylesbury, Bucks. HP19 3BU

Therm Tech Engineering Ltd, Vale Mill, Huddersfield Road, Mossley, Ashton under Lyne, Lancs. OL5 9JL

John Thurley Ltd, Ripon Road, Harrogate, Yorks. HG1 2BU

TI Richards & Ross Ltd, Neachells Lane, Wednesfield, Wolverhampton WV11 3QH

Titanium Fabricators Ltd, TFL House, Orgreave Crescent, Handsworth, Sheffield S13 9LQ

Tolltrek Ltd, Priory House, Friar Street, Droitwich, WR9 8ED

Trace Heat Pumps Ltd, Unit 6, Industrial Estate, West Witham, Essex CM8 3TQ

Trans Vac Process Equipment Ltd, Stonebroom Industrial Estate, Stonebroom, Derby DE5 6LQ

Trendpam Engineering Ltd, 29 Nork Way, Banstaad, Surrey

Turbinservice International Ltd, Units 23/24 Bracken Hill, SW Industrial Estate, Peterlee, Co. Durham

Twin Industries Agencies Ltd, Stoneyard Works, Park Street, Camberley, Surrey GU15 3PB

United Air Coil Ltd, Northdown Road, St. Peters, Broadstairs, Kent

United Air Specialists (UK) Ltd, 15 Waterloo Place, Leamington Spa, Warwick CV32 5LA

Wabco Westinghouse (UK) Ltd, 30 Burners Lane, Kiln Farm Ind. Est., Milton Keynes MK11 3AT

Watermiser Ltd, Tower Works, Stoneygate Road, New Milns, Ayrshire KA16 9AJ

Water Saver Systems Division, Shrewsbury Tool and Die Co. Ltd, Harlescott Lane, Shrewsbury, Salop SY1 3AS

Water Technology Ltd, 1 Church Square, Oldbury, Birmingham B69 4DX

Fred Watkins (Eng.) Ltd, Sling Engineering Works, Coleford, Glos.

Watt, Joule and Therm Ltd, Coventry Point, Coventry CV1 1EA

Weaverbrook Ltd, Granville Chambers, 2 Radford Street, Stone, Staffs. ST15 8DA

Weir Heat Exchange Ltd, 149 Newlands Road, Cathcart, Glasgow G44 4EX

J. A. Welch (Plant and Vessel) Ltd, Stalco Works, Livingstone Road, Stratford, London E15 2LP

Welding Technical Service Ltd, Hurst Mill, 107 Pershore Road S., Kings Norton, Birmingham B30 3EN

Wellman Furnaces Ltd, Cornwall Road, Smethwick, Warley, W. Midlands B66 2LB

Wellman Sales Ltd, 8/20 City Road East, Manchester M15 4PJ

Whessoe Heavy Engineering Ltd, 40 Broadway, London SW1H OBR

Widnes Foundry and Engineering Co. Ltd, Lugsdale Road, Widnes, Cheshire WA8 6DA

Winson Eagle Machinery Ltd, Aqueduct Road, Raikes Lane, Bolton BL3 1RP

Worthington-Simpson Ltd, PO Box 17, Lowfield Works, Newark, Notts. NG24 3EN

R. Wright and Son (Engineers) Ltd, Church Broughton Road, Foston, Derby DE6 5PW

York Division, Borg Warner Ltd. 715 North Circular Road, London NW2 7AV

John Zinc Co. Ltd, Acrewood Way, St. Abans, Herts.

W S Atkins and Partners, Woodcote Grove, Ashley Road, Epsom, Surrey KT18 5BW

Computer Aided Design Centre, Madingley Road, Cambridge CB3 0HB

HTFS, Heat Transfer and Fluid Flow Service, Harwell Laboratory, Oxon OX11 0RA

National Engineering Laboratory, East
Kilbride, Glasgow G75 0QU
(NIFES) National Industrial Fuel Efficiency
Service Ltd, Orchard House, 14 Great
Smith Street, London SW1
Scicon Computer Services Ltd, Brick Close,
Kiln Farm, Milton Keynes MK11 3EJ

Simulation Sciences Inc, Regent House,
Heaton Lane, Stockport SK4 1BS

Yugoslavia

Duro Dakevic, Njegoseva br. 1, Slavonski
Brod.

Bibliographic guide to the literature of heat exchangers

INTRODUCTION

A rich and diverse literature exists in the field of heat exchangers, which is constantly being expanded. In the last few years new impetus has arisen from increased attention to energy conservation and the recovery of useful energy from thermal effluents.

The profusion of literature is likely to bewilder newcomers to the field and baffle users who have no aspirations to design or build their own heat exchangers. This book was written to help both these groups. For one group the book serves as the vehicle whereby readers can get up to speed before tackling the professional literature. For the other it provides a distillation of the subject sufficient in itself for many purposes but serving also as an introduction to a deeper study. For both groups a brief review of the extensive literature is an appropriate ending for the volume.

CLASSIFICATION OF THE LITERATURE

Different sources of information about heat exchangers may be classified as follows:

1. Books
2. Periodicals
3. Engineering-society transactions and conference proceedings
4. Continuing-education courses
5. Government reports and publications
6. National standards
7. Trade-association publications

8. Manufacturers' literature
9. Contract research reports

BOOKS

Basic Textbooks

Many books have been written on the science of heat transfer. Classic texts that come readily to mind include:

Jakob, M.: "Heat Transfer," vol. 1, 2, Wiley, New York, 1949.
McAdams, W. H.: "Heat Transmission," 3d ed., McGraw-Hill, New York, 1954.

Current texts in use for college courses in heat transfer include:

Bayley, F. J., J. M. Owen, and A. B. Turner: "Heat Transfer," Nelson, London, 1972.
Chapman, A. J.: "Heat Transfer, 3d ed., Macmillan, New York, 1974.
Gebhart, B.: "Heat Transfer," 2d ed., McGraw-Hill, New York, 1971.
Holman, J. P.: "Heat Transfer," 5th ed., McGraw-Hill, New York, 1981.
Incropera, F. P., and D. P. DeWitt: "Fundamentals of Heat Transfer," Wiley, New York, 1981.
Karlekar, B. V., and R. M. Desmond: "Engineering Heat Transfer," West, St. Paul, Minn., 1977.
Kreith, F., and W. Z. Black: "Basic Heat Transfer," Harper & Row, New York, 1980.
Lienhard, J. H.: "A Heat Transfer Textbook," Prentice-Hall, Englewood Cliffs, N.J., 1981.
Ozişik, M. N.: "Basic Heat Transfer," McGraw-Hill, New York, 1976.
Parker, J. D., J. H. Boggs, and E. F. Blick: "Introduction to Fluid Mechanics and Heat Transfer," Addison-Wesley, Reading, Mass., 1974.
Rohsenow, W. M., and H. Choi: "Heat, Mass and Momentum Transfer," Prentice-Hall, Englewood Cliffs, N.J., 1961.
Simonson, J. R.: "Engineering Heat Transfer," Macmillan, London, 1975.
Thomas, L.: "Fundamentals of Heat Transfer," Prentice-Hall, Englewood Cliffs, N.J., 1980.
Welty, J. R.: "Engineering Heat Transfer," Wiley, New York, 1974.

Many mechanical-engineering courses in the U.K. and those following the British model include lectures on heat transfer in the applied thermodynamics courses. This pattern is followed in an excellent, widely used text that contains a closing section on basic heat transfer:

Rogers, G. F., and Y. R. Mayhew: "Engineering Thermodynamics Work and Heat Transfer," 2d ed., Longmans, London, 1967.

Reference Books

Reference books on heat exchangers and heat transfer include:

Afgan, N., and E. Schlunder (eds.): "Heat Exchangers: Design and Theory Sourcebook," Hemisphere, Washington, D.C., 1974.

Anderson, D. A., J. C. Tannehill, and R. B. Pletcher: "Computational Fluid Mechanics and Heat Transfer," Hemisphere, Washington, D.C., 1984.

ASHRAE: "ASHRAE Handbook," vols. 1–4, American Society of Heating, Refrigerating and Air Conditioning Engineers, New York.

Bergles, A. E., J. G. Collier, J. M. Delhaye, G. F. Hewitt, and F. mayinger: "Two-Phase Flow and Heat Transfer in the Power and Process Industries," Hemisphere, Washington, D.C., 1981.

Bliem, C., et al.: "Ceramic Heat Exchanger Concepts and Materials Technology," Noyes, Park Ridge, N.J., 1985.

Butterworth, D., and G. F. Hewitt: "Two-Phase Flow and Heat Transfer," Oxford University Press, Oxford, 1977.

Chisholm, D. (ed.): "Heat Exchanger Technology," Elsevier, New York, 1988.

Deng, S. (ed.): "Heat Transfer Enhancement and Energy Conservation," Hemisphere, Washington, D.C., 1990.

Eckert, E. R. G., and R. M. Drake: "Analysis of Heat and Mass Transfer," Hemisphere, Washington, D.C., 1987.

Edwards, D. D., "Radiation Heat Transfer Notes," Hemisphere, Washington, D.C., 1988.

Fraas, A. P., and M. N. Osizik: "Heat Exchanger Design," Wiley, New York, 1965.

General Electric: "General Electric Heat Transfer and Fluid Flow Data Books," General Electric Co., Schenectady, N.Y.

Gupta, J. P., "Working with Heat Exchangers: Questions and Answers," Hemisphere, Washington, D.C., 1990.

Hetstroni, G.: "Handbook of Multiphase Systems," Hemisphere, Washington, D.C., 1982.

Hewitt, G. F.: "Measurement of Two-Phase Flow Parameters," Academic Pres, New York, 1978.

Hewitt, G. F. (ed.): "Hemisphere Handbook of Heat Exchanger Design," Hemisphere, Washington, D.C., 1990.

Jaluria, Y., and K. E. Torrance: "Computational Heat Transfer," Hemisphere, Washington, D.C., 1986.

Kakaç, S., A. E. Bergles, and F. Mayinger: "Heat Exchangers: Thermal-Hydraulic Fundamentals and Design," Hemisphere, Washington, D.C., 1981.

Kakaç, S., and D. B. Spalding: "Turbulent Forced Convection in Channels and Bundles: Theory and Applications to Heat Exchangers and Nuclear Reactors," vols. 1 & 2, Hemisphere, Washington, D.C., 1979.

Kakaç, S., and Yener, Y. "Heat Conduction," 2nd ed., Hemisphere, Washington, D.C., 1985.

Kakaç, S., et al. (eds.): "Two-Phase Flow Heat Exchangers: Thermal-Hydraulic Fundamentals and Design," Kluwer Academic, Norwell, Mass., 1988.

Kays, W. M., and A. L. London: "Compact Heat Exchangers," McGraw-Hill, New York, 1984.

Kern, D. Q.: "Process Heat Transfer," McGraw-Hill, New York, 1950 (the classic and perhaps the single most important book written on heat exchangers).

Kim, J. H., S. T. Ro, and T. S. Lee: "Heat Transfer: Korea–U.S. Seminar on Thermal Engineering and High Technology," Hemisphere, Washington, D.C. 1988.

Kline, S. J., and N. H. Afgan (eds.): "Near Wall Turbulence," Hemisphere, Washington, D.C., 1990.

Kraus, A. D., and A. Bar-Cohen: "Thermal Analysis and Control of Electronic Equipment," Hemisphere, Washington, D.C., 1983.

Kutateladze, S. S., and A. I. Leontiev: "Heat, Mass Transfer, and Friction in Turbulent Boundary Layers," Hemisphere, Washington, D.C., 1990.

Manzoor, M.: "Heat Flow Through Extended Surface Heat Exchangers," (Lecture Notes in Engineering, vol. 5), Springer-Verlag, 1983.

Markatos, N. C. (ed.): "Computer Simulation for Fluid Flow, Heat and Mass Transfer, and Combustion in Reciprocating Engines," Hemisphere, Washington, D.C., 1989.

Martynenko, O. G., and A. A. Zukauskas (eds.): "Convective Heat Transfer," Hemisphere, Washington, D.C., 1989.

Martynenko, O. G., and A. A. Zukauskas (eds.): "High Temperature Heat Transfer," Hemisphere, Washington, D.C., 1990.

Miropolsky, Z. L., and R. I. Soziev (eds.): "Fluid Dynamics and Heat Transfer in Superconducting Equipment," Hemisphere, Washington, D.C., 1990.

Mori, Y., A. E. Sheindlin, and N. Afgan (eds.): "High Temperature Heat Exchangers," Hemisphere, Washington, D.C., 1986.

Nogotov, E. F.: "Applications of Numerical Heat Transfer," Hemisphere, Washington, D.C., 1978.

Osizik, M. N., "Heat Conduction," Wiley-Interscience, New York, 1979.

Palen, J. W. (ed.): "Heat Exchanger Sourcebook," Hemisphere, Washington, D.C., 1986.

Patankar, S. V.: "Numerical Heat Transfer and Fluid Flow," Hemisphere, Washington, D.C., 1980.

Saunders, E. A.: "Heat Exchangers: Construction and Thermal Design," Wiley, New York, 1988.

Schlünder, E. U. (ed.): "Heat Exchanger Design Handbook," Hemisphere, Washington, D.C., 1983.

Shah, R. K., and A. L. London: "Laminar Flow Forced Convection in Ducts," Academic Press, new York, 1978.

Shah, R. K., E. C. Subbarao, and R. A. Mashelkar: "Heat Transfer Equipment Design," Hemisphere, Washington, D.C., 1988.

Sheindlin, A. E. (ed.): "High Temperature Equipment," Hemisphere, Washington, D.C., 1986.

Siegel, R., and J. R. Howell: "Thermal Radiation Heat Transfer," 2nd ed., Hemisphere, Washington, D.C., 1981.

Simonson, J. R.: "Engineering Heat Transfer," 2nd ed., Hemisphere, Washington, D.C., 1988.

Singh, K. P., and A. I. Soler: "Mechanical Design of Heat Exchangers and Pressure Vessel Components," Arcturus, New York, 1984.

Spalding, D. B. (ed.): "Heat and Mass Transfer in Gasoline and Diesel Engines," Hemisphere, Washington, D.C., 1989.

Taborek, J., G. F. Hewitt, and N. H. Afgan: "Heat Exchangers: Theory and Practice," Hemisphere, Washington, D.C., 1983.

Thome, J. R.: "Enhanced Boiling Heat Transfer," Hemisphere, Washington, D.C., 1990.

Tien, C. L., V. P. Carey, and J. K. Ferrell: "Heat Transfer 1986: Proceedings of the Eighth International Heat Transfer Conference," (six volumes), Hemisphere, Washington, D.C., 1986.

van Stralen, S., and R. Cole: "Boiling Phenomena," vols. 1 & 2, Hemisphere, Washington, D.C., 1979.

Vilemas, J., B. Cesna, and V. Survila: "Heat Transfer in Gas-Cooled Annular Channels," Hemisphere, Washington, D.C., 1987.

Wang, B. (ed.): "Heat Transfer Science and Technology 1988," Hemisphere, Washington, D.C., 1989.

Zukauskas, A.: "High Performance Single Phase Heat Exchangers," Hemisphere, Washington, D.C., 1989.

Zukauskas, A., and R. Ulinskas: "Heat Transfer in Banks of Tubes in Crossflow," Hemisphere, Washington, D.C., 1988.

Zukauskas, A., R. Ulinskas, and V. Katinas: "Fluid Dynamics and Flow-Induced Vibrations of Tube Banks," Hemisphere, Washington, D.C., 1988.

Many other specialist monographs are available. Two annually published series are:

"Progress in Heat and Mass Transfer," Pergamon Press, New York.
"Advances in Heat Transfer," Academic Press, New York.

Readers are urged to consult the mechanical, chemical, and nuclear engineering, energetics, heat-transfer, multiphase phenomena, and thermofluids sections of the principal publishers' lists, the relevant sections of a directory of books in print, and the catalog section of the library or seek the help and advice of a technical librarian.

PERIODICALS

The best technical periodical about heat exchangers is the quarterly *Heat Transfer Engineering,* edited by Kenneth J. Bell and published by Hemisphere. It is devoted in large part to heat exchangers and matters related thereto. The associate editors include top experts in the field, and I consider this journal the clear leader in this special-interest field.

Hemisphere publishes other specialty journals, frequently containing articles of interest to heat-exchanger engineers. These journals, with their editors' names in parentheses, include:

Bulletin of the International Centre for Heat and Mass Transfer (N. Afgan)
Experimental Heat Transfer (G. F. Hewitt and C. L. Tien)
International Archives of Heat and Mass Transfer (N. Afgan)
IVTAN Reviews (A. E. Sheindlin)
Numerical Heat Transfer (W. J. Minkowycz)
Canadian Journal of Chemical Engineering (N. Epstein)
International Communications in Heat and Mass Transfer (J. P. Hartnett and W. J. Minkowycz)
International Journal of Heat and Mass Transfer (J. P. Hartnett)
International Journal of Multiphase Flow (G. Hetsroni)
The Chemical Engineer (P. Varey)
Chemical Engineering Progress (A. Bisio)
Heat Transfer—Japanese Research (T. F. Irving and K. Suzuki)
Heat Transfer—Soviet Research
Journal of Heat Transfer (Transactions of the ASME) (K. T. Yang)

Other journals devoted entirely to heat transfer and often including articles of interest about heat exchangers are the following (the publishers are given in parentheses):

International Journal of Heat and Mass Transfer (Pergamon, Oxford, U.K.)
Journal of Heat Transfer (American Society of Mechanical Engineers, New York)
Journal of Engineering for Power (American Society of Mechanical Engineers, New York)
Journal of Heat Recovery Systems (Pergamon, Oxford, U.K.)

Many engineering periodicals embrace a broader field of interest but frequently include articles or have special issues about heat exchangers and process heat-transfer equipment. These include:

Chemical Engineering (McGraw-Hill, New York)
Industrial and Engineering Chemistry, Process Design and Development (American Chemical Society, Washington, D.C.)
Hydrocarbon Processing (Gulf Publishing, Houston)
Chemical Engineering Progress (American Institute of Chemical Engineers, New York)
Mechanical Engineering (American Society of Mechanical Engineers, New York)
Chemical Engineering Journal (Elsevier Sequoia, Lausanne, Switzerland)
Oil and Gas Journal (Petroleum Publishing, Tulsa)

The Chemical Engineer (Institution of Chemical Engineers, London)
Chartered Mechanical Engineer (Institution of Mechanical Engineering, London)
Journal of Mechanical Engineering Science (Institution of Mechanical Engineers, London)
ASHRAE Journal (American Society of Heating, Refrigerating and Air Conditioning Engineers, New York)

ENGINEERING–SOCIETY TRANSACTIONS AND CONFERENCE PROCEEDINGS

The transactions or proceedings of the engineering institutions and societies in the United States, England, West Germany, Japan, and other countries are published annually. They are incomparable collections of engineering papers embracing all aspects of the field and will be found in any good technical reference library. In some cases the collection extends back over a century. Nowadays the number of technical papers presented at the various meetings of the societies has grown so great that only a few selected papers are published in the proceedings. Papers that are presented at a meeting but not included in the annual transactions are sometimes published with other papers on the same topic in a specialist volume. This is an increasingly popular practice, for it provides at low cost a compilation of specialist material about a topic.

Similar specialist volumes are also published as the proceedings of conferences held to consider designated topics (e.g., the symposia volumes published by the ASME Heat Transfer Division). These meetings are usually sponsored by the professional engineering societies and institutions, sometimes acting in concert. Reference to the list of publications of the principal mechanical and chemical engineering societies will quickly provide an extensive list of topics covered by these conference proceedings. Many of the volumes contain material related to heat exchangers and process heat-transfer equipment.

We have emphasized mechanical and chemical engineering societies, but others too are very much involved. There are specialist societies on welding, metals and materials, corrosion, air conditioning, heating and refrigeration, nuclear systems, control, computer simulation, and systems. All publish their annual transactions, specialist volumes, and conference proceedings.

In addition, there are numerous international conferences on special topics. The International Heat Transfer Conference is held every four years in a different locale. The eighth conference was held in San Francisco, California, in 1986, and the ninth is being held in Jerusalem, Israel, in 1990. The proceedings of these conferences are published in seven or eight volumes and contain many papers pertaining to heat exchangers.

The International Centre for Heat and Mass Transfer (ICHMT) in Belgrade, Yugoslavia, organizes conferences and courses on special-interest topics, whose proceedings are published. The von Karman Institute for Fluid Mechanics in Brussels, Belgium, has an annual series of two-week courses on many aspects of fluid mechanics, some of which are related to heat exchangers. The notes for these courses are available from the institute and are a valuable supplementary source of specialist information.

In North America the National Heat Transfer Conference is organized jointly by the American Society of Mechanical Engineers and the American Institute of Chemical Engineers. The proceedings of this conference, in its twentieth year in 1981, provide an impressive data base of heat-transfer and heat-exchanger related material. Details of the proceedings, most of which are still available, may be found in the lists of publications of the ASME and AIChE.

CONTINUING-EDUCATION COURSES

Short courses of two to five days duration provide an effective forum for bringing newcomers to a field "up to speed" and for enhancing the expertise of those already working in the area. The engineering societies all have an organized program of continuing education. Several short courses related to heat transfer, heat exchangers, and process heat transfer equipment are offered by the ASME, AIChE, and ASHRAE.

Advanced courses on specialist topics are sometimes available. These are organized by a variety of universities, professional engineering societies, and commercial organizations.

GOVERNMENT REPORTS AND PUBLICATIONS

Government publications can be an excellent and rewarding source of information that is frequently not exploited to the full. Many appear to be unaware of the quality of information available and the procedures used to attain it. The following may be helpful.

British Government Reports

In Britain the distribution of government reports and publications is handled by Her Majesty's Stationery Office (HMSO). The headquarters is located at Holborn, London, but HMSO bookshops may be found in most of the larger cities. In smaller centers, commercial booksellers are designated as official government agents and distributors.

Catalogs of available British government publications are provided at no charge. These are divided into about 20 individual sections for the publications of the different ministries and departments of state.

Current publications are therefore easily obtained, but problems arise when one seeks information about out-of-print reports and documents. The best source for xerographic copies of past publications whose details are known is the National Science and Technology Library, Harrogate, Yorks. In more obscure cases where the full details are not known the author has found the libraries of the Institution of Mechanical Engineers and the Science Museum, South Kensington, London, to be most helpful.

United States Government Reports

The U.S. Government publishes an astonishing range of books, papers, and reports on virtually every topic. Copies of these documents can be obtained from three sources:

Superintendent of Documents (SupDocs), Government Printing Office, Washington, D.C.

National Technical Information Service (NTIS), U.S. Department of Commerce, Springfield, Virginia 22161

Defense Documentation Center (DDC), Defense Logistics Agency, Cameron Station, Alexandria, Virginia

Superintendent of Documents (SupDocs)

The Superintendent of Documents is the sales organization of the U.S. Government Printing Office (GPO), the federal printer and part of the legislative branch of the U.S. government. SupDocs sells only material printed by the GPO. This includes at any one time about 20,000 titles, on every topic of government and public interest.

New and current stock titles are announced in the monthly catalog and are distributed through 19 regional centers in the United States.

Various reports from government research laboratories are published by the GPO and distributed through SupDocs. They do not handle government contractor reports, nor is there any obligation to maintain stocks of particular items, so that frequently one's request is met with an "out-of-print" response. NTIS handles the same materials as SupDocs plus government contractor reports. In most instances NTIS does not have and therefore cannot supply copies of the original report from stock (as SupDocs will if it is available), but will print (photocopy) a single copy from an original on file. This is the essential difference between NTIS and SupDocs.

National Technical Information Service

NTIS handles the distribution of copies of all U.S. government unclassified, unrestricted reports in all fields.

Anyone may order anything from NTIS. The range of interest and the scope of available material staggers the imagination. As a consequence of the extent of the system, it is difficult to obtain quick response. One often receives information after the need for it has passed. One can opt for priority mail ($10 for 24-h turnaround) or personal collection at NTIS Springfield and the Washington, D.C., Information Center and Bookstore.

The system works best for regular users who have an NTIS deposit account established for payment. Customers can be billed but a hefty surcharge is involved. Payment can also be made by including an American Express credit-card number. Normal delivery is said to be 3 to 5 weeks.*

*In Canada, it tends to be 8 to 12 weeks.

To minimize delay it is necessary to quote the order number, for example, AD-A042 786/4ST. This is much more important than the title, author, source, etc. Failure to quote the order number will necessitate someone at NTIS handling the order manually (probably searching indexes and lists for the order number), and this adds many weeks to the delivery time.

Reports may be obtained as paper or microfiche copies. Sometimes the paper copies are actual editions of the report as delivered by the contractor to the sponsor, but normally they are photocopies. The quality ranges all the way from as good as the original to unreadable. Microfiche copies are pieces of photographic film containing images of up to 8 to 12 pages of text. One needs a special machine, a microfiche reader, to be able to read what is recorded. The machines generally display the text on an illuminated screen. Microfiche is much favored by librarians and information specialists because of the reduced storage requirements, and by the photographic companies selling the film and reader machines. It is not favored by users wishing to get information, however. One is not able to make copies of pages that are of particular interest. Nor is it possible to read microfiche on a plane or train. Even in libraries or offices where the readers are available, they are tedious to use and difficult to read. One's concentration is easily lost, and one's eyes rapidly tire.

NTISearches

In addition to simply filling orders for copies of reports, NTIS provides another important service. They will generate bibliographies on specified topics of reports that they have on file. The NTIS data base includes over 800,000 document and data records covering U.S. government-sponsored research from 1964. (Some 250 research reports are added daily!) Therefore, whatever one's interest, it is likely that there is a report or two already on file that would be worth looking at. Several hundred bibliographies have already been assembled and are listed in the NTISearch Subject Index, obtainable on request from NTIS. These existing searches are available at a cost of $25. If an existing bibliography on the topic of interest is not available, NTIS will undertake a custom search of their data base. The cost is a moderate $100 for up to 100 reports, increasing incrementally to $200 for up to 500 reports—an incredible value for anyone used to manual library research.

In 1979 the list of published searches contained several entries under "Heat Exchangers," which, surprisingly, were for such exotica as heat pipes, rather than common or garden-variety heat exchangers. It appeared there simply had not been an NTISearch prepared for heat exchangers. Therefore a search was commissioned; starting from mid-1979 and working backward, it yielded references to 500 reports by 1971. This NTISearch will be listed in future editions of the subject index. Presumably the data base contains references to many other reports on heat exchangers contributed before 1971, but these have not been classified so far as is known.

In addition to NTIS, there are other computerized data bases available for search purposes.

Defense Documentation Center

The Defense Documentation Center (DDC) supplies copies of reports on defense- and security-related matters at all levels of security classification to approved requesters. Only approved "DDC users" will likely have success in requesting material from DDC. One becomes a DDC user by working for a U.S. government agency or government contractor on defense-related matters.

Canadian Government Reports

The distribution of Canadian government publications is handled by the Department of Supply and Services, with headquarters in Hull, Quebec. There is a government bookstore in Ottawa, and various booksellers in the larger cities are designated as government sales agents. Catalogs of the various publications in print are available in English and French. Xerographic copies of out-of-print materials may be obtained from the National Research Council Technical Information Service, Ottawa.

NATIONAL STANDARDS

Every major country has an organization with legal or quasilegal responsibility for the drafting and introduction of national standards. Such organizations include the American National Standards Institute, the British Standards Institution, and the Canadian Standards Association.

The essence of their activity is the preparation and publication of national standards. All these organizations make lists of their publications available. Most have standards or codes of practice relating to heat exchangers and process heat-transfer equipment. Adherence to these standards may be required by the national regulatory agency, the client, or the client's insurance company. It is essential, therefore, that the standards literature be available and well known to users, designers, and manufacturers. Further information on this topic may be found in Chap. 10.

TRADE-ASSOCIATION PUBLICATIONS

Trade-association publications are similar in many respects to publications of the national standards organizations. The publications include standards and recommended practices, written to define minimum quality levels. The documents are prepared by representative committees of experts or users in the industry.

For heat exchangers the best known trade-association publications are the Standards of the Tubular Equipment Manufacturers Association, New York, and the Standards of the American Petroleum Institute, New York. There are many others for boilers, fired heaters, condensers, steam equipment, refrigerators, coolers, and heaters of various kinds. Further information is contained in Chap. 10, and a comprehensive list of codes and standards is given in the handbook of the American Society of Heating, Refrigerating and Air Conditioning Engineers.

MANUFACTURERS' LITERATURE

Important sources of information are the publications of the manufacturers of heat exchangers and related equipment. Quite frequently manufacturers' literature is the best and most readily available for special-interest topics. Items usually include reprints of papers concerning their equipment presented at conferences of the engineering societies, and handbooks of design data and applications, installation, repair, and operations procedures. There are, in addition, the useful glossy brochures cataloging the range of equipment available, brief specifications, and recommendations for use. Many manufacturers regularly publish technical periodicals describing applications of their products.

Manufacturers' literature is characterized by positive comments and hyperbole about their equipment. It would be unrealistic to expect anything else; nevertheless, one must recognize this and seek an appreciation of the warts as well as the roses. The warts of one system are often discussed in the literature of the manufacturer of a competing system. To obtain a balanced view, one must therefore consult the various sources available. Fortunately the manufacturers of heavy industrial equipment including heat exchangers are not given to the excessive hyperbole found in consumer-product literature, and much factual, dependable information is available from them.

CONTRACT RESEARCH REPORTS

There are numerous organizations, particularly in the United States, willing to undertake contract research on virtually anything including heat exchangers. However, in this specialist field two organizations completely dominate the scene. Between them they have accumulated the greatest concentration of proprietary material in matters relating to heat transfer and heat exchangers.

Heat Transfer and Fluid Flow Service (HTFS)

The Heat Transfer and Fluid Flow Service (HTFS) is operated jointly by the Harwell Laboratory of the U.K. Atomic Energy Authority and the National Engineering Laboratory of the U.K. Department of Industry. The service provides design methods, computer programs, and research facilities in heat transfer and related fluid-flow applications.

Heat Transfer Research Inc. (HTRI)

Heat Transfer Research Inc. (HTRI) is a cooperative, nonprofit corporation organized in 1962 by users, designers, and manufacturers of heat-exchange equipment to carry out applied research in heat transfer and fluid flow. The organization is located on the scenic grounds of the C. F. Braun Co. at Alhambra, Calif., near Los Angeles.

SUMMARY

This appendix provides a brief review of the principal sources of information in the heat-exchanger field. The single most important resource is the good technical library found in the central libraries of large metropolitan areas or universities with faculties of engineering. It is essential to understand how to use the catalog system in the library. Time spent on instruction in this aspect of library usage will be repaid many times over by time saved in locating the material required.

Many libraries operate information-retrieval systems using computer searches of data and information banks. One supplies key words describing the topic of interest, and in short order and at very low cost is presented with a package of titles and abstracts of publications in the field whose descriptors contain the key words provided in the original request. Sometimes one is embarrassed by the plethora of available information. As an example the one key word "robot" recently yielded over 900 references from a single data bank.

Index